...But If a Zombie
Apocalypse *Did* Occur

CONTRIBUTIONS TO ZOMBIE STUDIES

White Zombie: Anatomy of a Horror Film. Gary D. Rhodes. 2001

The Zombie Movie Encyclopedia. Peter Dendle. 2001

*American Zombie Gothic: The Rise and Fall (and Rise)
of the Walking Dead in Popular Culture.* Kyle William Bishop. 2010

*Back from the Dead: Remakes of the Romero
Zombie Films as Markers of Their Times.* Kevin J. Wetmore, Jr. 2011

*Generation Zombie: Essays on the Living Dead
in Modern Culture.* Edited by Stephanie Boluk and Wylie Lenz. 2011

*Race, Oppression and the Zombie: Essays on Cross-Cultural Appropriations
of the Caribbean Tradition.* Edited by Christopher M. Moreman
and Cory James Rushton. 2011

Zombies Are Us: Essays on the Humanity of the Walking Dead.
Edited by Christopher M. Moreman and Cory James Rushton. 2011

The Zombie Movie Encyclopedia, Volume 2: 2000–2010. Peter Dendle. 2012

Great Zombies in History. Edited by Joe Sergi. 2013 (graphic novel)

Unraveling Resident Evil: *Essays on the Complex Universe
of the Games and Films.* Edited by Nadine Farghaly. 2014

"We're All Infected": Essays on AMC's The Walking Dead
and the Fate of the Human. Edited by Dawn Keetley. 2014

Zombies and Sexuality: Essays on Desire and the Walking Dead.
Edited by Shaka McGlotten and Steve Jones. 2014

...But If a Zombie Apocalypse Did *Occur: Essays on Medical, Military,
Governmental, Ethical, Economic and Other Implications.*
Edited by Amy L. Thompson and Antonio S. Thompson. 2015

*How Zombies Conquered Popular Culture: The Multifarious Walking
Dead in the 21st Century.* Kyle William Bishop. 2015

...But If a Zombie Apocalypse *Did* Occur

Essays on Medical, Military, Governmental, Ethical, Economic and Other Implications

Edited by Amy L. Thompson
and Antonio S. Thompson

Foreword by Wade Davis

CONTRIBUTIONS TO ZOMBIE STUDIES

McFarland & Company, Inc., Publishers
Jefferson, North Carolina

LIBRARY OF CONGRESS CATALOGUING-IN-PUBLICATION DATA

...But if a zombie apocalypse did occur : essays on medical, military, governmental, ethical, economic and other implications / edited by Amy L. Thompson and Antonio S. Thompson ; foreword by Wade Davis.

pages cm. — (Contributions to zombie studies)
Includes bibliographical references and index.

ISBN 978-0-7864-7550-6 (softcover : acid free paper) ∞
ISBN 978-1-4766-2090-9 (ebook)

1. Zombie films—History and criticism. 2. Zombies in popular culture. 3. Emergency management. 4. Survival. 5. Zombies in literature. 6. Zombies. I. Thompson, Amy L., editor. II. Thompson, Antonio S. (Antonio Scott), 1975– editor.

PN1995.9.Z63B88 2015
791.43'675—dc23   2015019294

BRITISH LIBRARY CATALOGUING DATA ARE AVAILABLE

Cover image © iStock/Thinkstock

Printed in the United States of America

McFarland & Company, Inc., Publishers
Box 611, Jefferson, North Carolina 28640
www.mcfarlandpub.com

Table of Contents

Acknowledgments

Amy: This project has been a huge endeavor and I have so many people to thank for its success. This feat could not have been accomplished without the hard work of our contributors. They worked tirelessly to complete quality pieces and I can't thank them enough for their dedication and patience throughout the editing process.

I must thanks my parents, Forest and Shirley Ingram, who encouraged me to see my goals through to the end and my uncle David Beach, who taught me to never give up no matter how difficult the challenge.

I would never have considered taking on a project of this magnitude without the encouragement and support of my co-editor and love of my life, Antonio. His expertise in writing showed throughout the process. It is difficult to find a project that a biologist and historian can work on together and it has been fun to have the opportunity to work with him in an academic capacity. He has been my biggest supporter throughout this project and in my life. Thank you!

Antonio: I would first like to thank all of our contributors for providing excellent essays, without them this edited collection literally could not have been finished. They are all respected academics in various disciplines and this project allowed us to gather this diverse talent into this creative endeavor. They each patiently read through our edits and suggestions as we worked on the various drafts. I am glad to have been able to work with all of you and thank you again for your help bringing our zombie work to life.

Finally, I would like to thank my family. My parents, Paul and Robbin Thompson, are partly responsible for this and may not even know it. When I was a child they rented movies on the weekend, many of them horror. My parents were great at movie trivia and instilled in me a love for movies and horror movies that has lasted a lifetime. I would also like to thank my children, Madeline, Julian, and Sophia. They have been patient with me over the last few years as I've finished several projects. My wife and co-editor, Amy Thompson, deserves the biggest thanks. She not only encourages me to finish projects, but she is also a creative sounding board, a partner in all things, and, more importantly, she has put up with me all these years.

Foreword

WADE DAVIS

The subject of the Haitian zombie is one that has fascinated—indeed titillated—foreign observers for decades. The common image is one of a corpse in tattered rags, trailing remnants of necrotic flesh as it rises from the cemetery in a state of trance-like animation, entirely subservient and beholden to the nefarious authority of some unknown master. This macabre figure has entered Western popular and academic culture in an unusual variety of forms, as this curious and fascinating book surely demonstrates.

My interest in zombies was prompted by the discovery of the first case that was believed to be scientifically verifiable. The putative victim, Clairvius Narcisse, had been pronounced dead in 1962 at an American-directed hospital in Haiti that kept excellent records. His death had been witnessed by two physicians, both American-trained, one an American, and by his sister. In 1980 Narcisse returned to his family's village, where he told a chilling tale of having been victimized by a *bokor*, a negative priest or sorcerer.

Lamarque Douyon, then Haiti's leading psychiatrist, conducted a thorough investigation and, in collaboration with the late Nathan Kline, winner of two Lasker Awards and a pioneer in the field of psychopharmacology, concluded that Narcisse had indeed been misdiagnosed dead. Their attention then focused on reports of a folk preparation, mentioned frequently in the popular and ethnographic literature, which was said to induce a state of apparent death so profound as to fool a Western-trained physician. The existence of this preparation was acknowledged in the penal code of the country and penalties for its use were severe. Intrigued by the medical potential of such a substance, Kline in 1982 contacted Harvard's Botanical Museum and I was dispatched to Haiti to seek the formula and collect raw samples. The assignment, initially but a fortnight, would in the end consume four years.

Although each bokor had a unique formula, the consistent ingredients, aside from human remains, were species of marine fish belonging to the order

1

Tetraodontiformes. The viscera and skin of these fish contain tetrodotoxin, a nerve poison roughly a thousand times stronger than cyanide: a lethal dose of the pure toxin would balance on the head of a pin. Exposure to the poison causes metabolic rates to fall dramatically. The pulse becomes imperceptible and peripheral paralysis is total. Though unable to move, the victim remains fully conscious until the moment of actual death.

In Japan tetrodotoxin-containing fish are a delicacy, and the biomedical and popular literature contains numerous accounts of individuals being mis-diagnosed dead, nailed into coffins alive or by folk tradition laid out by their graves for several days until known to be dead. These accounts confirmed that the sorcerers in Haiti had indeed found in their environment a natural product that could induce a state of apparent death and evidently had done so many times in the past in a quite different cultural context. That Narcisse's symptoms were consistent with the known effects of tetrodotoxication suggested the possibility that he had been exposed to the poison.

While the formula of the preparation took the zombie phenomenon from the phantasmagoric into the realm of the plausible, it by no means solved the essential mystery. For the Vodounist the *poudre zombi*, the zombie powder, is seen as but a support for the magical force of the sorcerer, and it is this power, not a poison, that creates the zombie.

In Vodoun there are two kinds of death: those that are natural, acts of God beyond the reach of sorcery, and those that are unnatural, mediated by the bokor. Only those who die an unnatural death may be claimed as a zombie. The poison is an effective way to induce such a death, but the performance of a magical rite is what actually creates a zombie.

The bokor gains power over the victim by capturing the *ti bon ange*, the little good angel, which is the component of the soul that creates character, personality and willpower. A zombie appears cataleptic precisely because it has no ti bon ange. Trapped in purgatory, the body is but an empty vessel. The notion of external forces taking control of the individual, and thus breaking the sacred cycle of life, death and rebirth that allows human beings to give rise to the gods, terrifies the Vodounist. The fear in Haiti is not of zombies, but rather of becoming one.

Understanding zombification from the perspective of the believer led to yet another revelation. Though there is no doubt that tetrodotoxin can induce apparent death, levels of the poison in the fish vary greatly, and at certain times of the year, as much as half the population may contain none of the toxin at all. Any particular batch of the folk preparation may range from being truly lethal to being completely inert. So what happens if the potion does not work? What insulates the sorcerer from being exposed as a fraud?

The answer is the belief system itself. The bokor does not have to account for his failures. If he administers a powder that has no effect, he can claim that his magic was deflected by the intervention of a benevolent priest. If, on the other hand, the victim actually dies, the bokor can suggest that the death was a call from God and beyond the reach of his sorcery. A bokor's failed attempts do not count, only his successes. Even if the poison was effective but once in dozens of attempts, the outcome would support the reputation earned by the zombie phenomenon. Its power as a concept depends not on how often it occurs but rather on the fact that it can and apparently has occurred.

But how and why is someone chosen to become the victim of a bokor's sorcery? From testimony of family and villagers it was clear that Narcisse had been a pariah at the time of his demise. Some claimed he had been brought before a tribunal to be judged.

In my time in Haiti I often crossed paths with the Bizango, a notorious secret society much feared by the Haitian elite. The Bizango were said to dominate the rural peasant society, constituting a force parallel to the Vodoun temples headed by the priests. Several contacts maintained that the Bizango controlled the zombie powders. The origins of the Bizango, as in the case of so many Haitian institutions, could be traced in direct lineage to West Africa, where to this day secret societies remain the most powerful arbiter of social and political life. They function as judicial tribunals, apply sanctions, and to punish those who violate the codes of their communities, they administer poisons.

Intrigued by the possibility that the Bizango might play a similar role in Haiti, I focused the last months of my research exclusively on the secret societies, and with the help of key contacts, I was able to undergo preliminary training as an initiate. What emerged after several months of study was a clear sense that beyond the ritual activities, the secret societies constitute in Haiti a true and effective political force that protects community resources, particularly land, even as they define the power boundaries of the villages. Sorcery and poisons are their traditional weapons, and within the Bizango, there is a complex judicial process by which those who violate the codes of the society may be punished. Zombification is the ultimate sanction. Clairvius Narcisse, it seems, was no innocent victim. His condition had been deserved, and his fate sealed, by his own misdeeds.

Looking back at this research after nearly three decades, I am reminded of something that Nathan Kline told me soon after I agreed to take on the assignment. "The purpose of science," he said, "is not to discern absolute truth but, rather, to generate better ways of thinking about phenomena." Although I went to Haiti to seek the scientific basis of a social mystery, my work in the end explored the psychological, spiritual, political and cultural dimensions of a chemical possibility. Ultimately it was impossible to prove that Clairvius Narcisse had received a dose of the poison, or for that matter that he had been buried alive. But his case, provocative as it was, obliged the scientific world to take seriously a folk tradition that had historically been invoked, often in an explicitly racist manner, to denigrate an entire culture and their religious worldview. Thirty years on, the link between the toxic powder and zombification remains compelling and the hypothesis still stands, even as Vodoun has become recognized as one of the more remarkable religious traditions ever brought into being by the human spirit and imagination.

But if the ultimate goal of my research was to make sense out of sensation, and lay to rest the phantasmagoric notions that had long been associated with the Haitian zombie, I surely failed to deliver. Today, more than ever, the macabre image of old remains the template of the popular imagination. This in itself is curious, given all that is actually known about the phenomenon. This stubborn obsession accounts for the attraction of this wonderfully eclectic collection of essays. They range from studies of the human brain to reflections on the nature of death and dying, from the contemplation of the deep impulses of religious faith to the examination of the psychological, political and economic responses of large populations faced with apocalyptic natural disasters. These essays do not concern the Haitian zombie as much as the metaphorical resonance of the very notion of zombies, a term that has been invoked in any number of peculiar and unexpected ways over the years.

An individual, for example, who feels particularly listless and apathetic, may refer to himself or herself as "feeling like a zombie." Such a sensation might result from indulgence the night before in one too many zombies, an iced cocktail notorious for its alcohol content. Had this person been drinking in a bar whose jukebox was some decades out of date, she or he might well have listened to one of the many hit songs of a famous rock band of the 1960s, The Zombies. A more sober individual might have stayed home and read a book, *Zombies Gone Wild* (Pierre 1978), or taken in a play, *The Zombie* (Kelly 1983). The choice of late night films on television would be almost limitless: *White Zombie* (1932), *King of the Zombies* (1941), *I Walked with a Zombie* (1943), *Zombies on Broadway* (1945), *Valley of the Zombies* (1946), *Zombies of the Stratosphere* (1953) and more recently *The Serpent and the Rainbow*, based very loosely on my book of the same name (1987).

The academic literature has for many years shown an odd fascination with the topic, as is evident in the casual and gratuitous use of the word itself. Witness, for example, two papers in descriptive linguistics: Nancy A. Johnson's "Zombies and Other Problems: Theory and Method in Research on Bilingualism" and Jerrold J. Katz's "Interpretive Semantics Meets the Zombies: A Discussion of the Controversy About Deep Structure." Or, in a journal of social psychiatry, Joshua Bierer's paper entitled simply "Zombi," in which the term is co-opted to define a "man who as a child has neither experienced nor been trained in the three ingredients of emotional life: attention, love and affection." Finally the word appears in the philosophical literature in a series of articles that again have nothing to do with the correct etymological and cultural meaning of the term (Locke "Zombies, Schizophrenics, and Purely Physical Objects"; Kirk "Zombies v. Materialists").

At one level the essays in this book continue this tradition as if a celebration of appropriation. Yet because of the quality of the contributions and the care taken by the editors in bringing these diverse narratives together, all beneath the shadow of the zombie if you will, there is a provocative unity to the collection that I think every discerning reader will very much appreciate. It is certainly fascinating, for example, to consider what the pop obsession with a "zombie apocalypse" might tell us about the potential reaction of the American people to a moment of real danger and true national peril. If nothing else the "zombie apocalypse" speaks to a herd mentality and a culture of paranoia, irrational thinking and scientific illiteracy that is both highly disturbing and all too real. Just consider the wave of fear that swept the nation in the wake of 9/11. The reaction led to terrible consequences, egregious political and military decisions that will haunt the world for a very long time.

America was at one time a land of courage and grit. It is today a place where the media whips up a constant frenzy of fear and the public seems to revel in it. One shudders to think how an American public capable of believing in a "zombie apocalypse" might respond to a serious challenge to its survival, an actual natural or manmade disaster of apocalyptic scale. New Orleans in the wake of Hurricane Katrina most certainly did not inspire. This is something worth studying and writing about, and if even the most tangential reference to the zombie phenomenon draws attention to the work, so much for the better, as this wonderful book reveals.

BIBLIOGRAPHY

Bierer, Joshua. "Zombi." *International Journal of Social Psychiatry* 22.3 (1976): 200–201.
Johnson, Nancy A. "Zombies and Other Problems: Theory and Method in Research on Bilingualism." *Language Learning: A Journal of Applied Linguistics* 24 (1974): 105–33.
Katz, Jerrold J. "Interpretive Semantics Meets the Zombies: A Discussion of the Controversy About Deep Structure." *Foundation of Language* 9.4 (1973): 549–96.
Kelley, Tim. *The Zombie.* New York: Samuel French, 1983.
Kirk, Robert. "Zombies v. Materialists." *Aristotelian Society Supplementary* 48 (1974): 135–52.
Locke, Don. "Zombies, Schizophrenics, and Purely Physical Objects." *Mind* 85 (1976): 97–99.
Pierre, Romulus. *Les zombis en furie.* Port-au-Prince: Ateliers Fardin, 1978.

Wade Davis holds the Leadership Chair in Cultures and Ecosystems at Risk at the University of British Columbia, where he is a professor of anthropology in the Liu Institute for Global Issues. He has served as explorer-in-residence at the National Geographic Society and is the author of 17 books, including Into the Silence, *winner of the 2012 Samuel Johnson Prize. His many film credits include the documentary series* Light at the Edge of the World.

Introduction

Amy L. Thompson *and*
Antonio S. Thompson

In *...But If a Zombie Apocalypse* Did *Occur* we have combined discipline specific expertise with popular culture to study real-world implications of a zombie apocalypse. Through the lens of zombies we examine the human condition. The zombie apocalypse thus serves as a vehicle to discuss probability of life, behavior, and reaction during an undead invasion or any major traumatic or world ending event. We do this by pulling in real world scenarios, historical evidence, and theoretical examples, as well as drawing on popular culture.

The roots of the zombie and the zombie apocalypse are grounded in mythology and religion. The idea of the dead eating the living can be traced back to the ancient Mesopotamian poem *The Epic of Gilgamesh*. In this poem an angry goddess threatens to "...bring up the dead to consume the living..."[1] There are also concerns about life after death and reanimation in modern religions. One aspect of the Voodoo religion is the belief that people can become "zombie" slaves by being poisoned by others. Even the Christian religion has beliefs of resurrection and an end of times revelation where the dead are raised and judged along with the living.

Zombies originated in mythology and religion and transcended into popular culture. There are a few seminal points that define the zombie in popular culture. William Seabrook's 1930 novel *The Magic Island* examined Voodoo and may have been among the first works to use the word zombie.[2] The 1932 movie *White Zombie*, based on Seabrook's book, was the first film depiction of the voodoo zombie. Reanimated corpses first appeared in 1968 in the genre defining film *Night of the Living Dead* by George Romero.[3] *The Walking Dead* started as a comic book in 2003 and then transitioned to television in 2010 on AMC, becoming one of basic cable's most watched shows.

The facination with zombies is evident from George Romero's *Night of the Living Dead* to other movies like *28 Days Later*, and even *Shaun of the Dead*. Literature has followed suit. In two non-fiction books, *The Serpent and the Rainbow* and *Passage of Darkness: The Ethnobiology of a Haitian Zombie*, ethnobiologist Wade Davis discussed the voodoo zombie and the poison used in zombification. The former was turned into a fictional movie based on the book. Robert Kirkman's writing in *The Walking Dead* comic book was developed into the popular AMC television series of the same name. Zombies have also infested classical literature with the recent *Pride and Prejudice and Zombies*.

A major question throughout this book emerges asking does the fact support the fiction? Whether one is concerned about the fictional zombie or an EMP pulse that reduces mankind to an earlier technological period, or a major disease that depopulates society, or a world ending war; does the preparation and popular culture seem to match the academic theories and historical realities of when these types of things have happened in the past? For instance, this is not the first time that humans have considered creating governments from scratch. It is not the first time that mankind has wondered if the mechanisms to control the spread of disease or manage disasters are adequate and, in fact, this book will demonstrate times when they have not been. While people worried about destructive wars, the world has faced and survived catastrophic war. The world has lived through major diseases and although outbreaks seem to be out of control, past experience shows us that this will not be the end of man.

Disasters big and small highlight the reliance on modern technology and fear of being cut off or poorly prepared for disasters. Take recent events, such as a relatively minor power outage in September of 2014 on the campus of Austin Peay State University. The lack of electricity prevented classes from meeting due to the fact that many of them were typically held in interior rooms with no light and no air conditioning during a very humid and warm Tennessee summer. Computers could not be used and even for a period of time the campus-operated phones were down. Food services could not serve food, including to students who were on meal plans depending on this and many of whom lacked transportation to go to off-campus dining. Even that option was limited by the extent of the power outage throughout that part of Clarksville. Eventually those who had access to email on their phones were able to get a message from the University stating that classes were finally cancelled. This small scale situation illuminates some of the concerns that a larger disaster might entail.

This book examines some of those concerns with each professional explaining how his or her discipline specific zombie topic relates to the real world. After a thorough examination of the zombie as it relates to popular culture, the history of apocalyptic events will be presented. Next, the work of the Centers for Disease Control and Prevention (CDC) will be evaluated as it relates to a zombie apocalypse followed by the expected response of the military. No study of an apocalypse or other disaster would be complete without looking at the importance of communication channels or the work of medical professionals. Questions such as what constitutes a zombie will also be explored. The work of engineers will be highlighted as will the work of agencies such as FEMA and Homeland Security. Typical apocalyptic governments will be examined along with legal questions that might arise in disaster situations. The sociological impact of grouping and the psychological stress of the situation will be discussed. Ethical questions will be presented as will questions of gender roles and economic impact. Through studies such as these, individuals will have a better understanding of lessons that can be learned from fiction and past events. These lessons can then be applied to future situations thus allowing readers to know what to expect and how to prepare for a zombie apocalypse or any other similar type of disaster.

NOTES

1. *The Epic of Gilgamesh*, translated and edited by Andrew George (London: Penguin, 1999), 51.

2. William Seabrook, *Magic Island* (Whitefish, MT: Kessinger, 2003).

3. *Night of the Living Dead*, written by George Romero and John A. Russo, directed by George Romero, Image Ten, 1968.

The Rise of the Zombie in Popular Culture

James F. Thompson

Scary stories are likely to be as old as storytelling itself. Many people enjoy a good scare in a safe environment. What sorts of frights does a person find in a work of entertainment? Fearful things may be grounded in reality or they may be drawn from within the realm of the supernatural; they may be derived from a human source or from something else. If the scary stimulus has elements of the human, it may be called a Bogeyman. The Bogeyman takes many forms; some natural and others supernatural. Most incarnations of a Bogeyman have become familiar staples in the horror genre of popular culture.

The most pervasive Bogeymen in modern popular culture are represented by one or more novels of international fame which define their icon. For the vampire, there is Bram Stoker's *Dracula* (1897). For the mad scientist, there is Mary Shelley's *Frankenstein; or, the Modern Prometheus* (1818) and Robert Louis Stevenson's *Strange Case of Dr. Jekyll and Mr. Hyde* (1886). The mummy novels include *Romance of a Mummy* (1857) by Théophile Gautier, two Arthur Conan Doyle short stories, "The Ring of Thoth" (1890) and "Lot 249" (1892), and *The Jewel of the Seven Stars* (1903) by Bram Stoker. The werewolf has the slightly less acclaimed *The Werewolf of Paris* (1933) by Guy Endore, though, like all these literary landmarks, Endore's novel was preceded by a variety of lesser known earlier folktales, legends, short stories and more obscure novels.

Another of the mythic "human monsters" is the ghoul, a flesh eater which was sometimes seen as one of the undead.[1] There is no classic novel to establish the ghoul as an icon, though they make appearances in William Beckford's Orientalist novel *Vathek* (1786), Richard Francis Burton's (among other translations of) *The Book of the Thousand Nights and a Night* (1885) and mentions in poems by Lord Byron, "The Giaour" (1813), and Edgar Allan Poe, "The Bells" (1848). Between the world wars, ghouls appeared in short stories in *Weird Tales* and similar pulps from authors such as Clark Ashton Smith and H.P. Lovecraft. There have been fewer films to focus on the ghoul as distinct from the vampire or the zombie. Boris Karloff appears in *The Ghoul* (1933, director T. Hayes Hunter), a British production that was a disappointment to critics and audiences alike. A decade later, George Zucco is the mad scientist transforming his assistant, David Bruce, into *The Mad Ghoul* (1943, director James P. Hogan) with assistance from a gas concocted by the Mayans. Bosley Crowther took *The Mad Ghoul* to task in his *New York Times* review from December 11, 1943:

> Ghoul Daze: Most of the ghouls we've met in horror films have been more or less scatter-brained, so there's really nothing out of the ordinary about the one in Universal's "The Mad Ghoul," which came clomping and goggle-eyeing into the Rialto yesterday. He is just another poor unfortunate who has been turned into a walking fiend by another maniacal scientist who has discovered some peculiar witching-gas. And the nature of his madness is no different from that of any ghoul who has gone clawing around among cadavers and generally making a nuisance of himself. As a matter of fact, we would call him a definitely second-rate ghoul. And if anyone is privileged to be crazy, it's us poor folks who have to look at such things.[2]

The zombie, the modern reinterpretation of the ancient ghoul, is a recent addition to the pantheon of premiere perils. Similar to the ghoul, there is no seminal literary incarnation for the zombie genre. One early contender is William Buehler Seabrook's *The Magic Island* (1929) and another is Zora Neale Hurston's *Tell My Horse* (1938), but neither is fiction. Rather they are the accounts of two travelers to Haiti who wished to pursue the legend of the zombie in the Vodou ("voodoo") religion and in the alleged powers of rural shamans, "bokors," to raise and enslave the dead. Seabrook's tale was resurrected by Lancer Books in 1968, timed to allow the reprint to ride the coattails of the success of George Romero's *The Night of the Living Dead* (1968). These bokor traditions may also be traced back to similar beliefs common to some West African tribes where the bokor captures and commands the zombi astral spirit, a portion of the soul, rather than the body. Similar West African gods or snake gods are the zumbi, ndzumbi, nzumbe, and nzambi. Decades later, Canadian ethnobotanist E. Wade Davis visited Haiti in pursuit of the zombie legend. Davis's hypothesis was that natural agents such as pufferfish tetrodotoxin, or Jimson weed ("loco weed," *Datura stramonium*), might be the active agents in the "zombie powder" used by the bokor to subjugate his victims. Davis's book on his investigations, *The Serpent and the Rainbow* (1986), became a best seller and the basis for *The Serpent and the Rainbow* (1988), directed by Wes Craven.

While there is no major literary foundation for the zombie mythos, one may look to art and myth for possible inspirations for the flesh-eating zombie. The Greek myth is that of Cronos/Kronos (Saturn in Roman myth), one of the first generation of titans who attacked and overthrew his own father, Uranus. Cronos then ruled during "The Golden Age."[3] Cronos learned from his parents, Uranus and Gaia, that he too would be overthrown by a child, so he subsequently consumed his first five children, the first of the Olympian gods, to prevent the prophecy's fulfillment. Nevertheless, Rhea, their mother, conspired with her mother-in-law to deceive Cronos and save the sixth son, Zeus. In adulthood, Zeus forced Cronos to disgorge his siblings, and they led a revolt, the Titanomachy, the end result of which was the overthrow of Cronos, the imprisonment of many of the titans, and the establishment of the reign of the Olympian gods, led by Zeus. Saturn devouring his children has been depicted many times in Western art. A key painting is *Saturn Devouring His Son* (1636) by Peter Paul Reubens. The same heinous act is depicted in the sculpture *Saturn Devouring One of His Children* attributed to Simon Hutrelle (~1700).[4] The Reubens painting is thought to have inspired Francisco Goya's most horrific of his series of fourteen private murals, *The Black Paintings*, painted on the walls of his home between 1819 and 1823. Goya did not name the works, assuming they would never be shown. This particular one is referred to as *Saturn, Saturn Devouring One of His Sons*, or by similar names today. Where Reubens' painting is classical and "refined," Goya's is more expressionistic, gory and grotesque. If one knows the painting, one cannot help but think of this work when viewing zombie attacks in films.[5]

To complete this circle, Seabrook's *The Magic Island* (1929), along with Kenneth Webb's 1932 New York stage play, *Zombie*, inspired *White Zombie* (1932).[6] This picture lacked the stature of star Bela Lugosi's earlier roles in *Dracula* (1931, directors Todd Browning and Karl Freund) and *Murders in the Rue Morgue* (1932, director Robert Florey), but marked the birth of the classical zombie in horror film history. Other low-budget films followed *White Zombie* to American theater screens, including *Ouanga* (1935, director George Terwilliger), *Revolt of the Zombies* (1936, director Victor Halperin), ostensibly a follow up to *White Zombie*, *Four Shall Die* (1940, directors William Beaudine and Leo C. Popkkin), *The Ghost Breakers* (1940, director George Marshall), one of the earliest zombie comedies, and *King of the Zombies* (1941, director Jean Yarbrough).[7] Director Jacques Tourneur's *I Walked with a Zombie* (1943), produced by Val Lewton for RKO (Radio-Keith-Orpheum) Pictures, now viewed as a minor classic of the zombie genre, was not received well in its initial release:

> But to this spectator, at least, it proved to be a dull, disgusting exaggeration of an unhealthy, abnormal concept of life. If the Hays office feels it has a duty to protect the morals of movie-goers by protesting the use of such expressions as "hell"

and "damn" in purposeful dramas like "In Which We Serve" and "We Are the Marines," then how much more important is its duty to safeguard the youth of the land from the sort of stuff and nonsense that their minds will absorb from viewing "I Walked with a Zombie."[8]

With no literary anchor nor classic film archetypes nor defining performances by a particular actor, no Bela Lugosi, Boris Karloff, John Barrymore, Frederick March, Spencer Tracy (the latter three portraying Dr. Jekyll and Mr. Hyde) or Lon Chaney, Sr., or Lon Chaney, Jr., the "zombie" brand began to drift. Zombies served a comic purpose in Bob Hope's *The Ghost Breakers* (1940). Bela Lugosi starred in *Voodoo Man* (1944, director William Beaudine) as a doctor hoping to use voodoo rituals to restore his dead wife to life. The lone "zombie" of *Valley of the Zombies* (1946, director Philip Ford), though a reanimated corpse, is a blood drinker, not a flesh-eating ghoul. In *Zombies of the Stratosphere* (1952, director Fred C. Brannon), best known for an early appearance of Leonard Nimoy, the "zombies" are Martian invaders. *The Ghoul* (1957, director Freddie Francis) features Peter Cushing as a doctor with a son who has become a cannibal during their stay in India, but whose status as alive or undead is unresolved, though there is a mysterious connection to Hindu mythology. This film and a small number of similar British horror films from the postwar era have a sublimated theme that suggests regret for the declining British Empire and its colonial excesses.

Among the first of the zombie films to shift the causal agent for the zombie from the supernatural to the "mad" scientist is *Bowery at Midnight* (1942, director Wallace Fox) with Bela Lugosi as the malevolent criminal mastermind converting his reanimated corpses to an afterlife of crime. *Bowery at Midnight* was followed soon after by Lon Chaney, Jr., in the *Indestructible Man* (1956, director Jack Pollexfen). *Creature with the Atom Brain* (1955, director Edward L. Cahn) catapulted the zombie into the atomic age. Low-budget zombie pictures continued to trickle out of the poverty row and B picture studios, perhaps turning a small profit on their minimal investments, but doing little to strike a stirring chord with the movie going audience. When *Plan 9 from Outer Space* (1958, director Edward D. Wood, Jr.) first circulated, it was just another cheap thrill, merely another bad sci-fi zombie movie. *Plan 9 from Outer Space* did not reach iconic status as one of, if not the "worst movie ever made," until director Ed Wood, Jr., received a Golden Turkey Award posthumously for Worst Director Ever in 1980.[9]

The fifties saw the occasional return to the threat of the supernatural zombie with obscure films such as *Zombies of Mora Tau* (1957, director Edward L. Cahn) and *Voodoo Island* (1957, director Reginald Le Borg) starring Boris Karloff. The sixties were a time of redefinitions and revolutions which saw cultural movements relating to the generation gap and youth counterculture, the anti–Vietnam War movement, and the Civil Rights movements, not just for African Americans, but for Hispanics and Chicanos, women, homosexuals, the mentally ill and the disabled, most under the umbrella of the political New Left. Of all the liberal films of that decade, none captured the spirit of this changing time more than *To Kill a Mockingbird* (1962, director Robert Mulligan) set in Depression-era Alabama. Gregory Peck plays the strong father and idealistic lawyer, Atticus Finch. What brings *To Kill a Mockingbird* to mind in an essay on zombie tales? It is the striking denouement of *To Kill a Mockingbird*. Black farmer Tom Robinson (Brock Peters) has wrongly been convicted of rape. Rather than bide his time to await the outcome of the appeal promised by attorney Finch, Tom breaks out of jail and runs for his life. By this point in the film, most audience members have become convinced not only of Robinson's innocence, but of his profound humanity. Robinson is a good man, a kind man, a noble man, the equal of Gregory's Peck's Atticus Finch. When the report of his death comes to the Finch family, the cinema audience members are as devastated by Robinson's death as were the sympathetic characters in the film, the Finch family, the humane elements of the white community and the entirety of the black community. Just six years later, George Romero would establish a similar black protagonist, Ben (Duane Jones), the man who unites the humans under siege in the rural farmhouse during the zombie apocalypse in the now classic *Night of the Living Dead* (1968). In the film's final moments, in the dawn after the final pyrrhic victory against the last zombie assault which breeches the farmhouse's defenses, Ben is unexpectedly and accidentally shot dead by the law officers, locals, and a national guard posse attempting to destroy the zombies in a mopping up operation. This unexpected, undeserved death always disturbs each new audience. Romero clearly wanted to emphasize the heroic humanity of his black protagonist so that the shock of his death would carry a symbolic message of injustice straight to the hearts of his audience, exactly the same effect achieved by director Robert Mulligan (and author Harper Lee) in *To Kill a Mockingbird*.

Nineteen sixty-eight was a watershed year for American cinema. The thirty-plus-year reign of the Hays Code of voluntary film censorship was over. Jack Valenti of the Motion Picture Association of America instituted a replacement film rating system (originally G, M, R, and X) which emboldened filmmakers to stretch the boundaries of what was acceptable to contemporary audiences. The top grossing films of 1968 included Stanley Kubrick's *2001: A Space Odyssey*, Franco Zefferelli's *Romeo and Juliet*, Roman Polanski's *Rosemary's Baby*, Franklin J. Schaffner's *Planet of the Apes*, George Romero's *Night of the Living Dead*, Anthony Harvey's *The Lion in Winter*, Mark Rydell's *The Fox*, Ralph Nelson's *Charly*, Richard Fleischer's *The Boston Strangler*, and Christian Marquand's *Candy*. Most films in this list shocked audiences with frank adult attitudes toward sexual matters, but a minority took a more adult attitude toward violence. Most in this list also have dark endings and a significant minority have fantastic elements. One might argue that many of them embody a worldview like that of film noir in which authorities cannot be trusted, friends betray friends, and character arcs do not necessarily end with positive outcomes. It is interesting to find Romero's low-budget indie with its cast of unknowns in such good company. *Night of the Living Dead* revitalized (re-un-vitalized?) the zombie genre. While *Night of the Living Dead* may not be the first zombie film to describe a global catastrophe of zombie transformations or to substitute a pseudoscientific for a supernatural mechanism of zombie transformation, it was the seminal film that served as the model for so many which followed. *Night of the Living Dead* also had elements of dark humor that have influenced many of its successors.

George Romero is certainly the Jonathan Swift of zombie film auteurs. Swift proposed that the starving Irish eat their babies and Romero merely generalized the recommended cannibalism to modern America and its social ills. Romero's films are always laced with social satire. His *Dawn of the Dead* (1978) was set primarily in a large, well-stocked shopping mall in Monroeville, Pennsylvania. At the time it was one of the largest malls in the nation and Romero was clearly suggesting mindless consumerism and the desire to go to the mall and shop as unconscious motivations for the undead surrounding the mall in the film. This is demonstrated by the following exchange of dialogue between Francine and Stephen: "What are they doing? Why do they come here?" "Some kind of instinct. Memory of what they used to do. This was an important place in their lives."[10] There was no gun store in the actual Monroeville mall; Romero was prescient in anticipating the subsequent proliferation of retail gun stores in suburban shopping areas.

Meanwhile, John A. Russo, Romero's co-screenwriter, was taking the living dead franchise in a different direction. Where Romero continued to focus on genuine terror and political satire, Russo preferred the humorous aspects of the zombie apocalypse. Russo collaborated with Dan O'Bannon on a very satisfying comic sequel, *The Return of the Living Dead* (1985, director Dan O'Bannon). O'Bannon had been screenwriter of the cult favorite *Dark Star* (1974, director John Carpenter), the low-budget independent film whose script was reworked into *Alien* (1979, director Ridley Scott). Many fans recall it for the moment when a zombie enters an ambulance after the EMTs have been devoured and uses the radio to say, "Send ... more ... paramedics." A lesser comic sequel, *The Return of the Living Dead II* (1988), followed with a different writer and director, Ken Wiederhorn, and the return of cast member James Karen in a new role. As a franchise, the *Return* cycle ended on a high note with a completely serious tragic love story in *The Return of the Living Dead III* (1993, director Brian Yuzna).

Zombie Honeymoon (2004, director David Gebroe), *Boy Eats Girl* (2005, director Stephen Bradley), *Make Out with Violence* (2008, directors the "Deagol brothers") and *Warm Bodies* (2013, director Jonathan Levine) also belong in the small but fascinating subgenre where the theme is of a normal human in love with a zombie. *Make Out with Violence* has had limited distribution, but has won several awards on the festival circuit. In *The Return of the Living Dead III* and *Make Out with Violence*, the zombie is the girl; in the other three, the zombie is the boy. *Zombie Honeymoon* and *Make Out with Violence* end badly for the couple. In *Boy Eats Girl*, there is only a single zombie, the boy resurrected by his mother who improperly uses a necromantic spell from a forbidden book she has stolen from the Catholic church where she works. The romance ends happily because a supernatural snake bite can reverse the evil spell. In *Warm Bodies* the couple lives happily afterward, because the emotion of love cures the zombie state for some stricken individuals. In the wider background story of *Warm Bodies*, the surviving humans and those zombies "cured" by love combine forces to destroy the incurable insatiable zombies, the "bonies," of the apocalypse. *Warm Bodies* gives the most overt expression of the social symbolism inherent in the zombie love story trope. The zombie boy has only vague memories of his normal life and can only recall the letter "R" as the first initial of his name. The young heroine is Julie. The obvious reference is to the conflict surrounding the young lovers from an Italian folk tale tragedy made famous in Shakespeare's play *Romeo and Juliet*. This same conflict, young lovers from different backgrounds facing potentially tragic consequences, has been expressed even more frequently in vampire tales, most recently in the *Twilight* series of novels and subsequent films featuring vampires and werewolves from author and producer Stephenie Meyer.

George Romero also had more plans for the "living dead." His third feature on the subject was *Day of the Dead* (1985), probably the weakest in the entire Romero cycle in entertainment value. Here Romero once again takes his morbid mythology in a new direction. Military scientists attempt to convert a captive zombie, Bub, into a controllable automaton. Only a very few zombie films since have followed that direction, looking to tame the zombies and put them to work. The humorous coda at the end of *Shaun of the Dead* (2004, director Edgar Wright) shows the recovered English society with tamed zombies doing menial work. *Fido* (2006, director Andrew Currie) is set in a recovered American society where electronic collar-pacified zombies are rented or purchased from the Zomcom corporation. By the end of *Fido*, the central family's wife is happier in the company of her zombie, Fido, than she had been with her late husband.

Romero turned his attention to other projects for twenty years, but eventually found inspiration for a new and effective installment in the series, *Land of the Dead* (2005). Following on his theme with Bub, that some zombies retain some intelligence and some self-identity, we meet "Big Daddy" (Eugene Clark, a former Canadian football player), a zombie who leads a "coordinated" attack on the remaining human fortress, Fiddler's Green, in what had once been the center of some major urban center on a river, perhaps a nod to Romero's hometown of Pittsburgh. Once again, Romero's satirical view of American society is front and center, as the economic haves and have-nots in Fiddler's Green mirror the inequality of wealth in the United States in the years of the national real estate bubble before the financial crisis and subsequent Great Recession of 2007 and 2008.

Shortly thereafter, Romero directed a lower budget but still effective entry in the cycle, *Diary of the Dead* (2007). *Diary of the Dead*, though given a contemporary setting, was actually a parallel story to that which Romero told in *Night of the Living Dead*, following a group of individuals encountering the zombie apocalypse at its inception. *Diary of the Dead* was followed by his last installment to date, *Survival of the Dead* (2009), which had only minimal theatrical release, primarily playing on the festival circuit before being dumped to the DVD and video on-demand market. In *Survival of the Dead*, some of the protagonists attempt to restore life to "living dead" loved ones while simultaneously fighting off attacking zombies.

George Romero studiously avoids the term zombie in his films, but it would be impossible to discuss the genre without discussing his contributions to it. In *Night of the Living Dead* (1968), a TV broadcast repeatedly describes the "living dead" as "ghouls," and one character refers to the "living dead" as zombies in the sequel, *Dawn of the Dead* (1978). Late in his career, Romero indicated it was a matter of personal creativity.

> In the first film, I never thought of them as zombies. I thought I had invented some new thing: The dead just didn't stay dead any more. And when everyone started to write about that film and call it important, I said, "Oops, maybe they are [zombies]." So in the second film I call them zombies.[11]

Another iteration of the zombie is found in the *Re-Animator* (1985, director Stuart Gordon) from the H.P. Lovecraft story "Herbert West, Re-Animator," screenplay by Dennis Paoli, William J. Norris and Stuart Gordon, and its two sequels, *Bride of Re-Animator* (1989, director Brian Yuzna) and *Beyond Re-Animator* (2003, director Brian Yuzna). The re-animated corpses from the Lovecraft universe represent a sort of hybrid between the ghoulish risen dead and the Frankenstein monster, but the threat they represent to normal humans, and the favorable response of the Lovecraft film fans, compels them to be included in this history. Another noteworthy science fiction zombie film from the eighties is *Night of the Comet* (1984, director/writer Thom E. Eberhardt) that successfully combined genuine scares with a dry sense of humor.

The East Coast "underground" exploitation studio, Troma Entertainment, founded by Lloyd Kaufman and Michael Herz in 1974, made the occasional foray into the zombie genre. Troma films are very low budget and usually humorous, or ridiculous, or pathetic, depending on one's taste. Among their first were *Curse of the Cannibal Confederates* (also known as *The Curse of the Screaming Dead*; 1982, director Tony Malanowski) and *Zombie Island Massacre* (1984, director John N. Carter). Troma Entertainment continued their zombie cycle with *Redneck Zombies* (1989, director Pericles Lewnes), *Space Zombie Bingo!!!* (1993, director George Ormrod), *Wiseguys vs. Zombies* (2003, director Adam Minarovich), *Zombiegeddon* (2003, director Chris Watson), *Pot Zombies* (2005, director Justin Powers), and *Poultrygeist: Night of the Chicken Dead* (2006, director Lloyd Kaufman). Troma Entertainment sees itself as a source of satirical social criticism:

> Film lovers have been starved for sustenance. The relentless diet of predictability and pretense Hollywood has been serving up just doesn't cut it. *Poultrygeist* is hearty food for thought. In *Poultrygeist*, Troma takes on the fast-food industry—skewering the soulless restaurateurs—in the world's first horror-comedy film to feature zombie chickens, American Indians and a bit of singing and dancing! It's *Poultrygeist!*[12]

Because *Night of the Living Dead* had long since lapsed into the public domain, writer and director James Riffel took the original Romero film and dubbed an entirely new dialogue track for the parody *Night of the Day of the Dawn of the Son of the Bride of the Return of the Terror* (1991). The creative well dug by screenwriters George Romero and John A. Russo in the creation of *Night of the Living Dead* has never run dry. A superb color recreation followed twenty-two years later, *Night of the Living Dead* (1990), from director Tom Savini, who had been a longtime make-up artist collaborator with Romero. The original screenplay plot and concepts have also been the springboard for a series of less faithful adaptations and borrowings including *Night of the Living Dead 3D* (2006, director Jeff Broadstreet), *Night of the Living Dead: Reanimated* (2009, director Mike Schneider), *Another Night of the Living Dead* (2011, director Alan Smithee), *Night of the Living Dead 3D: Re-Animation* (2012, director Jeff Broadstreet), and *Night of the Living Dead: Resurrection* (2012, director James Plumb).[13]

Episodes two and three of Romero's ongoing saga have also been successfully remade as *Dawn of the Dead* (2004, director Zack Snyder) and *Day of the Dead* (2008, director Steve Miner). The true classics of the science fiction, fantasy and horror genres are destined to be remade from time to time to introduce the enduring legends to new generations of viewers. Romero's undead are certainly a part of that tradition some forty-five years out from their introduction to the fan audience in 1968.

The nature of the family is often a subtext in zombie films, whether strictly horrific or combined with comedy. A new family coming together from disparate individuals to defend the group against some hostile circumstance is a classic plot device in cinema. The zombie genre got a huge box office boost from the very successful *Shaun of the Dead* (2004, director Edgar Wright) which cleverly infused elements of romantic comedy and family dynamics into the traditional lumbering zombie apocalypse. In *Shaun of the Dead*, protagonist Shaun (Simon Pegg) has to multi-task, defending himself and others from the zombie attacks and resolving a recent break up with his girlfriend, Liz (Kate Ashfield), while taking additional risks to rescue his mother, Barbara (Penelope Wilton), and his step-father, Philip (Bill Nighy), whom Shaun resents. Similar multi-generational obligations drive the plot in *Cockneys vs. Zombies* (2012, director Matthias Hoene) in which some amateurs are robbing a bank to get the funds to save their grandfather's retirement home from demolition at the hands of land developers when the zombie apocalypse intervenes and they must then expend considerable effort in making the detour to the retirement home before escaping from the zombie hordes. Another successful zombie comedy thriller was *Zombieland* (2009, director Ruben Fleischer) which shifts effortlessly between horror and romantic and family comedy as protagonist Columbus (Jesse Eisenberg) acquires a father figure, Tallahassee (Woody Harrelson), a potential bride, Wichita (Emma Stone), and a ward, Wichita's younger sister, Little Rock (Abigail Breslin), while fending off the zombie hordes. *Night of the Living Dead* (1968) depicts the formation of a new family coming together out of necessity in an abandoned farm house under siege by the "living dead." Within the new family context are two nuclear families, one a young couple, the other a middle aged married couple with a preteen daughter who has already been bitten and will "turn" in a climactic scene near the film's end.

Zombie literature was revivified when it invaded the world of metafiction beginning with *Pride and Prejudice and Zombies: The Classic Regency Romance— Now with Ultraviolent Zombie Mayhem!* by Jane Austen and Seth Grahame-Smith (2009). *Pride and Prejudice and Zombies* may not be the archetypal novel for the zombie genre, but it did spawn a significant cottage industry in which zombies are combined with a variety of classic works of fiction by authors from Jane Austen to Mark Twain as well as films combining zombies with historical figures, for example, the low-budget *Abraham Lincoln vs. Zombies* (2012, director Richard Schenkman) not to be confused with the major studio release, *Abraham Lincoln: Vampire Hunter* (2012, director Timur Bek-mambetov).

Up until this point in this essay, the zombies, the re-animated corpses, and the risen undead, have all had some fundamental behavioral similarities. All have been relatively dull, unintelligent, slow-moving, and dangerous primarily because of their numbers, not their individual destructive predatory prowess. That changed in the twenty-first century. The ancestral hulking walking dead did not become extinct, but its lineage split, and a different form was born, the fast-moving feral and ferocious cousin, usually the victim of science gone astray, and not necessarily a risen corpse, but rather a transformed human monster, but one still seen as in the "zombie" mold by the viewing audience.

This new fast moving zombie appeared in movies in the new millennium, although the rise of the rapid zombie began in the gaming world in the 1990s. The *Resident Evil* franchise began as a survival horror video game. It spawned a series of action horror films of the same name starring Milla Jovovich. While *Resident Evil* fans were coming out to support the first transition of their game to the big screen, a more inspired film drew even larger audiences. This was *28 Days Later* (2002, director Danny Boyle) and its equally successful sequel, *28 Weeks Later* (2007, director Juan Carlos Fresnadillo). *Planet Terror* (2007, director and writer Roberto Rodriguez) also combined humor and suspense in a localized zombie apocalypse.

There is now a candidate for the archetypal novel for the zombie genre, the serious metafiction *World War Z: An Oral History of the Zombie War* by Max Brooks, a bestseller from 2007 which followed Brooks' earlier satirical *The Zombie Survival Guide: Complete Protection from the Living Dead* (2003). *World War Z* has epic scope and incisive character portrayals. Readers find the novel consists of reports compiled from all across the globe from survivors of the zombie apocalypse, collected on behalf of the United Nations "Postwar Commission" by the fictional narrator, a UN agent. Many real-world events such as the Chinese authorities' delay in reporting Severe Acute Respiratory Syndrome (SARS) to world health authorities in 2002 are repurposed to drive or explain events of the fictional global zombie apocalypse in Brooks' novel. Hollywood transformed the novel into an effective and profitable summer blockbuster, *World War Z* (2013, director Marc Forster), but for all its fine qualities, the movie pales in comparison to its literary source, which could potentially be mined for many additional movies.

The shambling zombie, however, caught the public eye in the twenty-first century with *The Walking Dead*, a television drama series developed by Frank Darabont from an ongoing series of award-winning monthly comic books by writer Robert Kirkman, which premiered in 2003. The television series is currently basic cable television's highest-rated drama series, extending over five years (Season 1 [2010], Season 2 [2011–12], Season 3 [2012–13], and Season 4 [2013–14]). While some of *The Walking Dead*'s success may be no more than the good fortune of riding the crest of the larger wave of interest in the ghoul, zombie, walking dead phenomenon, the series deserves credit for quality character development, dramatic plots, and a focus on important questions, such as what constitutes family and what are the correct moral or ethical decisions to be made in crisis situations. Consider, for example, this quote from the deceased character Herschel Green (Scott Wilson):

> You step outside, you risk your life. You take a drink of water, you risk your life. And nowadays you breathe, and you risk your life. Every moment now … you don't have a choice. The only thing you can choose is what you're risking it for. Now, I can make these people feel better and hang on a little bit longer. I can save lives. And that's enough reason to risk mine.[14]

There is no doubt that today's viewing public is drawn to drama that engages intellectually, morally, and viscerally, as *The Walking Dead* does, even when packaged in extreme fantastic settings.

So why has the public's fascination with the zombie risen in the last century? In part, it is the general increase in public interest in horror themes. That question was addressed thoughtfully more than a quarter of a century ago by Stephen King in his *Danse Macabre* (1981), a non-fiction book about horror fiction in print, radio, film and comics, and the horror genre's influence on United States popular culture. King recognized that horror plots could restate normal social and cultural fears in the shroud of otherness, a fantastic and unreal threat, and once that threat has been resolved, whether the threat was conquered or victorious in the fiction, the consumer audience member obtains some catharsis and, perhaps, is less stressed by the normal social or cultural fears that drew the viewer or reader to the work in the first place. King's is a very useful analysis and a useful tool in the search for the sublimated real-world fears that drive the creation of works of horror in any genre.

A recent article revisits that same theme, "The Collective American Scream" by James Wolcott in *Vanity Fair* (January 2014). Wolcott speculates on the same questions King had explored. Wolcott spends considerable time with the zombie and with an academic book, *Theories of International Politics and Zombies* (2011) by Daniel W. Drezner, professor of international politics at The Fletcher School of Law and Diplomacy at Tufts University. Wolcott provides a summary of Drezner's analysis of the ways in which the zombie horror icon can be the substitute for real-world social cultural fears:

> But there's no disputing the paramount importance of zombies as postmodern multi-purpose signifiers that can be plugged into any pandemic-disaster scenario and survivalist manual. With their staggering gait and ravenous need to feed, "the mobile deceased"— as Daniel W. Drezner calls them in his provocative study *Theories of International Politics and Zombies*[15]—conjure images of famine, holocaust, plague, toxic waste, genetic mutation, unburied furies, borderless chaos, racial warfare, urban riots, suburban home invasions, soul-less consumerism (a shopping mall becomes the slaughterhouse in zombie master George Romero's 1978 *Dawn of the Dead*), the rampaging march of Marx's reserve army of unemployed workers in a post-industrial economy, and opaque projections of the Other: a Hieronymus Bosch buffet for the whole family.[16]

There may be a more fundamental reference supporting the current interest in the zombie icon. The zombie, especially the zombie *en masse*, wandering mindlessly through what had been familiar normal city streets and suburban neighborhoods, ruthlessly but unthinkingly attacking and devouring normal humans as they encounter them, incapable of speech and impossible to reason with, destroying the social fabric and replacing it with a deadly irreversible hopeless chaos, is the perfect sublimated symbol for the increasing polarization in human society, particularly American society. So much of the analysis of real world issues is conveniently oversimplified into a conflict between "us" and "them," regardless of what underlies the social conflict, politics, race, class, religion, or sexual orientation. A recent Associated Press–GfK Public Affairs & Corporate Communication poll documents the reality of this thesis:

> Only one-third of Americans say most people can be trusted. Half felt that way in 1972, when the General Social Survey first asked the question.... An AP-GfK poll conducted last month found that Americans are suspicious of each other in everyday encounters. Less than one-third expressed a lot of trust in clerks who swipe their credit cards, drivers on the road, or people they meet when traveling.... Even the rancor and gridlock in politics might stem from the effects of an increasingly distrustful citizenry, said April K. Clark, a Purdue University political scientist and public opinion researcher.... Trust has declined as the gap between the nation's rich and poor gapes ever wider, Uslaner says, and more and more Americans feel shut out. "They've lost their sense of a shared fate. Tellingly, trust rises with wealth. People who believe the world is a good place and it's going to get better and you can help make it better, they will be trusting," Uslaner said. "If you believe it's dark and driven by outside forces you can't control, you will be a mistruster."[17]

Much of human society has lost its belief in compromise, in respecting and accepting differences in points of view or peaceable methods for resolving problems. The culture wars become transformed in fantastic media entertainments into the zombie wars. There is such comfort in the simplicity of knowing all one needs to do is to target the brain of the "other," destroy that "other's" brain, and the problem would be resolved; no diplomacy, no traditional ethical considerations, no rational analysis, no compromise required. George Romero warned us all of the extreme polarization in *Night of the Living Dead* in 1968 through his Johnny (Russell Streiner): "They're coming to get you, Barbra, there's one of them now!"

Notes

1. "Ghoul," www.tribe.net, posted August 23, 2007, http://tribes.tribe.net/b9b544af-89e5-4aa7-8dec-c917f83c3bd7/thread/09133c0d-f6cd-4e93-920c-7bdcea6c40c5.

2. Bosley Crowther, "The Mad Ghoul (1943) THE SCREEN; Ghoul Daze," *New York Times*, December 11, 1943, http://movies.nytimes.com/movie/review?res=9C05E7DF1238E33BBC4952DFB4678388659EDE.

3. "Kronos," *Theoi Project*, accessed September 9, 2014, http://www.theoi.com/Titan/TitanKronos.html.

4. *Saturn Devouring One of His Children*, J. Paul Getty Museum, accessed September 9, 2014, http://www.getty.edu/art/gettyguide/artObjectDetails?artobj=1302.

5. *Saturn Devouring His Son, 1820–23 by Francisco Goya, Francisco Goya: Paintings, Biography, Quotes*, accessed September 9, 2014, http://www.goya.net/saturn-devouring-his-son.jsp.

6. Jeff Stafford, "White Zombie," Turner Classic Movies, accessed September 9, 2014, http://www.tcm.com/this-month/article.html?isPreview=&id=208674|549&name=White-Zombie.

7. "Zombie Movie List by Year," *Zombie Zone News*, accessed September 9, 2014, http://www.zombiezonenews.com/zombie-movies-list/zombie-movie-list-by-release-date/.

8. Thomas M. Pryor, "I Walked with a Zombie (1943) at the Rialto," *New York Times*, April 22, 1943, http://movies.nytimes.com/movie/review?res=9B0CEEDB1730E53BBC4A51DFB2668388659EDE.

9. Matt Patches, "How Plan 9 from Outer Space earned, and lost, the title of worst movie of All Time," *The Dissolve*, June 19, 2014, http://thedissolve.com/features/movie-of-the-week/623-how-plan-9-from-outer-space-earned-and-lost-the-ti/.

10. *Dawn of the Dead*, written and directed by George Romero, 1978.

11. Dave McGinn, "George A. Romero keeps his zombie obsession alive," *The Globe and Mail*, August 19 2010, http://www.theglobeandmail.com/arts/film/george-a-romero-keeps-his-zombie-obsession-alive/article1377686/.

12. *Poultrygeist: Night of the Chicken Dead*, written by Gabriel Friedman, Daniel Bova, and Lloyd Kaufman and directed by Lloyd Kaufman, 2006.

13. *Night of the Living Dead: Reanimated*, directed by Mike Schneider, 2009. ["*Night of the Living Dead: Reanimated* features the work of various artists, animators, and filmmakers from around the globe. The mixed media featured includes puppetry, CGI, hand-drawn animation, illustration, acrylics, claymation, and even 'animated' tattoos, just to name a few. This mass-collaboration approach is less about remaking Romero's film and more about viewing the classic through an experimental lens. Instead of trying to alter Image Ten's work, NOTLD:R seeks to showcase the responses that artists from around the world have had to this landmark film." http://www.imdb.com/title/tt1520368/plotsummary?ref_=tt_ov_pl)].

14. "Isolation," written by Robert Kirkman and directed by Daniel Sackheim, third episode of Season 4 of American Movie Classics' *The Walking Dead*, originally aired on October 27, 2013, http://walkingdead.wikia.com/wiki/Hershel_Greene_%28TV_Series%29.

15. Daniel W. Drezner, *Theories of International Politics and Zombies* (Princeton: Princeton University Press, 2011).

16. James Wolcott, "Hollywood: The Collective American Scream," *Vanity Fair*, January 2014, http://www.vanityfair.com/hollywood/2014/01/horror-films-american-interest.

17. Connie Cass, "In God we trust, maybe, but not each other," AP-GfK Poll (Associated Press–GfK Public Affairs & Corporate Communications), November 30, 2013, http://ap-gfkpoll.com/featured/our-latest-poll-findings-24.

Bibliography

Cass, Connie. "In God we trust, maybe, but not each other." *AP–GfK Poll* (Associated

Press–GfK Public Affairs & Corporate Communications), November 30, 2013, http://ap-gfkpoll.com/featured/our-latest-poll-findings-24.

Crowther, Bosley. "The Mad Ghoul (1943) THE SCREEN; Ghoul Daze." *New York Times,* December 11, 1943. http://movies.nytimes.com/movie/review?res=9C05E7 DF1238E33BBC4952DFB4678388659EDE.

Dawn of the Dead. Written and directed by George Romero, Laurel Group, 1978.

Drezner, Daniel W. *Theories of International Politics and Zombies.* Princeton: Princeton University Press, 2011.

"Ghoul." www.tribe.net, August 23, 2007. http://tribes.tribe.net/b9b544af-89e5-4aa7-8dec-c917f83c3bd7/thread/09133c0d-f6cd-4e93-920c-7bdcea6c40c5.

"Isolation." Written by Robert Kirkman and directed by Daniel Sackheim, third episode of Season 4 of American Movie Classics' *The Walking Dead,* originally aired on October 27, 2013. http://walkingdead.wikia.com/wiki/Hershel_Greene_%28TV_Series%29.

"Kronos." *Theoi Project.* Accessed September 9, 2014, http://www.theoi.com/Titan/TitanKronos.html.

McGinn, Dave. "George A. Romero keeps his zombie obsession alive." *The Globe and Mail,* August 19, 2010. http://www.theglobeandmail.com/arts/film/george-a-romero-keeps-his-zombie-obsession-alive/article1377686/.

Night of the Living Dead: Reanimated. Directed by Mike Schneider, Neoflux Productions, 2009.

Patches, Matt. "How Plan 9 From Outer Space earned, and lost, the title of worst movie of All Time." *The Dissolve,* June 19, 2014. http://thedissolve.com/features/movie-of-the-week/623-how-plan-9-from-outer-space-earned-and-lost-the-ti/.

Poultrygeist: Night of the Chicken Dead. Written by Gabriel Friedman, Daniel Bova, and Lloyd Kaufman and directed by Lloyd Kaufman, Poultry Productions LLC and Troma Entertainment, 2006.

Pryor, Thomas M. "I Walked with a Zombie (1943) at the Rialto." *New York Times,* April 22, 1943. http://movies.nytimes.com/movie/review?res=9B0CEEDB1730 E53BBC4A51DFB2668388659EDE.

"*Saturn Devouring His Son, 1820–23* by Francisco Goya." *Francisco Goya: Paintings, Biography, Quotes,* accessed September 9, 2014, http://www.goya.net/saturn-devouring-his-son.jsp.

Saturn Devouring One of His Children, J. Paul Getty Museum, accessed September 9, 2014, http://www.getty.edu/art/gettyguide/artObjectDetails?artobj=1302.

Stafford, Jeff. "White Zombie." Turner Classic Movies, accessed September 9, 2014, http://www.tcm.com/this-month/article.html?isPreview=&id=208674|549& name=White-Zombie.

Wolcott, James. "Hollywood: The Collective American Scream." *Vanity Fair,* January 2014. http://www.vanityfair.com/hollywood/2014/01/horror-films-american-interest.

"Zombie Movie List by Year." *Zombie Zone News,* accessed September 9, 2014, http://www.zombiezonenews.com/zombie-movies-list/zombie-movie-list-by-release-date/.

Other Apocalypses
*Historical Perspectives
on Mass Destruction*

NICK PROCTOR

Recent health crises, terrorist strikes, natural disasters, and wars have made people familiar with the idea of sudden, unexpected devastation. Media coverage of these events focuses upon images of destruction, which serve as harbingers of doom. Although coverage is usually sensationalist, such events remain worthy of study because they are useful in refining our understanding of public health policy, risk management, emergency response techniques, and the use of military force. Studying these events provides useful data for considering various *tactical* responses to zombies as an isolated phenomenon, but these recent disasters are not particularly useful in understanding the challenges posed by a zombie apocalypse as a *strategic* and *existential* menace to humanity as a species.

First their scale is comparatively limited. For example, while global in reach and potentially devastating, the 2002–2003 outbreak of Severe Acute Respiratory Syndrome (SARS) resulted in fewer than 1000 deaths. In addition, while recent natural disasters and armed conflicts wreaked massive destruction, they are, when cast into a global context, rather limited in scope. Similarly, the 2011 Tōhoku earthquake and tsunami resulted in over 15,000 deaths, but all of the fatalities were in a single country: Japan.[1] As a result, regardless of the intense suffering of those caught up in these events, the vast majority of humanity remained unaffected. This facilitated international relief efforts, which greatly reduced death and suffering.

A zombie apocalypse is different. It is not an "outbreak" or a natural disaster. It is an *apocalypse*—a simultaneous global cataclysm featuring mass killing and social collapse that occurs everywhere. There is no safe haven. Even well into the progression of a zombie apocalypse, those who consider themselves secure are almost invariably proven wrong. In military terms, this resembles a particularly determined insurgency, albeit on an unprecedented geographical scale. In terms of disease, it can be compared to a pandemic with an especially intense initial outbreak and an unprecedented universal geographical distribution. In a zombie apocalypse, there is rarely a "Patient Zero"; instead there are millions of them.

The second defining characteristic of a zombie apocalypse is the particular nature of the threat. Driven by their insatiable desire for human flesh, zombies pursue their victims without thought of self-preservation. This relentlessness distinguishes them from human adversaries and adds to the terror of the apocalypse. There is no reasoning with zombies. They never tire. They give no quarter. Consequently, zombies are similar to a disease, but unlike a pathogen, they are much more mobile. Furthermore, given their capacity for vocal signaling (usually manifested as moaning), they possess significantly more ability to coordinate than simple organisms.

The final distinguishing feature of the zombie threat is the nature of their method of recruitment. It is quick, irresistible, and total. People who are bitten (but not devoured) become new zombies very quickly. This rapid augmentation of the initial threat makes zombies an especially terrifying enemy. Their ability to bring anyone to their side almost instantaneously is essential to the apocalyptic quality of the zombie onslaught. Since everyone is a potential zombie, you are as likely to be "part of the problem" as you are "part of the solution."[2]

Consequently, while recent events provide useful instruction in how a society might deal with some of the tactical and operational challenges posed by a zombie apocalypse, they do not provide much insight at the *strategic* level. This is because a zombie apocalypse presents a profound existential threat to the entire species.

Short of pre-historic asteroid strikes, few such events exist, but four historical episodes do include sufficiently intense local destruction to provide particularly fruitful avenues of inquiry. In terms of warfare, the Mongol Invasions of the 13th, 14th, and 15th centuries and World War II (1937–1945) brought widespread destruction and death to millions by human hands. Diseases have killed far more, but even alongside mass killers like smallpox and influenza, the Black Death of the 14th century was shocking in its level of devastation. Finally, the Columbian Exchange in the Americas, which took place from the 15th to the 19th centuries, combined conquest and disease in ways that shattered societies and destroyed whole peoples. These events are quite varied, but they are all worth studying when contemplating how our society might confront and ultimately survive a zombie apocalypse.

The Mongol Invasions

When compared to the death toll of recent catastrophes, the casualty figures of past conflicts stagger the imagination. They represent suffering on a scale that far outweighs all of the conflicts of the 21st century combined. Yet, even with numerous episodes of bloodletting on a monumental scale, two military conflicts rise above all the others, the Mongol Invasions and World War II.

The former resulted in the deaths of 30 to 60 million people. While this number was eclipsed by World War II in terms of overall body count, it has no equal in terms of what it represents as a portion of the global population. Taken altogether, the depredations of the Mongols led to the deaths of somewhere between 7.5 percent and 17.1 percent of the population of the entire world. Northern China, Eastern Iran, and Central Asia suffered something that, as historian David Morgan puts it, "must have seemed to approximate very nearly to attempted genocide."[3] The geographical dimension of the Mongol Invasions was equally enormous, encompassing most of Eurasia. Indeed, under the rule of Genghis Khan's successors, some Mongols contemplated conquering the entire world.[4]

Led by a succession of capable commanders, the Mongol army resembled a zombie horde in many ways. In the field, they ruthlessly destroyed opposing armies and in conquest they often slaughtered hosts of civilians. To some, they appeared to be an unstoppable force. A survivor of the Mongol sack of the Central Asian city of Bukhara provided a terse description of their onslaught: "They came, they sapped, they burnt, they slew, they plundered, and they departed."[5]

Several of the strategies employed by the Mongols in the expansion of their empire resemble zombie methods of recruitment. They readily incorporated local populations into their war machine. Other steppe peoples provided additional cavalry while more sedentary populations provided infantry and siege troops. In addition, they sometimes conscripted enemy civilians as "arrow fodder" when assaulting fortifications. As Matthew Paris, an English monk, explained in his *Chronica Majora,* "If by chance they did spare any who begged their lives, they compelled them, as slaves of the lowest condition, to fight in front of them against their own kindred."[6] So, in terms of scope, remorselessness, and recruitment, the Mongol horde shared a number of qualities with that of the zombies.

While an examination of the Mongol Invasions may provide instruction about the capability and impact of a powerful and ruthless military capable of continent-spanning operations, drawing parallels between the Mongols and zombies is difficult for a number of reasons. The first is the Mongol reliance upon horses. The abilities of Mongols to ride long distances, shoot from the saddle, and engage in coordinated battlefield maneuvers were keys to their success; however, this reliance upon horses also meant that they generally limited their operations to areas in which they could find sufficient pasture for their large herds.

These strengths and weaknesses are clearly distinct from those of zombies, which cannot exploit horses as a source of transportation. Presumably, they might eat them as is the case in the first episode of AMC's adaptation of *The Walking Dead*.[7] Consequently, while tireless, zombies are unable to move as quickly as the Mongols. More importantly, even though zombies do appear to possess some rudimentary communication ability, they seem to lack group coordination, which is useful for tactical advantage and absolutely essential at the operational or strategic level of operations. Additionally, in spite of their reputation for indiscriminate slaughter, Mongols generally sought to conquer areas with their populations intact. As historian Timothy May notes, "the Mongols were not interested in ruling a desert devoid of life."[8]

Zombies do appear to possess certain advantages that the Mongols lacked, but these are double-edged. For example, Mongol rule was regularly challenged by local rebellion. Zombies do not have this problem due to the irreversible quality of zombie recruitment, but in practice this quality stiffens the resolve of those combating the zombies. The Mongols might spare you. The zombies will not.

In addition, while the process of zombie recruitment is demoralizing to their opponents, it appears to be largely incidental to zombie efforts to devour the flesh of the living. As a result, zombies often cease their attacks in order to eat their victims. This allows possible recruits to escape. Furthermore, since zombies often eat their victims whole, many of those who fall into their clutches are consumed rather than recruited. In addition, when people become zombies, they appear to lose all of their technical skills.

A unified and undifferentiated horde prevents infighting. Zombies occasionally scrabble with one another over chunks of flesh, but they are united in their ultimate goal of devouring every living thing on the planet. Mongols lacked this single-mindedness, and this proved to be a political problem. In terms of leadership, the Mongol Empire and the khanates that succeeded it all depended upon systems of dynastic succession, which led to periodic pauses in Mongol operations and eventually led to the fracturing of their empire.[9] Zombies do not appear to rely upon hierarchical leadership, which eliminates this weakness. Once again, what appears to be a strength for zombies may actually be a weakness because the lack of a zombie hierarchy greatly hampers their ability to coordinate forces strategically. Instead of sweeping forward in a carefully orchestrated effort, zombies shamble about. They are driven by stimulus response rather than coordinated planning.

The Second World War

Much shorter in duration than the Mongol Conquests, the Second World War was, nevertheless, exceptionally bloody. Given the geographically sprawling nature of the conflict, which may be considered to include the Sino-Japanese War (1937–45), it is difficult to estimate the total number of deaths; historians' estimates fall between 40 and 72 million people killed. While this number does represent a smaller proportion of the global population than the Mongol Conquests, the fact that all of this destruction took place over the course of less than a decade is astonishing.

This intensity of destruction was possible because of advances in military technology, the creation of the modern nation state, and the development of totalitarian ideologies. Together, these factors brought destruction to an even wider geographical expanse than the Mongol Conquests and involved—at least peripherally—almost all of the world's different peoples.

Once drawn into the conflict, all nations proved remarkably capable of turning civilians into soldiers. Their ability to conscript millions of their citizens led to the global scale of the war. Through the use of propaganda, military training, and public education the belligerent powers all instilled a willingness to kill members of opposing societies indiscriminately and without mercy. This was particularly pronounced in totalitarian states like Germany, Japan, and the Soviet Union. This meant that, as was the case with the Mongols, the militaries that perpetrated the vast majority of the death and destruction in these wars were remorseless and willing to kill anyone who lay between them and victory. New technologies often facilitated long-distance mass killing, but death continued to be dispensed at close range by highly-indoctrinated troops.[10]

As the war continued and intensified into a true "total war," the rising body count made ever greater levels of destruction politically possible. For example, at the outset of the conflict, most militaries and governments considered the strategic bombing of civilian centers an immoral act that violated the rules of war, but by 1945, bombing entire cities from high altitude, rather than targeting specific buildings, had become nearly unremarked upon. The atomic bombings of Hiroshima and Nagasaki dramatically illustrate the rapid evolution of doctrine towards indiscriminate mass destruction.[11]

Thus, in several ways, including conscription, indoctrination, and the development of military doctrine endorsing the mass killing of civilians, the ways in which World War II was fought bears some resemblance to zombie recruitment. Furthermore, no other conflict has been waged over such a geographically immense area, which suggests a close parallel to a zombie apocalypse.

As horrific and vast as this war was, its most severe effects were focused on specific regions, while large parts of the world suffered comparatively light casualties. Some of this was due to geographical isolation—oceans insulated the Americas—but it was also due to politics. Nations like Switzerland and Sweden neighbored areas ravaged by war, but they escaped unscathed because of their neutrality. This makes elements of the parallel forced. Zombies are not respecters of borders. In addition, while the war caused suffering on an unprecedented scale, these casualty figures amount to between 1.7 percent and 3.1 percent of the world population at the time, which is considerably less than the devastation wrought by the Mongol Conquests, and far below that experienced in most popular depictions of a zombie apocalypse.[12]

Therefore, when considering the Second World War as a model, one must focus on the areas that suffered most severely: China, the Soviet Union, Germany, and Eastern Europe. These areas all suffered immense destruction, but even in these areas, belligerents fought with goals typical of nation states; they sought conquest rather than devastation. Millions of people in these areas were murdered, starved, and enslaved, but even in areas overrun by death squads, fought over by armies deploying modern weapons, and controlled by nations that exhibited little concern about civilian deaths, most people survived. Some even continued to resist their conquerors through clandestine means, which clearly could not be the case with zombies. The "pretend we're zombies too" tactic may work in some situations for a short time, but it would be impossible for a sizable human population to move among zombies without detection in the long term. Consider Bill Murray's golf outing in *Zombieland* the exception that proves the rule.[13]

Given the level of destruction in the Second World War, one is struck by the fact that most societies caught up in the war survived. Every polity engaged in the war underwent substantial upheaval, but human organizations—particular nations and ethnic groups—proved remarkably resilient. This is cold comfort when confronting a zombie apocalypse, but it also suggests that one must explore other, even more devastating historical models of mass destruction.

The Black Death

Despite the destruction wrought by war, historically, far more people have been killed by disease. Indeed, many of the casualties of military conflicts are the victims of disease brought on by the stresses of war rather than to direct military action. The history of pandemics, then, would seem to be promising area of study.[14] Of all the lethal diseases that have ravaged humanity, in terms of massive body count, nothing equals the Black Death of the

14th century. The most devastating plague of all time, it caused between 75 million and 200 million deaths worldwide. Originating in China or Central Asia, it quickly advanced across South Asia, North Africa, and on to most of Europe. While the initial epidemic receded in the mid–14th century, these areas continued to suffer from the disease, which became a persistent menace for several centuries. Death rates due to the initial onslaught in Europe are now often estimated at something around 50 to 60 percent. Clearly, this is death on a scale unequaled by war alone.[15]

Even though it took years for the Black Death to move across Asia, Africa, and Europe, it advanced with what was great speed for the time. This means that like other catastrophes, it was not a truly global phenomenon, however, since 14th century communications tended to flow by word of mouth or hand-delivered letter, news of the plague was sometimes outpaced by the progress of the disease itself. This meant that, in some sense, each community endured its own Black Death experience; it became a sort of *rolling* global disaster. Consequently, most communities suffered the initial outbreak in a manner similar to a zombie apocalypse: a sudden, unexpected, and unexplainable onslaught of death. Furthermore, although large areas of the world (most notably, the Americas) did not suffer from the Black Death at this time, these areas did not contribute significantly to the rebuilding and repopulation of areas ravaged by the disease. No outside aid came to areas afflicted by plague.

Biologically speaking, the plague bacillus *Yersinia pestis* is driven only by the need to reproduce. It spreads through a variety of means. In the case of the septicemic and bubonic varieties it is carried by infected fleas or the bodily fluids of infected rodents. The pneumonic plague is even more insidious as it can be contracted by inhaling water droplets expelled by the lungs of the infected.[16] Even though the disease usually progresses rapidly towards death, humans often serve as vectors of the disease, but their actions indirectly aid the disease by hosting fleas or by transporting rodents as much as through coughing infected water droplets. Thus, in a manner similar to the growth of a zombie horde, spreading infection increases the strength of the destructive force of the plague.

As is the case with most depictions of a zombie apocalypse, the social stresses and depopulation that resulted from the sudden onslaught of an unprecedented, deadly threat led to the destabilization of the economy.[17] This led to behaviors that further destabilized society. Italian poet Giovanni Boccaccio described various reactions to the disease in his *Decameron,* which resemble popular understandings of the likely actions of people confronted with zombies. A common reaction described by Boccaccio seems a likely response: flight. As he explains, "They said there was no better medicine against the plague than to escape from it. Moved by this argument and caring for nothing except themselves, a large number of men and women abandoned their city, houses, families and possessions in order to go elsewhere."

Some combined flight with isolationism. Struck by the rapid advance of the disease, "they congregated and shut themselves up in houses where no one had been sick, partaking moderately of the best food and the finest wine, avoiding excess in other ways as well, trying their best not speak of or hear any news about the death and illness outside, occupying themselves with music and whatever other pleasures they had available."

Isolation was not always accompanied by moderation. As Boccaccio goes on to explain, "Others were of the opposite opinion. They believed that drinking a good deal, enjoying themselves, going about singing and having fun, satisfying all their appetites as much as they could, laughing and joking was sure medicine for any illness."

Those who isolated themselves often failed to keep the disease at bay through quarantine, but given the lack of understanding about how the disease was transmitted, it was a reasonable response—provided one had sufficient supplies and defenses. In many ways, attempts to survive by avoiding the disease met greater success than those who took forceful action against it. These counterproductive efforts featured scapegoating and zealous expressions of faith. In Western Europe, the fear and confusion sown by the plague often led to attacks against Jews, and in some areas, this led to what historian Robert Gottfried calls "a concerted and well-organized anti–Semitism that bordered on genocide." Other radical movements sought divine intercession through organized group self-flagellation.[18]

Less vicious but no more effective, some people, driven to desperation and totally lacking an understanding about how the plague was spread, "went about carrying flowers, fragrant herbs or various spices which they often held to their noses, assuming that the best thing for the brain was to comfort it with such odors, since the air was filled with the stench of dead bodies and illness and medicine."[19] Given the straightforwardness of the zombie threat, it seems unlikely that many people would turn to herbal sachets as potential solutions to the problem, but flight, isolationism, scapegoating, and religious zeal all seem likely based on historical evidence.

In addition to allowing one to see this wide variety of responses to an apparent apocalypse, the Black Death is useful as a test of social resilience. Somewhat surprisingly, despite the massive death toll and the disruption of political, cultural, and economic institutions, societies ravaged by the Black Death did not collapse. Most institutions temporarily suspended operations, but their authority was usually quickly restored. Even in the midst of the plague, courts met, churches held services, taxes were collected, and wars continued to be fought.[20]

This reflects an observation made by economist Jack Hirshleifer that the experience of populations "suddenly struck by disaster does not include the wild, asocial behaviour described by the more lurid popular writers on such themes." As he explains, "panic does not ordinarily occur. Survivors first reorient and extricate themselves, and then their families. Some, even when seriously injured themselves, assist others." He qualified this altruism by noting, "If there is reason to fear another hazard (explosion, spreading fire, renewed bombing, etc.), there may be hasty flight." He acknowledges that some people slip into non-responsive shock in response to a crisis, but overall people take charge and they do so by cooperating with other survivors rather than by selfishly murdering one another. As he explains, "all this is rational behavior."[21] Groups are more survivable than individuals.

This tendency towards community survival is reflected in the history of the Black Death. J.N. Hays, a scholar of disease, points out that in Northern Italy, "governments began taking a more active role in resisting disease, and their public health bureaucracies, quarantines, and isolation facilities may ultimately have diverted the second pandemic from their borders."[22] Most officials did not flee to the hills or board themselves up in their palazzos. Instead, they acted upon an ongoing desire to contend with the enormous challenges confronting them.

Popular representations of zombies frequently describe a collapse of public services, food and water supplies, public heath, and law and order. This is invariably accompanied by a downward spiral into a Hobbesian nightmare of hyper-individuality, vigilantism, and violence. The history of the Black Death suggests that this is another manifestation of the "lurid popular writers" derided by Hirshleifer. Yet, even in these depictions one can see the resilience of government and social organizations. Consider the arrival of the military at the end of *Shaun of the Dead* or the zombie-hunting vigilantes that appear in the conclusion of *Night of the Living Dead*.[23]

In all of these cases, those who survived the initial wave of mass destruction began to develop countermeasures against the zombie threat. Even though some people initially struggle with the concept of zombies, the vast majority quickly adapt because the nature of the threat is so straightforward. Diseases mutate. Human combatants modify their behavior. Zombies, however, are so intent on devouring humanity that the nature of their threat remains essentially unchanged. Their relentless bloodthirstiness makes their actions predictable and understandable. When a zombie is trying to eat your face, it clearly identifies itself as a hostile force. When it "stays dead" as the result of a headshot or incineration, future tactics become clear. The possibility of reactive zombies with learned behavior is chilling, but it is also quite unusual.[24] They are vicious, yet simple adversaries.

The Columbian Exchange

The most devastating wars and plagues in history inflicted horrific damage on human populations, but considering all the factors described above it remains difficult to use them to understand how societies and individuals might function during a zombie *apocalypse*. An important part of this problem is due to the nature of zombies, but it also relates to the level of devastation popularly associated with a zombie apocalypse. Most films, books, and graphic novels feature the reduction of civilization to handfuls of survivors.

Historically, this level of death has rarely been reached on a large scale. Given the capacity for destruction possessed by organized, industrialized, and competently led nation states during the Second World War and the mass death resulting from poor sanitation and misunderstood diseases in the medieval era, it is difficult for historians to imagine how such a situation might develop—particularly in the modern era.

The process by which the onslaught of zombies causes society to collapse is often elided in popular depictions of zombies. The steps by which the initial rise of the zombies leads to the extermination of most of humanity are rarely explored. A popular narrative arc follows a protagonist who simply *sleeps* through these critical stages. Both *The Walking Dead* and *28 Days Later* begin with protagonists waking from comas well after civilization has fallen.[25]

If we put the mechanics of apocalypse aside and focus on the impact of the destruction itself, there may still be events in human history that *combine* the onset of multiple deadly diseases with ruthless military aggression to reach a level of social destabilization that actually resembles that of a zombie apocalypse. The largest of these is the European conquest of the Americas. Even though it occurred over centuries, it provides significant insight into the social stresses that accompany the massive casualties inflicted by an array of insatiable and unprecedented threats.[26]

The combination of disease and military operations in this centuries-long conquest, first by several western European nations and then by the independent nations that succeeded them, led to the enslavement, colonization, and destruction of numerous peoples. The specific experience of different indigenous populations to European invasions varies significantly and, in some cases, lacks substantial historical documentation, but several examples show levels of death and destruction that far exceed those of the Black Death. In a number of cases, they essentially resulted in the total destruction of entire peoples.

During the conquest, smallpox, measles, influenza, and typhus (along with other diseases) accounted for millions of deaths. These diseases regularly wracked Europe, but when brought to the Americas, the devastation was much worse because of the lack of immunities (acquired or genetic) among indigenous populations; this meant that pathogens struck with great severity. Disease spread rapidly and in some cases it preceded European explorers and settlers. As a result, many areas were depopulated before Europeans even arrived.

As anthropologist Jared Diamond explains, "The winners of past wars were not always the armies with the best generals and weapons, but were often merely those bearing the nastiest germs to transmit to their enemies." He adds, "Far more Native Americans died in bed from Eurasian germs than on the battlefield from European guns and swords."[27]

This host of diseases proved devastating, and, of course, Europeans also brought war. Horses, firearms, and steel weapons provided important advantages, but in the long term their greatest strength was numbers. Indigenous populations, reeling from repeated eruptions of disease, resisted mightily, but usually were eventually overcome. Although rarely explicitly genocidal in intent, European wars of conquest were nonetheless often genocidal in effect. In many cases, the precipitous drop in indigenous populations as a result of the combination of war and disease led to the collapse of many native societies. For example, central Mexico had a population of around 20 million when the conquistador Hernán Cortés arrived in 1519. As Diamond notes, a century later, the population had "plummeted to about 1.6 million."[28]

This begins to resemble the level of devastation in a zombie apocalypse, but the parallels are weak in term of the nature of the threats. As was the case with the Mongol Invasions, the Second World War, and the Black Death, the problems of historical specificity make comparisons difficult to draw. Conquistadors, smallpox, and the U.S. cavalry do not have a great deal in common with one another, let alone with zombies. Furthermore, as is the case with the Mongol Invasions and World War II, the key perpetrators are human. Thus, they lack the utter ruthlessness of zombies. While the vast majority of Europeans advocated the cultural assimilation of native peoples, most opposed outright genocide.

Despite these shortcomings, if one steps back from the details, the Columbian Exchange remains useful when assessing impact because, in the end, the zombie-like Europeans, motivated by greed, racism, Manifest Destiny, and Christian evangelicalism (among other things) essentially win. In the Americas, only a few small areas survive in which there is no European influence. Almost all the indigenous societies that survived have done so by rebuilding themselves along new cultural and social lines. Certainly, some elements of pre–Columbian culture have endured, but the vast majority of native societies have been forced to adapt to the presence of European peoples and institutions; much of the pre–Columbian world has been lost forever. So, when faced with this most zombie-like of historical threats, many native peoples survived, but only by almost totally altering their cultures, societies, politics, economies, and ways of life.

Commonalities

How might these historical experiences inform our understanding of the destructive potential of a zombie apocalypse? Despite the difficulties in drawing precise historical parallels, there are definitely some commonalities.

All of these historical events feature relentless and lethal agents that are more than willing to kill millions of people in order to achieve their objectives. In every case, this led to widespread, indiscriminate death. Wealth, gender, and class do not appear to offer any particular protection when this sort of mass destruction is at hand. The general trend appears to be: when disaster strikes, everyone is at risk. When confronting a threat of this type, appeasement and isolationism do not work; anticipating some sort of preferential treatment from the threatening force—human, pathogen, or zombie—is a good way to die.

When faced with this sort of opponent, it is imperative to fully mobilize your population in order to face the threat. In World War II, the Allied Powers eventually defeated the Axis because they shifted to a strategy of total war. During the Black Death, ruthlessly quarantined cities appear to have experienced lower death rates.[29] Even in the case of the conquest of the Americas, indigenous groups that dedicated themselves to resistance were able to defeat the forces arrayed against them—sometimes.[30]

History also shows that in terms of mobilizing society, there is no equal to the organization, resources, and social unity provided by modern nation states. Consequently, strengthening your affiliation with one in the midst of a zombie apocalypse offers a number of advantages. First, it offers community. Given the level of devastation one expects, the opportunity to build and maintain social alliances with strangers is extremely important. Furthermore, given the organizational resources available to governments, they probably have access to food, water, and medical care. In addition, governments—with their militaries and research facilities—likely possess the deepest well of resources for fighting the zombie menace. After all, the compiler of *World War Z* works for the United Nations Postwar Commission.[31]

Some may be reluctant to affiliate with governments because they fear tyranny and the use of force, but these cases show that in times of great crisis, coercion has utility. Quarantine laws may seem unethical to those who are caught on the wrong side of them, but in terms of maximizing the number of survivors, they are effective. Conscription may violate one's rights, but given the need for the full mobilization of society's resources to face the zombie threat, the potential for this sort of coercion also offers a reasonable tradeoff of rights for security. For a meditative consideration of this tradeoff consider the plight of the "sweeper" character Mark Spitz in Colson Whitehead's novel *Zone One*.[32]

Isolated families and communities did survive all of the historical episodes described above, but this owed more to luck the than anything else. Small communities, however tightly-knit, will also find it extremely difficult to defend themselves against a zombie onslaught. Similarly, despite its romantic allure, the "Lone Wolf" approach to resistance is even more unwise. It may enable a particularly gifted individual to survive an initial onslaught, but the need for supplies, shelter, sleep, and companionship will eventually erode the ability of even the most determined survivor to "go it alone." Even the omni-competent protagonist in J.L. Bourne's *Day by Day Armageddon* needs a friend.[33]

The immensity of suffering, the large body count, and the travails of survival will certainly alter the shape of the societies that endure, but in every historical case, humanity proved remarkably resilient in the face of disaster. Even when confronting what appeared to be the end of the world, people did their best to adapt to the challenges and they did so as communities. So, in addition to increasing chances of survival, affiliating with societies increases the speed of eventual recovery.[34]

NOTES

1. "March 11, 2011 Japan Earthquake and Tsunami," National Geophysical Data Center, National Oceanic and Atmospheric Administration, 9 March 2012, www.ngdc. noaa.gov/hazard/honshu_11mar2011.shtml, accessed 12 July 2014.

2. Philip Munz, Ioan Hudea, Joe Imad, Robert J. Smith, "When Zombies Attack! Mathematical Modelling of an Outbreak of Zombie Infection," *Infectious Disease Modelling Research Progress*, J.M. Tchuenche and C. Chiyaka, eds. (New York: Nova Science, 2009), 133–150.

3. David Morgan, *The Mongols* (Cambridge, MA: Blackwell, 1986), 74.

4. Morgan, 17–25, 84–85.

5. Timothy May, *The Mongol Art of War* (Yardley, PA: Westholme, 2007), 4.

6. May, 4, 12–16.

7. "Days Gone Bye," *The Walking Dead*, AMC, 31 Oct. 2010.

8. Quotation from May 119. See also: Morgan, 65, 82, 93, 137–138.

9. Morgan, 82.

10. Richard Rhodes, *Masters of Death: The SS-Einsatzgruppen and the Invention of the Holocaust* (New York: Vintage, 2003); Christopher R. Browning, *Ordinary Men: Reserve Police Battalion 101 and the Final Solution in Poland* (New York: HarperCollins, 1992).

11. Herman Knell, *To Destroy a City: Strategic Bombing and Its Human Consequences in World War II* (Cambridge, MA: DaCapo Press, 2003); A.C. Grayling, *Among the Dead Cities: The History and Moral Legacy of the WWII Bombing of Civilians in Germany and Japan* (New York: Walker, 2006).

12. Scholarship on the Second World War is vast, but several new one volume

surveys of the conflict provide an excellent place to start. Andrew Roberts, *The Storm of War: A New History of the Second World War* (New York: Harper Perennial, 2012); Anthony Beevor, *The Second World War* (New York: Little, Brown, 2012); Max Hastings, *Inferno: The World at War, 1939–1945* (New York: Vintage, 2011).

13. Timothy Snyder, *Bloodlands: Europe Between Hitler and Stalin* (New York: Basic Books, 2010); S.C.M. Paine, *The Sino-Japanese War of 1894–1895: Perceptions, Power, and Primacy* (Cambridge: Cambridge University Press, 2003); *Zombieland*, Ruben Fleischer, dir., Columbia Pictures, 2009.

14. Robert S. Gottfried, *The Black Death: Natural and Human Disaster in Medieval Europe* (New York: Free Press, 1983); J.N. Hays, *Epidemics and Pandemics: Their Impacts on Human History* (Santa Barbara: ABC-CLIO, 2005); John Kelly, *The Great Mortality: An Intimate History of the Black Death, the Most Devastating Plague of All Time* (New York: HarperCollins, 2005).

15. Ole J. Benedictow, *The Black Death 1346–1353: The Complete History* (Woodbridge: Boydell Press, 2004); John Aberth, *Plagues in World History* (Lanham, MD: Rowman & Littlefield, 2011), 33–61.

16. Kelly, 11–27. Centers for Disease Control website on Plague "Ecology and Transmission," www.cdc.gov/plague/transmission/index.html, accessed 15 Feb. 2013.

17. Gottfried, 51, 58, 91, 135–137.

18. Gottfried, 52, 73–74, 87–88. Quotation from 73.

19. All Boccaccio quotations are taken from Giovanni Boccaccio, *The Decameron*, Introduction, David Burr, trans., www.history.vt.edu/Burr/Boccaccio.html, accessed 15 Feb. 2013. For analysis of Boccaccio's description of the Black Death, see Gottfried, 47, 78–80.

20. Gottfried, 50, 94–97, 103, 135–138, 151–152; William H. McNeill, *Plagues and Peoples* (New York: Anchor Press, 1976), 186–191.

21. Jack Hirshleifer, *Economic Behaviour in Adversity* (Chicago: University of Chicago Press, 1987), 9–10.

22. Hays, 50; Sheldon Watts, *Epidemics and History: Disease, Power and Imperialism* (New Haven: Yale University Press, 1997), 15–25, 37.

23. *Night of the Living Dead*, George Romero, dir., Image Ten, 1968; *Shaun of the Dead*, Edgar Wright, dir., Universal Pictures, 2004.

24. Angela Watercutter, "Zombie Neuroscientist Explains the Ant-Like Behavior of *World War Z's* Running Dead," *Wired*, 9 Nov. 2012; *Land of the Dead*, dir. George Romero, Universal Pictures, 2005.

25. *28 Days Later*, Danny Boyle, dir., DNA Films, 2003. Robert Kirkman, *The Walking Dead: Compendium One*, illus. Charlie Adlard, Cliff Rathburn, and Tony Moore (Berkeley, CA: Image Comics, 2009).

26. Alfred W. Crosby, *Ecological Imperialism: The Biological Expansion of Europe, 900–1900* (Cambridge: Cambridge University Press, 1986); Alfred W. Crosby, Jr., *The Columbian Exchange: Biological and Cultural Consequences of 1492* (Westport, CT: Greenwood Press, 1972); Jared Diamond, *Guns, Germs, and Steel: The Fates of Human Societies* (New York: W.W. Norton, 1997); McNeill, 199–234; Watts, 84–93, 99–102.

27. Diamond, 197, 210.

28. Diamond, 210.

29. Wendy Orent, *Plague: The Mysterious Past and Terrifying Future of the World's Most Dangerous Disease* (New York: Free Press, 2004), 232–233.

30. For example, see S.C. Gwyne, *Empire of the Summer Moon: Quanah Parker and the Rise and Fall of the Comanches, the Most Powerful Indian Tribe in American History* (New York: Scribner, 2011).

31. Max Brooks, *World War Z: An Oral History of the Zombie War* (New York: Crown, 2006).

32. Colson Whitehead, *Zone One* (New York: Anchor Books, 2012).

33. J.L. Bourne, *Day by Day Armageddon* (New York: Pocket Books, 2009).

34. Lawrence J. Vale and Thomas Campanella, eds., *The Resilient City: How Modern Cities Recover from Disaster* (Oxford: Oxford University Press, 2005); Kevin Rosario, *The Culture of Calamity: Disaster and the Making of Modern America* (Chicago: University of Chicago Press, 2007).

Bibliography

Aberth, John. *Plagues in World History.* Lanham, MD: Rowman & Littlefield, 2011.

Beevor, Anthony. *The Second World War.* New York: Little, Brown, 2012.

Benedictow, Ole J. *The Black Death 1346–1353: The Complete History.* Woodbridge: Boydell Press, 2004.

Boccaccio, Giovanni. *The Decameron,* Introduction. David Burr, trans. www.history. vt.edu/Burr/Boccaccio.html. February 15, 2013.

Bourne, J.L. *Day by Day Armageddon.* New York: Pocket Books, 2009.

Brooks, Max. *World War Z: An Oral History of the Zombie War.* New York: Crown, 2006.

Browning, Christopher R. *Ordinary Men: Reserve Police Battalion 101 and the Final Solution in Poland.* New York: HarperCollins, 1992.

Centers for Disease Control. "Plague Ecology and Transmission." www.cdc.gov/plague/transmission/index.html. February 2, 2013.

Crosby, Alfred W. *Ecological Imperialism: The Biological Expansion of Europe, 900–1900.* Cambridge: Cambridge University Press, 1986.

Crosby, Alfred W., Jr. *The Columbian Exchange: Biological and Cultural Consequences of 1492.* Westport, CT: Greenwood Press, 1972.

"Days Gone Bye." *The Walking Dead.* AMC. October 31, 2010.

Diamond, Jared. *Guns, Germs, and Steel: The Fates of Human Societies.* New York: W.W. Norton, 1997.

Gottfried, Robert S. *The Black Death: Natural and Human Disaster in Medieval Europe.* New York: The Free Press, 1983.

Grayling, A.C. *Among the Dead Cities: The History and Moral Legacy of the WWII Bombing of Civilians in Germany and Japan.* New York: Walker, 2006.

Gwyne, S.C. *Empire of the Summer Moon: Quanah Parker and the Rise and Fall of the Comanches, the Most Powerful Indian Tribe in American History.* New York: Scribner, 2011.

Hastings, Max. *Inferno: The World at War, 1939–1945.* New York: Vintage, 2011.

Hays, J.N. *Epidemics and Pandemics: Their Impacts on Human History.* Santa Barbara: ABC-CLIO, 2005.

Hirshleifer, Jack *Economic Behaviour in Adversity.* Chicago: University of Chicago Press, 1987.

Kelly, John. *The Great Mortality: An Intimate History of the Black Death, the Most Devastating Plague of All Time.* New York: HarperCollins, 2005.

Kirkman, Robert. *The Walking Dead: Compendium One.* Illustrated by Charlie Adlard, Cliff Rathburn, and Tony Moore. Berkeley, CA: Image Comics, 2009.

Knell, Herman. *To Destroy a City: Strategic Bombing and Its Human Consequences in World War II.* Cambridge, MA: DaCapo Press, 2003.

Land of the Dead. Directed by George Romero. Universal Pictures. 2005.

"March 11, 2011 Japan Earthquake and Tsunami." National Geophysical Data Center,

National Oceanic and Atmospheric Administration, March 9, 2012. www.ngdc. noaa.gov/hazard/honshu_11mar2011.shtml. July 12, 2014.

May, Timothy. *The Mongol Art of War.* Yardley, PA: Westholme, 2007.

McNeill, William H. *Plagues and Peoples.* New York: Anchor Press, 1976.

Morgan, David. *The Mongols.* Cambridge, MA: Blackwell, 1986.

Munz, Philip, Ioan Hudea, Joe Imad, and Robert J. Smith. "When Zombies Attack! Mathematical Modelling of an Outbreak of Zombie Infection." *Infectious Disease Modelling Research Progress,* eds.: J.M. Tchuenche and C. Chiyaka. New York: Nova Science, 2009, 133–150.

Night of the Living Dead. Directed by George Romero. Image Ten, 1968.

Orent, Wendy. *Plague: The Mysterious Past and Terrifying Future of the World's Most Dangerous Disease.* New York: Free Press, 2004.

Paine, S.C.M. *The Sino-Japanese War of 1894–1895: Perceptions, Power, and Primacy.* Cambridge: Cambridge University Press, 2003.

Rhodes, Richard. *Masters of Death: The SS-Einsatzgruppen and the Invention of the Holocaust.* New York: Vintage, 2003.

Roberts, Andrew. *The Storm of War: A New History of the Second World War.* New York: Harper Perennial, 2012.

Rosario, Kevin. *The Culture of Calamity: Disaster and the Making of Modern America.* Chicago: University of Chicago Press, 2007.

Shaun of the Dead. Directed by Edgar Wright. Universal Pictures, 2004.

Snyder, Timothy. *Bloodlands: Europe Between Hitler and Stalin.* New York: Basic Books, 2010.

28 Days Later Directed by Danny Boyle. DNA Films, 2003.

Vale, Lawrence J., and Thomas Campanella, eds. *The Resilient City: How Modern Cities Recover from Disaster.* Oxford: Oxford University Press, 2005.

Watercutter, Angela. "Zombie Neuroscientist Explains the Ant-Like Behavior of *World War Z's* Running Dead." *Wired,* November 9, 2012.

Watts, Sheldon. *Epidemics and History: Disease, Power and Imperialism.* New Haven: Yale University Press, 1997.

Whitehead, Colson. *Zone One.* New York: Anchor Books, 2012.

Zombieland. Directed by Ruben Fleischer. Columbia Pictures, 2009.

Looking to the CDC
and WHO for Answers

AMY L. THOMPSON

People look to the Centers for Disease Control and Prevention (CDC) in the United States and the World Health Organization (WHO) headquartered in Geneva, Switzerland, to provide the answers to numerous different biomedical questions, especially during times of crisis, such as disease outbreaks like Ebola or H1N1. People do this in an unwavering, almost faith-like manner with most citizens having no real reason to believe that the CDC or WHO will fail to deliver the best and most useful scientific information. In reality, Americans put so much belief in the CDC that when the organization announced its "Preparedness 101: Zombie Apocalypse" campaign in 2011, some U.S. citizens actually believed that the zombie threat was real. This coupled with the fact that the CDC website was so overloaded that it was inaccessible made some believe that a zombie apocalypse was in the early stages. A writer for the *Huffington Post* stated that "the CDC's awareness campaign may have been so effective it's rendered itself ineffective."[1] In 2012 several incidents involved individuals behaving in zombie like ways.[2] Examples included a man eating the organs of his housemate in Maryland and another man gnawing on a homeless man's face in Miami. People again became paranoid over the possibility of a zombie apocalypse and the CDC issued a statement that zombies are not real.[3] While citizens may seem irrational for believing in zombies, necessitating the issuance of a CDC statement to dispel this myth, it proves an important point. People trust the CDC because of its reputation to provide sound advice and warnings, in addition to vaccines, treatments, and cures. Several popular culture zombie references even include science and the work of organizations such as the CDC and the WHO. Zombies and the zombie apocalypse may destroy the world's infrastructure and even the CDC building, as seen on the AMC zombie series *The Walking Dead*, but man's faith in science remains strong. While many Americans are aware of the CDC's location in Atlanta, Georgia, most really do not understand the abilities or limitations of this organization or science in general. Does Frank Borelli, the author of the book *Surviving the Zombies: Things the CDC Didn't Know*, have a better understanding of the limitations of the CDC, as the name of his book implies or are citizens (real and imaginary) correct in their belief that the CDC and WHO are the organizations to look to when it comes to understanding infectious disease and stopping illness from spreading?

Placing Trust in the CDC Based on Its Successes

The Centers for Disease Control and Prevention (CDC) originated in 1946 as the Communicable Disease Center, with its mission to control malaria. Malaria is caused by the *Plasmodium* parasite found in infected mosquitoes and transferred to the red blood cells of humans through a mosquito bite. Three short years after the Communicable Disease Center's inception, malaria was no longer a public health risk in the United States, attesting to the ability of this new agency's effectiveness. This early success in removing the threat of malaria for most Americans likely instilled trust in and support for the organization. This is especially important since malaria continues to be a threat outside the U.S. and is especially prevalent on the continents of South America, Africa, and Asia, although the disease is also in Mexico in North America, even in areas of Mexico bordering Texas.[4]

Following its initial success with malaria, the Communicable Disease Center continued to expand and started working with rabies, polio, and cholera, establishing best practices to control these diseases. Rabies in the United States, for example, is now typically only seen in wildlife, such as bats and raccoons, with humans and pets most often contracting the virus through bites by infected and most often wild animals.[5] This is in stark contrast to India, where citizens know little about rabies, stray animals are common, and when children are bitten, little if anything is done to treat the wounds.[6] In the United States, the rabies virus, part of the *Rhabodviridae* family, which also includes several bat viruses, is largely controlled in domestic animals through annual vaccination and exposure to the virus though bites can be treated via a series of shots.[7] Most Americans have been educated on what to look for with potential rabies infections and even children are taught by their parents to avoid unknown animals or those that are behaving strangely.

Some zombie literature suggests that the rabies virus might be a possible causative agent of zombie behavior. If this was the case, American citizens, educated by the CDC on identifying potential rabies infected creatures would most likely stay away from the strangely behaving individual, taking all means necessary to avoid being bitten. In this regard, the CDC has done a great job in its educational outreach.

The CDC additionally focused on polio during its early days. Poliomyelitis or polio, caused by the virus family *Picornaviridae*, is also controlled through vaccinations.[8] These vaccines are administered by medical personnel worldwide and are the result of the work of scientists Jonas Salk (inactivated polio vaccine in 1950s) and Albert Sabin (oral polio vaccine in 1960s). Young children are most susceptible to the polio virus through fecal-oral transmission or by consuming contaminated food or water, although anyone can contract the virus. The virus causes symptoms such as headache, fever, vomiting, fatigue, limb and neck pain, and even permanent paralysis.[9] Although no cure has been identified, polio has been eradicated in the U.S. since 1979, another testament to the CDC's work.[10] Worldwide efforts are in place to eradicate polio elsewhere, with a small number of wild polio virus cases still found primarily in the Middle East and several African nations.[11] The WHO projects that with continued vaccination efforts, polio can be eradicated worldwide by 2018. Clearly, the work of the CDC and WHO has dramatically impacted the prevalence of polio on national and international levels, again attesting to the knowledge and know-how of these agencies and the significant impact that science can have on humanity.

Cholera, caused by the bacterium *Vibrio cholera*, was another initial disease studied by the CDC. This organism is not a virus, but does spread most often through contaminated water like poliovirus and is transmitted through fecal-oral contact.[12] Cholera outbreaks are less of a problem in the U.S. due to the availability of clean, safe water, although some cases occur each year.[13] Worldwide, cholera outbreaks and deaths continue to happen especially in areas where water quality is compromised. Vaccines are available to help and the CDC and WHO continue to focus on reducing outbreaks.[14] Most Americans have no major reason to be concerned about cholera, but following a disaster situation, such as a zombie apocalypse, when water quality becomes questionable, cholera outbreaks could become a problem. Municipal power outages and lack of manpower to quality control test water could result in water that has dangerous levels of disease causing bacteria, potentially including the organism that causes cholera.

To communicate outbreaks, such as those caused by cholera or other public health concerns, the CDC created a National Surveillance Program in 1952 that is still in existence today. This program monitors diseases and potentially impactful medical problems and when needed releases early warnings. The CDC's *Morbidity and Mortality Weekly Report* (MMWR), which began in 1961, provides timely updates regarding emerging and ongoing issues. To ensure that the results released are the best possible, the CDC employs scientists that have the expertise and training to work with the most dangerous and deadly biological agents.[15] This work is done in expansive Biosafety Level (BSL) facilities, including BSL 4 facilities (on a scale from BSL 1–4), which are the highest level facility and require the most stringent safety rules and regulations. The CDC has set the standard for handling biological organisms with most researchers and clinical scientists following the BSL guidelines established by the CDC. This again testifies to the organization's impact on science throughout America and the world.

The scope of the CDC is not limited to infectious disease, however, as its work extends to other areas of public safety. It was the CDC that reported on the dangers of lead, which resulted in many buildings and houses being treated for lead-based paint. They also identified the link between aspirin and Reye syndrome leading to the warning against giving aspirin to children and ultimately resulting in the name change from baby aspirin to low dose aspirin. The surgeon general, the CDC's national spokesperson, told of the risks associated with smoking and tobacco use and citizens are reminded of these risks on each pack of cigarettes or tobacco product. Although the average smoker may not appreciate the warnings found on his or her tobacco products, he or she likely does not wonder about the legitimacy of the message provided by the surgeon general. Education and information leads to change. No parent questions whether the CDC was right in linking aspirin use in children to Reye syndrome or in describing the dangers of lead. They instead use ibuprofen or acetaminophen for their children and avoid products containing lead-based paint.

Although the organization's name changed from the Communicable Disease Center in 1967 to the National Communicable Disease Center and in 1970 to the current name, CDC, the focus has remained on the well-being of American citizens encouraging vaccination and discussing national health. Through its work with malaria, polio, rabies, and cholera, among other diseases, the CDC has established itself as a reputable and proactive organization that works to protect mankind. For this reason, Americans put their trust in the CDC and this trust has been earned and maintained through continued successes.

This trust placed in the CDC and on a worldwide scale the WHO is apparent in the zombie genre. In the AMC's *The Walking Dead*, the survivors believe that if they can just make it to the CDC, they will be saved.[16] This allows the characters to stay hopeful as they fight off zombies in their travels from rural Georgia to downtown Atlanta. Although amenities like hot showers, safe areas to sleep, and new books to read initially make the trying trip to the CDC building worthwhile for the group, this renewed faith is quickly lost upon finding that the strength of the organization rests on the shoulders of one unstable scientist, Edwin Jenner. Jenner's character is a nod to Edward Jenner, the scientist who developed the smallpox vaccine. Edwin Jenner does not live up to his big name, however, as the group learns that the power in the building is failing and that they are all destined to die if they do not exit the locked building and return to the zombie-laden streets from which they came. Jenner ultimately commits suicide by staying in the facility after activation of the CDC's high-impulse thermobaric (HIT) decontamination system, which results in an explosion that destroys the CDC in its entirety. While Jenner demonstrated questionable stability, the group did not doubt the validity of his zombie virus research or dispute Jenner's message delivered through group leader Rick Grimes, that everyone is infected with the zombie virus. The following seasons have not further explored this idea other than to show several dead, but not obviously bitten, people turn into zombies. No information has been provided regarding how everyone became infected, including the route of infection or if one who has never encountered a zombie or infected individual might in fact, not be infected. It was simply believed from that point on in the show that everyone is infected based on the trust that the group placed in science and the CDC, if not Edwin Jenner.

Scientific faith is also demonstrated by the psychopathic character, known as Philip Blake, aka The Governor, again in *The Walking Dead* series.[17] He believed that his resident scientist, Milton Mamet, of small Woodbury, Georgia, would discover how to communicate with zombies. This was most likely not to save mankind, but rather so that he could communicate with his own zombie daughter, Penny. Another character, Andrea, did not have faith in Milton's science and even pointed out potential flaws in his theory based on her own "field" research. Ultimately, all three characters died and zombie Penny was "killed" as well, leaving a big hole in this aspect of zombie research.

No further mention of science occurs until the show introduced Eugene Porter, a scientist making his way to Washington, D.C., because he has the answer to the zombie mystery.[18] Group members again look to science as the key to their future, never questioning the credentials of Eugene as it was explained that Eugene had been in contact with officials in Washington, D.C., via a satellite phone. In this sense Washington, D.C., becomes the new mecca in their science faith and the group will stop at nothing to reach this research haven. As was seen with Jenner and the CDC, blind faith in science again trumps rational skepticism.

Is there reason for man in real life to be skeptical of science? Certainly, the word of one researcher or research group or scientific study should not be sufficient to cause widespread panic or deviation from standard protocols. True scientific data is supported by facts and can be replicated by other appropriately trained scientists in other labs with the necessary equipment. Evidence to support the conclusions must be provided from a study population that is large enough from which to draw statistically significant conclusions. While the data of one study may be significant and impactful, in that it challenges what is already known or accepted, this should never lead to rash changes in how things are done. Rather one should ask more questions and demand further study. It is also important to place scientific data in the greater body of information already known. Finally, the credentials of the scientists conducting the studies must be examined.

Certainly, in a zombie apocalypse there are plenty of zombie test subjects, although materials to conduct and replicate experiments and qualified scientists to do this research may be less readily available, bringing into question any research coming out of a zombie apocalypse. What does one really know about the zombie data generated by Edwin Jenner as it relates to the data generated by others at the CDC or around the world? Is it possible that Milton's work only showed promise because it was keeping The Governor happy with him and bad data might lead to zombie Milton? Did anyone suspect Eugene Porter's "fight fire with fire cure" was a hoax for personal preservation?[19] Who really knows if Edwin Jenner or Milton Mamet are even reputable scientists or scientists at all? Does anyone check credentials during a zombie apocalypse?

Real World Implications

While the CDC and WHO work together with other international organizations, employ scientists with top credentials, and operate the best facilities with cutting edge resources, there are still limitations to what these organizations can achieve. The 2014 Ebola virus outbreak in several African countries and eventually resulting in cases in the United States demonstrated the coordinated worldwide effort to combat a disease, but also underscored the constraints of such organizations. Ebola continues to test these limitations with many questions about this virus remaining unanswered. For example, the manner in which the virus is transmitted to humans and its natural reservoir, likely the bat, is unknown. It has been suggested that exposure to infected body fluids is the main cause of virus spread prompting scientists and medical personnel to operate under this assumption. An established plan to control spread was not initially in place in areas where the infected numbers were high and developing a plan met with resistance as citizens feared consequences for reporting suspected Ebola cases, such as quarantine away from family and friends. To compound the Ebola problem, the lack of an approved vaccine to protect those not infected, including healthcare workers, and the absence of an approved and readily available drug to treat diseased patients caused containment issues and further deaths.

The need for an approved Ebola vaccine and treatment options left scientists and medical personnel scrambling to try to make use of what was available or already in the works. One drug, ZMapp, seemed to hold promise and was given to the first Americans infected. This drug, however, was in the early stages of clinical trials and prior to the major outbreak had not been tested in sick patients.[20] Due to very limited doses of this drug, many Ebola patients did not receive it. Vaccine development for Ebola proved equally problematic. The National Institutes of Health (NIH) and Food and Drug Administration (FDA) had been working on a vaccine against the Zaire and Sudan strains of Ebola for more than ten years.[21] The urgency of the outbreak situation pushed these two organizations to couple with drug company GlaxoSmithKline (GSK) to easily scale up production of the vaccine, should clinical trials prove effective. Another vaccine against the Zaire strain of Ebola studied by the NIH and a third vaccine created by the Public Health Agency of Canada also hold promise. Although the CDC did not develop these vaccines, the organization is involved in vaccine testing in Africa, the area of the world in most need of preventative medicine against Ebola. All three vaccines must be put through clinical trials to evaluate effectiveness and safety, however, delaying their potential widespread use for several months.

Clearly, the CDC, like similar organizations in the rest of the world including the WHO, was not ready for an Ebola outbreak of this magnitude. For much of the 2014 Ebola outbreak in West Africa the CDC referred healthcare workers to a more general recommendation flyer titled "Interim Guidance for Managing Patients with Suspected Viral Hemorrhagic Fever in U.S. Hospitals" dated May 19, 2005. Without specific guidelines, patients with the deadly Ebola virus would be treated the same as any other patient with viral hemorrhagic fever.[22] This is not to say that the guidelines were not helpful or applicable and they might even be useful during a zombie virus outbreak, but it is important to point out that they were not specific for Ebola. On August 1, 2014, the CDC, faced with the possibility of healthcare workers treating the Ebola virus in U.S. hospitals, released new guidance titled "Infection Prevention and Control Recommendations for Hospitalized Patients with Known or Suspected Ebola Hemorrhagic Fever in U.S. Hospitals." These guidelines required patient isolation, dedicated equipment that is disposable when possible, strict personal protection equipment requirements, and restrictions on aerosol generating procedures and procedures in general, especially those requiring needles. Environmental infection control measures outlined ways to clean hard surfaces, linens, and other potentially infected items and recommended standard precautions. Standard precautions are basic practices designed to prevent the spread of disease. The guidelines also placed limitations on visitors and visitor contact with others after coming into contact with an Ebola infected individual. These new 2014 Ebola guidelines differed significantly from the more generic guidance offered in 2005, attesting to the danger of this virus and renewing the CDC's vigilance in attempting to prevent and control the spread of Ebola.

On August 2, 2014, the new August 1, 2014, recommendations were put into place when U.S. physician Dr. Kent Brantly returned to the United States from charity medical work in West Africa infected with the Ebola virus. He was treated at Emory University in Atlanta, Georgia. Emory, just a few miles from the CDC, is one of four U.S. hospitals specially equipped with necessary biomedical personal protective equipment and appropriately trained staff to handle Ebola. The reality of this scary disease reached the U.S. from West Africa on a private airplane and all personnel involved in treating Dr. Brantly understood that mistakes could be deadly.

The morbid nature of the Ebola hemorrhagic fever outbreak has parallels to that depicted in a zombie outbreak. In Ebola, citizens are at risk for contracting the virus when treating the sick and dealing with the dead and the route of transmission is unknown, although handing body fluids is the suspected route. In a zombie virus outbreak, the virus in most cases is transmitted through a bite, although sometimes a scratch will also transmit the virus. In *The Walking Dead*, as previously mentioned, everyone is already infected, so upon death everyone turns into a zombie. In this regard, in Ebola and zombie virus outbreaks the route of transmission and the natural reservoir is largely unknown, although these can be speculated. Additionally, both Ebola and zombie virus infections begin with vague symptoms such as fever, headaches, and joint pain as well as gastrointestinal distress such as nausea, vomiting, and diarrhea. Finally, in both situations there may be no preventative vaccine or drug treatment that is readily available to all. Obviously, patients with Ebola hemorrhagic fever do not turn into zombies, but the uncertainty of an Ebola outbreak is also not that dissimilar from that depicted in zombie movies. This real life concern could be best seen in West Africa, where citizens were feeling especially nervous. Dr. Keiji Fukuda, WHO Assistant Director-General for Health Security, said, "We now have a lot of questions coming to WHO, and the level of anxiety is definitely high."[23]

Americans, far removed and little if at all impacted by Ebola, generally gave minimal attention to the Ebola risk. This opinion was initially mirrored by Tom Frieden, Director of the CDC, who told United States citizens, "The plain truth is that we can stop Ebola," later stating in the same speech, "I don't think it is in the cards that we would have widespread Ebola in this country."[24] This type of news was encouraging to Americans who put faith in the CDC and believed that Ebola or any other disease outbreak problem would be solved with little to no impact to the average citizen. A month later, however, Frieden told reporters, "This is not just a problem for West Africa, it's not just a problem for Africa. It's a problem for the world, and the world needs to respond.... Like it or not, we live in an interconnected world."[25] The CDC also issued warnings to U.S. colleges and universities to pay close attention to potential Ebola symptoms in students or others returning from Ebola stricken areas.[26] Warnings such as this suggested concern on the part of CDC for Ebola reaching the U.S. and the possibility for a greater impact than initially suspected.

This is not the first time that an infectious disease has caused concern in the United States. Researchers met with similar difficult challenges with the 2009 H1N1 influenza virus outbreak and in this case the average American was dramatically impacted. The initial problem with H1N1, like that since experienced with Ebola, was that the reservoir of this virus was unknown. The first two cases of H1N1 occurred in California in young females living more than 100 miles from one another and having no known connection. The novel virus was type influenza A and it was initially thought to be similar to viruses seen between 2005 and 2009 that infected swine as their reservoir. In the earlier cases there was a link that suggested the spread of the virus between swine and the infected humans and although no link seemed to exist for the new cases (or one of the older cases), early reports of the new virus described it as "swine origin influenza A virus."[27] As the virus spread among the populations, the general public referred to it as the swine flu, a name that stuck even after it was proven that the virus did not originate in swine. It was eventually discovered that the H1N1 virus contained genes from four different influenza viruses of swine, avian, and human origin. There certainly was not any anticipation of this type of flu by the CDC or similar authorities.

The H1N1 influenza outbreak led to greater than 284,000 deaths worldwide.[28] The CDC reported of pandemic status in June of 2009 and the World Health Organization ultimately raised the alert level to Phase 5 (the highest level, also raised during the 2014 Ebola outbreak in West Africa).[29] The U.S. began to stockpile antiviral agents, oseltamavir and zanamivir, to treat those infected and due to limited supplies, the CDC placed restrictions on who could receive these drugs. An intravenous antiviral drug was also available in limited quantities with its access controlled by the CDC as well. School districts closed in the hardest hit areas due to high numbers of absences and to control spread and some colleges and universities even shut down for several days. Signs posted in public locations warned of flu symptoms recommending staying home if sick and most knew someone who got the virus.

To further compound the problem, the yearly flu vaccine that many received as a preventative vaccine did not help with the 2009 H1N1 virus.[30] This meant that no vaccine existed during the 2009 outbreak that protected against H1N1. On top of this, the lack of readily available drugs kept infected individuals sick longer increasing the potential for spread of the virus to others. Additionally, those infected had to get the drugs (oseltamivir or zanamivir) within the first 48 hours or it was not as effective and due to limited doses not prescribed. Unlike other flu strains that had the greatest morbidity and mortality for the youngest and oldest patients as well as those immunocompromised, H1N1 impacted adults aged 18 to 64. This led to great concern in the United States as more and more people became sick. Eventually, the flu season came to a close, more doses of the antiviral drug were made available, and a vaccine protecting against the H1N1 strain was released. Although H1N1 still makes people sick, its devastating effects have never reached numbers like those experienced during the 2009–2010 flu season. With both Ebola and H1N1, the CDC was stretched to its limits with its emphasis being split between control and prevention.

CDC and WHO Dependence on Citizens to Prevent and Control Disease

In zombie lore fear of the creation of more zombies is front and center and the work of scientists becomes instrumental to worldwide survival. Scientists must determine the origin of the virus or natural reservoir, as well as the mode of transmission and then find a way to prevent further spread and cure those who are infected. This is a tall order, especially since during a zombie apocalypse or other similar disaster, resources and researchers may be very limited. In the movie *World War Z*, the world's future relied solely on the efforts of the WHO to solve the zombie disease plague. Gerry Lane, a former United Nations (UN) investigator, figures out that individuals who are sick are not susceptible to the plague. He travels to the WHO branch in Wales and along with the lead WHO doctor and a female soldier known as Segen a vaccine like treatment is developed and hope for a future is restored.

The restoration of hope in the movie *World War Z* depends on individuals taking the newly developed vaccine-like prophylactic (preventative) treatment.[31] In reality, even if a vaccine does exist, this does not guarantee that everyone will partake of the vaccine and thus outbreaks will continue to occur. For example, in 2014 even with H1N1 antigens being included in the annual influenza (flu) vaccine, H1N1 was very prominent in the 18- to 65-year-old age group.[32] Those who took the annual flu vaccine were protected, so these numbers originated from those who chose not to take the vaccine and had never previously been infected with H1N1.

Another similar situation is observed with the measles virus. A vaccine exists and is on the vaccination schedule in the United States, but not everyone takes this vaccine.[33] Although the measles virus is considered eliminated in the United States and Canada, outbreaks continue to occur, especially within religious groups that do not believe in taking vaccines. Most of the outbreaks are related to foreign travel and many can be traced back to an Orthodox Protestant group in the Netherlands, where there is an ongoing measles epidemic.[34] New reports from the CDC show that 2014 marked the highest number of measles cases in twenty years. This situation with measles could lead to a new measles epidemic in the United States.[35]

The measles virus causes a cough, runny nose, and watery eyes, and eventually causes a rash and high fever.[36] Most individuals may initially just think that they have a common cold, potentially infecting many others before the rash becomes prevalent or the high fever appears. This transmission could be prevented if those around the sick individual have been vaccinated. The problem arises when the measles-infected individual encounters someone who has not been vaccinated, such as a very young child. Although the symptoms of measles do somewhat resemble those seen in zombie virus-infected individuals about to turn into a zombie, it is unlikely that measles would ever lead to an outbreak of zombie apocalypse proportion, especially in the United States. This is because there is an effective vaccine and most Americans have received it. In fact, proof of the measles vaccination (often given as a mumps, measles, and rubella concoction) is required for public school in most states.

The massive emergence of the measles virus in the 2010s is, however, a great example of how vaccinations do not guarantee eradication. In fact, the World Health Organization reported 122,000 measles-related deaths worldwide in 2012, a growing trend that led to the culmination of the high number of cases in 2014.[37] A similar scenario might occur with a zombie outbreak, even if a vaccine was identified. New zombie outbreak cases would emerge in those not vaccinated and this would lead to the spread of the disease to more non-vaccinated individuals. Therefore during a zombie apocalypse, even if a preventative strategy was identified to stop further zombie virus spread, there likely would be continued outbreaks for some time. The only saving grace may be that most unlike with measles, a zombie virus infected individual turned zombie will probably not be mistaken for a person just having a cold.

Looking to the CDC for Answers

The CDC has a long history of success with controlling and preventing disease. Its excellent work done with malaria, rabies, polio, and cholera within the United States cannot be debated. When considering the number of individuals who die throughout the world from malaria alone, the early work of the organization stands out. The CDC has continued to make advances in science, issuing early warnings and generally protecting the public from infectious disease, as well as dangerous toxins including lead and nicotine. It emerges as one of the best agencies in the world for its work and its scientists have been credited with answering many of the world's more challenging questions. In recent years, the CDC has been faced with several dangerous and deadly outbreaks. The 2009 H1N1 outbreak tested the organization's resolve and the year 2014 proved to be a difficult one for the CDC with the major Ebola outbreak in Africa, the re-emergence of the 2009 H1N1 virus during flu season, and a drastic overall increase in measles cases. Even with these hurdles, the CDC fought back, continuing to educate the public on how best to avoid disease, working to understand the mechanisms that underlie these most dangerous infectious organisms, identifying novel ways to control disease spread, and researching treatment options for those already infected. Additionally, the CDC collaborates with other like-minded agencies, such the WHO worldwide and FDA and DOD within the United States, to further its mission to control and prevent disease spread.

This essay sought to determine whether the blind faith placed in the CDC and similar groups and science in general, as presented in zombie lore and observed in real life, is warranted. Diseases like Ebola and H1N1 demonstrate the limitations of such organizations. Even the CDC, with its advanced scientific prowess, cannot possibly anticipate all of the infectious organisms that might cause an outbreak or identify every environmental toxin that might lead to disease. Additionally, the CDC cannot make people follow their recommendations. There will always be smokers despite continued warnings of the toxins found in tobacco and there will always be individuals who choose not to stay home when they are sick, despite knowing the symptoms of an influenza infection. These choices made by a few individuals, however, may have great impact on the masses. Choosing not to vaccinate can result in a re-emergence of previously controlled viruses, such as was seen with measles.

Situations where individuals do not follow recommended guidelines cause the CDC to divert resources from new problems to older, previously solved problems. Additionally, viral and bacterial mutations, whereby older treatments or vaccines no longer work, also require the CDC to look at new versions of older organisms. Finally, the lack of approved and readily available vaccines and treatments can add to an already precarious situation. Both lack of available or effective vaccine and lack of available treatment doses proved problematic for disease control and prevention in the H1N1 outbreak in 2009 and during the 2014 Ebola outbreak. In both outbreaks, drug prototypes were available, but production of the needed amounts of drug was not possible. Without a means to prevent disease through vaccines or treat disease through effective drug therapy, the disease, in both outbreaks, spread.

This led to a large outbreak, but not to the level of apocalyptic proportions, shown in zombie lore. In zombie movies and disease related disaster shows depicting both organismal (viral or bacterial) mutation and lack of means to control these organisms, the end result is an outbreak (zombie or otherwise) that causes the illness and death of many. With no real way to regain control of the disease, an apocalyptic event occurs.

The CDC encounters situations involving emerging diseases on a regular basis that could in theory lead to an outbreak of apocalyptic proportion. Although they may discuss these diseases in their weekly report, much of the American public is often not even aware that a problem existed. In other cases, the event is so well contained that it impacted only a small subset of the population. This speaks to the CDC's focus on its mission to prevent and control disease. It also supports the notion that although the CDC collapsed in the zombie show, *The Walking Dead*, in real life Americans are in good hands where the CDC is concerned and their faith in this organization is well placed, well earned, and well deserved.

Notes

1. Meghan Neal, "Zombie Apocalypse? The CDC Describes How to be Prepared," *The Huffington Post*, July 18, 2011, http://www.huffingtonpost.com/2011/05/18/zombie-apocalypse-the-cdc_n_863900.html.

2. The Associated Press, "Maryland cannibal case: suspect accused of eating heart, brains ranted on Facebook," *New York Daily News*, June 1, 2012. http://www.nydailynews.com/news/national/maryland-cannibal-case-suspect-accused-eating-heart-brains-ranted-facebook-article-1.1088522.

3. Andy Campbell, "Zombie Apocalypse: CDC Denies Existence of Zombies Despite Cannibal Incidents," *The Huffington Post*, June 1, 2012, http://www.huffington post.com/2012/06/01/cdc-denies-zombies-existence_n_1562141.html.

4. "Malaria Facts," CDC, Centers for Disease Control and Prevention, Last modified March 26, 2014, http://www.cdc.gov/malaria/about/facts.html; and "Malaria: CDC Map Application," CDC, Centers for Disease Control and Prevention, last modified February 8, 2010, http://www.cdc.gov/malaria/map/.

5. "Rabies," CDC, Centers for Disease Control and Prevention, last Modified September 24, 2013, http://www.cdc.gov/rabies/.

6. Alakes Kumar Kole, Rammohan Roy, and Dalia Chanda Kole, "Human rabies in India: a problem needing more attention," *Bulletin of the World Health Organization* 92, no. 4 (2014), http://www.who.int/bulletin/volumes/92/4/14–136044/en/.

7. "The Rabies Virus," CDC, Centers for Disease Control and Prevention, last updated April 22, 2011, http://www.cdc.gov/rabies/transmission/virus.html.

8. "Global Health-Polio," CDC, Centers for Disease Control and Prevention, last updated January 6, 2012, http://www.cdc.gov/polio/why/index.htm.

9. "Media centre–Poliomyelitis," CDC, Centers for Disease Control and Prevention, last modified May 2014, http://www.who.int/mediacentre/factsheets/fs114/en/

10. "Global Health-Polio," CDC, Centers for Disease Control and Prevention, last updated January 6, 2012, http://www.cdc.gov/polio/why/index.htm.

11. "Polio eradication and endgame timeline," Global Polio Eradication Initiative, http://www.polioeradication.org/Dataandmonitoring/Polioeradicationtargets.aspx.

12. "Cholera-Vibrio cholera infection," CDC, Centers for Disease Control and Prevention, last modified October 18, 2013. http://www.cdc.gov/cholera/index.html.

13. "Cholera Fast Facts," *CNN*, May 25, 2014, http://www.cnn.com/2013/10/16/health/cholera-fast-facts/.

14. Ibid.

15. "Recognizing the Biosafety Levels," CDC, Centers for Disease Control and Prevention, http://www.cdc.gov/training/quicklearns/biosafety/.

16. *The Walking Dead*, Season 1, developed by Frank Darabont, based on the books by Robert Kirkman, Tony Moore, and Charlie Adlard, American Movie Classics, 2010–present, television series.

17. *The Walking Dead*, Season 3, ibid.

18. *The Walking Dead*, Season 4, ibid.

19. *The Walking Dead*, Season 5, ibid.

20. "Questions and Answers on Experimental Treatments and Vaccines for Ebola," CDC, Centers for Disease Control and Prevention, last updated August 29, 2014, http://www.cdc.gov/vhf/ebola/outbreaks/guinea/qa-experimental-treatments.html.

21. Liz Szabo, "U.S. Officials Announce Ebola Vaccine Trial Launch," *USA Today*, August 28, 2014, http://www.usatoday.com/story/news/nation/2014/08/28/ebola-vaccine-trial/14716833/.

22. "Interim Guidance for Managing Patients with Suspected Viral Hemorrhagic Fever in U.S. Hospitals," CDC, Centers for Disease Control and Prevention, May 19, 2005, http://www.cdc.gov/vhf/ebola/pdf/vhf-interim-guidance.pdf.

23. Megan Brooks, "WHO declares Ebola a global public health emergency," *Medscape Medical News*, August 8, 2014, http://www.medscape.com/viewarticle/829629.

24. Katie Zezima, "CDC director: Ebola is 'out of control' in West Africa but can be stopped," *Washington Post*, August 3, 2014. http://www.washingtonpost.com/blogs/post-politics/wp/2014/08/03/cdc-director-ebola-is-out-of-control-in-west-africa-but-can-be-stopped/.

25. Christine Mai-Duc, "'Window of opportunity' to control Ebola is closing, CDC director says," *LA Times*, September 2, 2014, http://www.latimes.com/world/africa/la-fg-africa-ebola-cdc-frieden-20140902-story.html.

26. Richard Pérez-Peña, " U.S. Health Agency Advises More Vigilance on Campuses: U.S. Colleges Advised to Tighten Ebola Precautions," *New York Times*, September 2, 2014, http://www.nytimes.com/2014/09/03/health/us-colleges-advised-to-tighten-ebola-precautions.html?_r=0.

27. "H1N1 Flu," CDC, Centers for Disease Control and Prevention, last updated August 3, 2010, http://www.cdc.gov/h1n1flu/cdcresponse.htm.

28. Robert Roos, "CDC estimate of global H1N1 pandemic deaths: 284,000," Center for Infectious Disease Research and Policy, June, 27, 2012, http://www.cidrap.umn.edu/news-perspective/2012/06/cdc-estimate-global-h1n1-pandemic-deaths-284000.

29. H1N1 Flu," CDC, Centers for Disease Control and Prevention, last updated August 3, 2010, http://www.cdc.gov/h1n1flu/cdcresponse.htm.

30. "Vaccine against 2009 H1N1 influenza virus," CDC, Centers for Disease Control and Prevention, http://www.cdc.gov/h1n1flu/vaccination/public/vaccination_qa_pub.htm.

31. *World War Z*, directed by Marc Forster, Paramount Pictures, 2013.

32. "Current Flu Situation," *Flu.gov*, http://www.flu.gov/about_the_flu/current_flu/index.html.

33. "Measles Vaccination," CDC, Centers for Disease Control and Prevention, last updated July 8, 2013, http://www.cdc.gov/measles/vaccination.html.

34. L. Mollema, G.P. Smits, G.A. Berbers, F.R. Van Der Klis, R.S. Van Binnendijk, H.E. De Melker, and S.J. Hahné, "High risk of a large measles outbreak despite 30 years of measles vaccination in The Netherlands," *Epidemiol Infect*, 142, no 5. (May 2014):1100–1108.

35. Robert Lowes, "Measles Vaccine Refusal Helps Make 2014 a Record Year," *Medscape Medical News*, May 29, 2014, http://www.medscape.com/viewarticle/825913.

36. "Measles Fact Sheet for Parents," CDC, Centers for Disease Control and Prevention, last updated July 8, 2013, http://www.cdc.gov/vaccines/vpd-vac/measles/fs-parents.html.

37. "Measles Fact Sheet Number 286," World Health Organization, last updated February 2014, http://www.who.int/mediacentre/factsheets/fs286/en/.

BIBLIOGRAPHY

Associated Press. "Maryland cannibal case: suspect accused of eating heart, brains ranted on Facebook." *New York Daily News*, June 1, 2012. http://www.nydailynews.com/news/national/maryland-cannibal-case-suspect-accused-eating-heart-brains-ranted-facebook-article-1.1088522.

Brooks, Megan. "WHO declares Ebola a global public health emergency." *Medscape Medical News*, August 8, 2014. http://www.medscape.com/viewarticle/829629.

Cable News Network (CNN). "Cholera Fast Facts." CNN, May 25, 2014. http://www.cnn.com/2013/10/16/health/cholera-fast-facts/.

Campbell, Andy. "Zombie Apocalypse: CDC Denies Existence of Zombies Despite Cannibal Incidents." *The Huffington Post*, June 1, 2012. http://www.huffington post.com/2012/06/01/cdc-denies-zombies-existence_n_1562141.html.

Centers for Disease Control and Prevention. "Cholera-Vibrio Cholera Infection." CDC, Centers for Disease Control and Prevention, October 18, 2013. http://www.cdc.gov/cholera/index.html.

_____. "Global Health-Polio." CDC, Centers for Disease Control and Prevention, January 6, 2012. http://www.cdc.gov/polio/why/index.htm.

_____. "H1N1 Flu." CDC, Centers for Disease Control and Prevention, August 3, 2010. http://www.cdc.gov/h1n1flu/cdcresponse.htm.

_____. "H1N1 Flu." CDC, Centers for Disease Control and Prevention, August 3, 2010. http://www.cdc.gov/h1n1flu/cdcresponse.htm.

_____. "Interim Guidance for Managing Patients with Suspected Viral Hemorrhagic Fever in U.S. Hospitals." CDC, Centers for Disease Control and Prevention, May 19, 2005. http://www.cdc.gov/vhf/ebola/pdf/vhf-interim-guidance.pdf.

_____. "Malaria: CDC Map Application." CDC, Centers for Disease Control and Prevention, February 8, 2010. http://www.cdc.gov/malaria/map/.

_____. "Malaria Facts." CDC, Centers for Disease Control and Prevention, March 26, 2014. http://www.cdc.gov/malaria/about/facts.html.

_____. "Measles Fact Sheet for Parents." CDC, Centers for Disease Control and Prevention, July 8, 2013. http://www.cdc.gov/vaccines/vpd-vac/measles/fs-parents.html.

_____. "Measles Vaccination." CDC, Centers for Disease Control and Prevention, July 8, 2013. http://www.cdc.gov/measles/vaccination.html.

_____. "Media Centre-Poliomyelitis." CDC, Centers for Disease Control and Prevention, May 2014. http://www.who.int/mediacentre/factsheets/fs114/en/.

_____. "Questions and Answers on Experimental Treatments and Vaccines for Ebola."

CDC, Centers for Disease Control and Prevention, August 29, 2014. http://www.cdc.gov/vhf/ebola/outbreaks/guinea/qa-experimental-treatments.html.

_____. "Rabies." CDC, Centers for Disease Control and Prevention, September 24, 2013. http://www.cdc.gov/rabies/.

_____. "Rabies." CDC, Centers for Disease Control and Prevention, September 24, 2013. http://www.cdc.gov/rabies/.

_____. "The Rabies Virus." CDC, Centers for Disease Control and Prevention, April 22, 2011. http://www.cdc.gov/rabies/transmission/virus.html.

_____. "Recognizing the Biosafety Levels." CDC, Centers for Disease Control and Prevention. http://www.cdc.gov/training/quicklearns/biosafety/.

_____. "Vaccine against 2009 H1N1 influenza virus." CDC, Centers for Disease Control and Prevention. http://www.cdc.gov/h1n1flu/vaccination/public/vaccination_qa_pub.htm.

"Current Flu Situation," *Flu.gov*. Last modified December 4, 2014. http://www.flu.gov/about_the_flu/current_flu/index.html.

Global Polio Eradication Initiative. "Polio Eradication and Endgame Timeline." *Global Polio Eradication Initiative*. http://www.polioeradication.org/Dataandmonitoring/Polioeradicationtargets.aspx.

Kole, Alakes Kumar, Rammohan Roy, and Dalia Chanda Kole. "Human Rabies in India: A Problem Needing More Attention." *Bulletin of the World Health Organization* 92, no. 4 (2014). http://www.who.int/bulletin/volumes/92/4/14-136044/en/.

Lowes, Robert. "Measles Vaccine Refusal Helps Make 2014 a Record Year." *Medscape Medical News*, May 29, 2014. http://www.medscape.com/viewarticle/825913.

Mai-Duc, Christine. "'Window of opportunity' to control Ebola is closing, CDC director says." *LA Times*, September 2, 2014. http://www.latimes.com/world/africa/la-fg-africa-ebola-cdc-frieden-20140902-story.html.

Mollema, L., G.P. Smits, G.A. Berbers, F.R. Van Der Klis, R.S. Van Binnendijk, H.E. De Melker, and S.J. Hahné. "High risk of a large measles outbreak despite 30 years of measles vaccination in The Netherlands." *Epidemiol Infect* 142, no 5. (May 2014):1100–1108.

Neal, Meghan. "Zombie Apocalypse? The CDC Describes How to Be Prepared." *The Huffington Post*, July 18, 2011. http://www.huffingtonpost.com/2011/05/18/zombie-apocalypse-the-cdc_n_863900.html.

Pérez-Peña, Richard. "U.S. Health Agency Advises More Vigilance on Campuses: U.S. Colleges Advised to Tighten Ebola Precautions." *New York Times*, September 2, 2014. http://www.nytimes.com/2014/09/03/health/us-colleges-advised-to-tighten-ebola-precautions.html?_r=0.

Roos, Robert. "CDC estimate of global H1N1 pandemic deaths: 284,000." Center for Infectious Disease Research and Policy, June, 27, 2012. http://www.cidrap.umn.edu/news-perspective/2012/06/cdc-estimate-global-h1n1-pandemic-deaths-284000.

Szabo, Liz. "U.S. Officials Announce Ebola Vaccine Trial Launch." *USA Today*, August 28, 2014. http://www.usatoday.com/story/news/nation/2014/08/28/ebola-vaccine-trial/14716833/.

The Walking Dead, Season 1. Developed by Frank Darabont, based on the books by Robert Kirkman, Tony Moore, and Charlie Adlard. American Movie Classics, 2010–present.

The Walking Dead, Season 3. Developed by Frank Darabont, based on the books by Robert Kirkman, Tony Moore, and Charlie Adlard. American Movie Classics, 2010–present.

The Walking Dead, Season 4. Developed by Frank Darabont, based on the books by Robert Kirkman, Tony Moore, and Charlie Adlard. American Movie Classics, 2010–present.

The Walking Dead, Season 5. Developed by Frank Darabont, based on the books by Robert Kirkman, Tony Moore, and Charlie Adlard. American Movie Classics, 2010–present.

World Health Organization. "Measles Fact Sheet Number 286." World Health Organization, February 2014. http://www.who.int/mediacentre/factsheets/fs286/en/.

World War Z. Directed by Marc Forster. Paramount Pictures, 2013.

Zezima, Katie. "CDC director: Ebola is 'out of control' in West Africa but can be stopped." *Washington Post,* August 3, 2014. http://www.washingtonpost.com/blogs/post-politics/wp/2014/08/03/cdc-director-ebola-is-out-of-control-in-west-africa-but-can-be-stopped.

A Matter of Timing

The U.S. Army Response to a Zombie Invasion

Jason W. Warren

Imagine the zombies crossing en masse over the fences around a lone farmhouse, where survivors of the zombie plague have huddled. Despite the best efforts of the inhabitants, as portrayed in season two of AMC's *The Walking Dead*, they were overrun by the flesh-eating horde (though some escaped). A much different scenario might have occurred had trained and properly equipped U.S. military forces been waiting on the other end of that fence.

The U.S. military is an all-purpose response force, which can counter various threats, including the outbreak of a zombie-like plague. Equipped with cutting-edge weaponry, and trained to the highest standard of proficiency in the world, the U.S. military is especially adept at fighting in high-intensity, conventional conflicts. The various capabilities of the U.S. Armed Forces notwithstanding, there are various systemic historical weaknesses for less "kinetic" operations, such as supporting civilian agencies during natural disasters. These shortcomings might hamper a military response to a zombie outbreak, which would place at risk the safety of millions of American citizens and civil order. The amount of time the military required to react to the initial zombie incursion would be the determining factor in its attempt to destroy the zombies and contain the crisis. If the military adequately countered the zombies in the initial hours of the outbreak, then there would be a high probability of containing the scourge. The U.S. military's response—particularly the Army's as it would bear the brunt of the operation—to Hurricane Katrina (2005) and the Los Angeles riots (1992), along with current doctrine and recent counterinsurgency operations in Afghanistan (2001–present) and Iraq (2003–2011; 2014–2015), serve as indicators of the level of success the organization would achieve during a zombie invasion. The U.S. Army's occupation of post–U.S. Civil War Texas also indicates how a strong military presence prevents civilian infighting and lawlessness during a zombie-like social and political crisis.

The U.S. Army would lead the effort against the zombies, although the other elements of U.S. national power would play a vital role, especially the other military services and the Coast Guard. Civilian agencies, such as the Department of Homeland Security, would work in conjunction with the military and perhaps act as lead agency to combat the zombie plague. The preponderance of power projection and its sustainment against the zombies, however, would fall to the Army. There is no substitute for the Army, particularly the Active Component (as opposed to the Reserve Component consisting of the National Guard and Reserves) because it is the only force *in the world* that can sustain itself logistically for prolonged periods of time away from its main cantonment areas without having to purchase or commandeer local supplies. The Navy relies on the necessarily limited hauls of its ships for re-supply, while the Air Force's ability to re-supply ground forces and its airfields depends on available aircraft, given a maintenance rotation, and space available inside the planes, sleep for its crews (flight restrictions occur with inadequate sleep), and high-quality jet fuel. The Marine Expeditionary Brigades (MEBs), the most tailored Marine unit for an initial zombie outbreak in one locale, can only re-supply itself for roughly 30 days (and 60 days as part of a larger Marine Expeditionary Force or MEF),[1] before relying ultimately on the Army. As the Marine Corps' primary mission is amphibious assaults and short-duration expeditionary missions, for which they are resourced, the service could not operate during a prolonged zombie plague without Army logistical support, as the Marine's long-term presence in Iraq required. During the zombie invasion, the Navy would, as portrayed in *World War Z*, maintain hospital ships and naval bases from which to evacuate citizens and personnel. It would also support the Marines' short-term logistical requirements. The Air Force (and Marine aircraft supporting Marine ground forces) would fly missions in support of the ground forces, but most of its combat airframes would be ineffective against zombie targets co-mingled with the population. Newly published *Field Manual (FM) 3-94, Theater, Army, Corps, and Division Operations* details the specific types of Army Support of Other Services (ASOS), including logistics, engineering, security, transportation, medical evacuation, missile defense, and other functions.[2]

Conversely, the Army has a well-resourced logistical tail for its forces and the other services, including durable trucks, boats, helicopters, and tracked vehicles, as well as the ability to integrate logistics into operations. It also maintains the capacity to augment its large-scale supply capability with contracts to civilian agencies or to obtain local resources with specially designated commander's funds. For instance, food or material could be purchased directly from civilian markets. The other military services, relying instead on high-tech platforms and systems, do not have similar capabilities for long-term action. Large numbers of the Active force and Army Reserve, once mobilized, would reinforce the National Guard's initial response to the zombie-plague. The first deployment might also include a ready Army active brigade on standby status, if authorities accurately discerned the threat. As occurred in the real-world instances described hereafter, the timeline for such a response would depend entirely on how rapidly and accurately the local authorities communicated the threat to state and federal officials.

As the lead response force, the Army would concentrate on a series of tasks to ultimately neutralize the zombies and restore civil authority in the effected region(s). The first Army objective would be to secure the site of the initial outbreak. This would ensure the virus' containment, which otherwise would complicate the response effort. The major problem posed by flesh-eating zombies is that the threat of violence is not an option, and armed Americans would have to slaughter the walking dead wherever they lurked, as zombie literature and movies such as *The Zombie Survival Guide: Complete Protection from the Living Dead* and the series *The Walking Dead* portray. A byproduct of the zombie plague is the breakdown of civil order leading to other crimes in infected areas. Given this menace, military forces in conjunction with local authorities would simultaneously provide humanitarian relief and protect lives and property. The military, augmented by local police forces, would protect vital infrastructure, such as power grids (especially nuclear facilities from which stolen waste could result in a terrorist incident), hospitals, civil administration buildings, water and waste treatment plants, jails, and road junctions. Federal agencies and local/state authorities would assist in guiding the Army's efforts to protect this critical infrastructure. The military would also have to remain alert to supranational threats, as rogue states and terrorists, if not larger powers, would consider monopolizing a security void to seek local advantages created by the home-deployment of U.S. military forces without fear of an armed U.S. response. All of these crucial tasks would depend on the timing and effectiveness of the initial military response; a slow response to the point of the initial zombie occurrence would necessitate the deployment of more forces and the enlargement of all operational requirements. It would also elevate the risk of foreign crises the longer that U.S. military forces deployed to stabilize the homeland. A recent example is Iraq's descent into chaos, when U.S. military withdrew after largely reducing the violence there with the support of Iraqi forces. A similar redeployment of U.S. forces from South Korea and Japan, for instance, would risk regional belligerents stepping into the power void.

The military response to Hurricane Katrina and the Los Angeles riots highlight likely scenarios for the Army's response to a zombie plague.[3] At first state and then federal authorities deployed the Army to rescue citizens and provide security and law enforcement. The initial response to both incidents relied on a National Guard deployment, which proved inadequate. The National Guard in the L.A. riots required time to mobilize from outlying areas, as unlike active duty troops, the part-time soldiers do not live in relatively centralized barracks and off-installation locations. This situation would be problematic if the zombie infection spread more rapidly than the Guard could mobilize and deploy into an infected area. Additionally, the zombies would likely interdict some personnel reporting for duty, disrupting unit chain-of-commands and hence effectiveness. In August 2005, Hurricane Katrina caused a similar situation. Although the states of Louisiana and Mississippi pre-deployed equipment and maintained active command and control centers prior to the storm's landfall, its severity undermined these preparations, delaying the Guard personnel and most of the first responders (police, ambulance, fire). The storm hindered these critical responders directly, rendering them victims, and hence exacerbating the states' inadequate initial response, and later the federal government's response. If local authorities face similar uncertainty for the zombie threat, and as a result it spreads rapidly, then a similar outcome is probable.

A major problem in the Army's response to both the riots and the storm centered on the inability of military forces to establish a viable command structure to "command and control" subordinate units. General David Petraeus, commander of U.S. forces in Iraq and Afghanistan, detailed how establishing an effective chain of command, focused on one leader atop a clearly delineated command structure, was one of his most difficult tasks as a commander in the Middle East. It was further complicated by having to integrate non-military diplomatic, intelligence, and foreign allied personnel.[4] Perceived threats to individual or agency power and a lack of interagency doctrine that would better integrate the bureaucracy caused the friction that Petraeus highlighted. The same situation plagued the Army's response to both Katrina and the L.A. riots, and would likely do so at the occurrence of zombies in search of human meat.

During the April 1992 L.A. riots, the initial National Guard responders, though somewhat delayed, integrated into the police force's planning for subduing the civil disturbance. The military chain-of-command did not challenge the civilian police's primacy. The police failed to deploy the Guard troops in a coherent structure that maintained the traditional chain-of-command, however, requiring the soldiers to act alone or with only a handful of peers within a police detachment. This failure of the police to observe the Army's traditional command apparatus, as detailed in squad, platoon, or company structure, caused the Guard troops to respond less effectively to the civil disturbance. The military only trains special operating forces to act individually or in very small elements, which was not the case with the National Guard soldiers.

When active duty troops mobilized and the president authorized them to deploy to Los Angeles, the chain-of-command issue became worse. At that point, the federal troops were legally in charge of subduing the riots, however, the regular Army forces lacked civil disturbance training and law enforcement know-how and subordinated themselves to police forces in many instances. This created an upside-down chain of command, where subordinates led those normally in charge. Federal troops should have maintained control of the police forces and integrated all agencies—local, state, and federal—into one effective response force, as federal authority trumps others levels of government once the president dictates a federal emergency. As it turned out, the riots had petered out largely of their own accord by the time the federal forces assumed command, and the winding down of the violence largely trumped the poor command and control structure. It cannot be ruled out, however, that the street gangs and individual hooligans dispersed with the knowledge that authorities were deploying Active Component forces into their neighborhoods.

The failure of the Army and federal interagency chain of command during Hurricane Katrina dwarfed that of the riots. State authorities, particularly Louisiana governor Kathleen Blanco's refusal to accept federal control of the response to the storm, and President George W. Bush's decision not to demand it, severely hampered the Army's effectiveness. This failure at the highest levels created a situation, especially in Louisiana, and to some extent in Mississippi, where the National Guard, and Active Army forces carried out their own, often overlapping, operations. This failure to coordinate diminished the timeliness and effectiveness of the relief effort, and in a longer duration zombie attack may well have jeopardized the mission. Where the operation did succeed, it was due to low-level leaders working out ad hoc agreements over command relationships. Coordinating the activities of local non-military responders and other agencies such as the Red Cross became all the more complicated due to the inability of state and federal forces to cooperate. On a practical level, the failure led to inefficiencies in the law enforcement response, such as inadvertently patrolling the same sectors. Without needed federal coordination, this could have led, for instance, to the redundant deployment of scarce aerial assets.

As the zombie threat would rapidly spread during the first few hours, a failure of local, state, and federal forces to cooperate and coordinate their efforts, especially by establishing a viable command structure, would derail the effort. As the active Army focuses its dwindling resources and training-time allotment to counterinsurgency operations in the Middle East and Southwest Asia, as well as prepares to rebuild its conventional-fighting capabilities focused on the Pacific region, the organization marginalizes preparation for civil disturbance and disaster relief. It also does not train adequately to respond to such crises with the National Guard or Reserve, let alone the interagency and non-governmental organizations (NGOs). The National Guard on the other hand, does respond to local disasters, and trains to some degree to do so, but the organization's inherent shortcomings—namely that it is a part-time force, lacking a high-level of efficiency—means that it is unlikely to perform to an adequate standard during a major crisis without federal support. These factors taken together, would likely result in an *initial* poor Army response to a zombie outbreak in terms of establishing command and control. As the time factor is the critical element in mission success in the event of a zombie outbreak, Army forces and the ability to coordinate with all levels of government and non-governmental agencies could become the causative factor of success or failure in containing its spread.

Other elements that would hinder an effective response emerged from both disasters, including the failure of useful doctrine and communication. Army doctrine for both the civil disturbance and disaster relief—and a zombie plague would involve both—caused a confused response. This was especially the case for Active Component units not accustomed to carrying out those operations. Army doctrine provides the "how" for everything from establishing command and control to performing duties related to the particular threat at hand. Although Army leaders sometimes discount or ignore doctrine, when faced with an unfamiliar situation, it is extremely likely that officers will consult doctrine for planning purposes. The Army's doctrine in both cases was vague and contradictory, and did not provide an adequate template for operations.

The lack of communication equipment, especially among National Guard and Reserve forces, exacerbated these other problems. Most non–Active Component troops are not provided the latest equipment, and the National Guard and Reserves normally provide scarce equipment only to units that are deploying overseas. In both disaster situations, Reserve Component forces lacked adequate communication equipment for command and control and to carry out simple tactics on the ground. As the Active Component maintained the newest equipment, the National Guard and Reserve forces during the response to Katrina, in many cases, could not communicate with each other and hence synchronize operations. With the threat of a zombie disease spreading rapidly, a failure to communicate in the early hours of a response would seriously hinder the ability to contain its spread.

The Army response to Katrina and the L.A. riots did reveal two positive aspects that would increase the chance of successful operations to counter zombies. The first was that some National Guard troops on both occasions had civilian law enforcement experience. This enabled these soldiers to better integrate into local police forces or act alone as local police. Active Component forces in New Orleans, the site of the worst law enforcement challenges, even ceded the law enforcement mission to the National Guard, so that it could better focus its resources on other areas of the disaster response. A zombie outbreak would bring with it the threat of disintegrating social order, calling for large-scale police measures, which the National Guard and Reserve troops could augment.

Recent battle experience in Iraq and Afghanistan also enabled a more efficient response during Katrina, allowing for battle-hardened veterans to overcome the failures of higher leaders to establish command and control and mission objectives. Fresh experience with counterinsurgency operations forced soldiers and Marines to become knowledgeable of politics, civil engineering, and law, and to employ these techniques in vague, challenging, and dangerous circumstances. Troops also had to distinguish rapidly between friends and foes, a skill that is vital to domestic threats, such as that posed by the walking dead. If the zombies attacked at a time of recent combat experience, then a similar Army response would be nearly assured.

Current Army doctrine detailing support for civilian authorities encompasses some of the experiences of the past-thirty years, but largely whitewashes the complexity of the situation and military responsibilities. This is the very doctrine that Army forces would rely on during a zombie attack. *Army Doctrine Publication (ADP) 3-28, Defense Support of Civilian Authorities*, counsels transferring command to civilian authorities at the first opportunity, as the situation on the ground permits.[5] This theory accords with traditional American values and law, as well as the preference of the armed forces to avoid irregular operations. The doctrine becomes problematic, however, in terms of operations, if commanders begin to look to transferring authority as soon as possible, instead of concentrating on providing adequate support or establishing permanent control of a situation. The goal should be the latter, which focusing on an "exit strategy" might overshadow. This attitude can also affect subordinate troops, who could take the mission less seriously.

The Army also has had concerns about *posse comitatus*, which prohibits unsanctioned Federal military support of local law enforcement. Active Component soldiers operate under a Title 10 legal statute that prohibits employment in states and local communities, unless ordered to by competent federal authorities (or in emergency circumstances). National Guard soldiers swear allegiance to state authority, and function according to Title 32 legal statute, unless activated to federal service by competent federal authority. Thus operationally, it is a more normal alignment to focus Title 32 forces on domestic law enforcement, while Title 10 forces serve in other capacities of support to civilian agencies. Over time, the Active Component Army has defined Title 10 restrictions more rigorously, imposing unnecessary restraints on its operations. The response of the Army's 7th Infantry Division along with local Marine units to the L.A. riots exemplified this trend, where its command feared violating *posse comitatus*, even though the President of the United States ordered the troops to end the riots, obviating further legal concerns. The Army Active component all but avoided military policing during its Katrina response, relying on National Guard units to perform that mission.

The Active Component's hesitation to perform law enforcement, even when authorized, stems from a post–Vietnam concern over its standing in the public opinion of Americans. Americans viewed military service extremely unfavorably in the 1970s after the Vietnam War, but the end of the draft and the development of an effective all-volunteer force that succeeded in the Gulf War (1991) changed these attitudes. The American public still holds the military in the highest regard when queried about major professions' contributions to "society's well-being," even after more than a decade of war with dubious results.[6] Most of the military's top leaders remember the post–Vietnam-era force, as they were junior officers during the tough transition to the all-volunteer military.[7] Any action, especially in the sphere of U.S. domestic politics that may jeopardize the volunteer military's hard earned reputation, appears to make these leaders balk. Using lethal force against Americans since the advent of the volunteer Army is the worst-case scenario for the brass of the Active Component, and they have hesitated to do so in practice, even with a clear legal mandate.

This reticence to use lethal force against the American citizenry greatly complicates the more straightforward tasks of Army civil support doctrine; the first of which is protecting lives. The Army seems to conceive of this task as aiding in evacuations from storm devastation or prior to storm-landfall. This is the task that most Active Component soldiers performed during Katrina, leaving law enforcement to the National Guard and Reserves. This was a luxury, as natural disasters, even powerful ones like Katrina, are limited in duration. A zombie apocalypse would be lengthier, occurring until authorities eliminated the zombies and the plague's means of transmission. The Reserve Component, unlike with Katrina and the L.A. riots, would be unable to perform such an intense, long-duration task without significant Active Component law enforcement support. Without the willingness to use lethal force, Active Army effectiveness in protecting lives and property could be limited. Shooting zombies might be easier for troops to handle in the near term, but using lethal force to prevent the breakdown of law and order, if not civil society in affected areas, would challenge the all-volunteer force's conception of itself as a champion of the citizenry. The use of deadly force in support of civil authority also applies to protecting property, which *Army Doctrine*

Publication (ADP), 3-28 specifies as a critical element of Army support to civilian authorities. Without the threat of force, protecting property becomes a meaningless objective, and the Army's prejudices would hamper the mission.

Friction, as influential military theorist Carl von Clausewitz defined the concept, between military objectives, battlefield intelligence, and real-time events, is inevitable in war. The nineteenth-century Prussian's writings form the basis for modern Western theories on the nature of warfare.[8] Successful military leaders and planners need to overcome friction in order to win. Military planners would need to undertake a mental transition to conceive properly of the threat posed by the walking-dead, and reduce friction by applying current doctrine and tactics and techniques, however flawed these remain. Without a clear-headed conception of the nature of the zombie threat, best approximated by the real-world military responses above, no plan would survive the reality of a monster-infested battlefield. This would demand a rapid clarification of the substance of zombie combat before more Americans became infected, while the planners reconceived the nature of war.

Although he allowed for the unpredictable nature of warfare once combatants initiated the conflict, Clausewitz's conclusion that war was a continuation of policy gave it a certain inherent rationale. In essence, war was a political and hence a logical action. This idea rationalized the seemingly chaotic nature of warfare, preempting a line of argument that the causes of war were more of an emotional and illogical complexion, although these irrational factors influence the *conduct* of war.

On what framework then, could military practitioners rely to understand and thus combat a zombie invasion? The zombie-warfare paradigm would be one detached from political reality and consisting of a purely irrational nature, corresponding with historian John Keegan's conception of the nature of war.[9] Keegan in *A History of Warfare* challenges Clausewitz's notion about war as an extension of policy, arguing for something deeper and darker that lurks in humankind. This new face of war would confront the men and women of the Armed Forces responsible for the fashioning of strategy and development of campaigns against this truly asymmetrical threat (asymmetrical threats are those that challenge weaknesses). Against an irrational yet deadly threat, only one idea could drive a new understanding of war—survival; the survival of as many Americans as possible, and the continued existence of the United States. The feasibility of accomplishing this objective would hinge on the ability of military leaders and planners to reconceive their intellectual framework of war. It is no small matter to adjust planning for an enemy that cannot be coerced or compelled, but simply must be eradicated. Zombie lore like the popular *The Walking Dead* series indicates such a scenario, where the flesh-eating zombies must be eradicated without hope for a "cure." Planning, derived from the new intellectual framework of warfare, would have to reflect this reality. A delay in conceiving of this new warfare and hence transitioning to a different paradigm to combat the zombies could lead to critical delays, where a wider swath of the United States succumbs to the zombie plague.

With survival as the overriding goal and purpose of the use of force against zombies, the war against the monsters would therefore adopt a decidedly straightforward purpose. With this singularity of mission, however, would come a whitewashing of moral and ethical standards that have been governed by the ancient laws of war dating back to the beginning of the Western epoch, as the Greek *poleis* emerged from the pre–Hellenic dark ages. The Greek city-states established markers declaring the laws of war on adjoining territorial boundaries. They also allowed truces for both sides to collect the dead. Then again, the dead in ancient Greece, unlike the zombie horde, had a way of remaining in that state. Thucydides attempted to relate how the endemic warfare of the years encompassing the Second Peloponnesian War— and issuing a warning to his readers—had deteriorated the morality of the Greeks. Horrific desecrations of Greek religious and other social traditions became commonplace, and slowly this led to the deterioration of the ethical fiber of Hellenic society.[10] The zombie scourge may accelerate this process in the United States because there can be no reciprocity with insentient beings. The survivors of the initial outbreak in *The Walking Dead* similarly struggled to adapt cultural norms, such as in the practice of burying the dead, particularly of those zombies that had been kin or friends.

Of course the great Hellenic war was not the only time in which prolonged, unrestrained warfare, undermined ethics in Western warfare. The Roman philosopher Augustine drafted rules of war that passed through the centuries. Jurist Hugo Grotius, who lived through the bloody upheaval of the Thirty Years' War, and men like him, updated Augustine's ideas and bequeathed legal interpretations to subsequent generations. Eventually, "modern" Europeans drawing upon their ideas designed the Geneva and Hague Conventions in the early twentieth century. This codification of Western ideals did not prevent the mass slaughter of last century's world wars, nor did the musings of the earlier philosophers always prevent the atrocities of contemporary conflicts. Some wars were more constrained and chivalrous, however, abiding by many of these rules.[11] The war against the zombies could have no chivalry with the immutable feature of the enemy, and perhaps this would damage the psyches of American soldiers and civilian responders.

The panoply of zombie literature and movies depict how society's moral standards would degenerate during an outbreak. Moral considerations are also affected by soldiers' indiscriminate ability to eliminate the monster horde without the possibility of reciprocity or a hope for peace. This might negatively affect a soldier's character, as his tolerance for humanity may decline with his license to kill the dead without consideration of another recourse. As a result, the military would have to plan for psychological care for some of its members at the conclusion of the campaign against the walking-dead, in a way that differed from fighting a living enemy.[12]

A decline in societal values and principles, calling for a robust Army occupation to prevent further bloodshed or disorder would accompany a zombie-outbreak that spread regionally. The need for domestic Army occupation forces is not unprecedented in American history, though Army officials resist the implementation of military personnel in governance, law enforcement, domestic courts-martials, and other occupation duties. Reconstruction in the post–U.S. Civil War Confederate South and Southwest regions serve as an example of what an Army occupation might entail. The legacy of Reconstruction-era Texas, which arguably lasted from 1865 through the Civil Rights era, is emblematic as a mixed success story of such occupation duty.

The commander of Union forces, General U.S. Grant ordered General Philip Sheridan to occupy Texas in the immediate post–Civil War period, as Confederate forces surrendered throughout the South. Unlike the other Confederate States, Texas had never suffered major invasion and thus retained a more recalcitrant population. The state also maintained active Confederate forces and government, both of which also required Union forces to intercede. The ensuing invasion ("invasion" as the Union had not clearly reasserted its authority over the state) was large-scale with more than 50,000 Union troops invading the rebel territory by land and sea, and included black troops. Black soldiers helped to signify the end of slavery in all corners of the former Confederacy. Although standing forces did not oppose the invasion, the appearance of "bluecoats" stirred localized violence in some cases. The overwhelming invasion force, however, ensured that there was no organized resistance as the Confederate infrastructure imploded. The invasion functioned to intimidate French interference in Mexico as well, and contributed to the withdrawal of French forces from Texas' southern neighbor. Comparatively untested French troops could not hope to defeat the battle-hardened Union Army, with the Americans maintaining much shorter supply lines from Texas than the French line-of-communication across the Caribbean Sea and Atlantic Ocean.

Union forces maintained an active-role in Texas politics during Reconstruction, ensuring a Radical Republican government. The Union Army in Texas sought to maintain the political achievements of federal victory, which translated into active interference in the domestic political process. This included appointing Radical Republican officials and actively suppressing former Confederate officials and sympathizers, as well as guaranteeing black suffrage and social and economic participation in the state. The Texas State Police (temporarily replacing the politically questionable Texas Rangers) formed to ensure these Radical Republican ideals. The measures were successful from a policy standpoint, but served largely to alienate the white population, which promptly subverted the Republican platform during the Grant administration's return to "normalcy." Once the U.S. Army returned to a peacetime footing as a frontier constabulary force, Texans rolled back Radical Republican gains and subverted racial equality for another century.[13]

One might imagine a similar U.S. Army role in a region of the U.S. that suffers from social discord and chaos post-zombie apocalypse. Perhaps the zombie incursion would be severe enough to subvert local government resulting in chaos, politically, socially, and economically. The U.S. Army, though hesitant to perform such domestic occupation duty, could serve in a similar way to the Union Army during the period of Reconstruction in Texas. As in other areas of the Reconstruction South and Southwest, Army duty might include interference in domestic politics, the appointment of officials, and the maintenance of military courts to control an unruly population. Given American attitudes toward military forces' prolonged involvement in domestic affairs, and the perceptions of senior U.S. military leaders toward domestic support, U.S. military forces would return control to civilian officials at the first opportunity. Judging from the Texas occupation of 1865 to 1870, they would do so even if the political situation remained less than stable.

Having considered how the military has responded recently to civil disturbances, and how current doctrine and concepts would apply to a walking-dead scenario, what would an actual operation entail, if the plague occurred in New York City? On the ground, in addition to local law enforcement support (which in NYC is similar to a small army), the New York National Guard and Reserve initially would activate under Title 32, with the New York governor as commander. If the Guard's response to the crisis proved inadequate, then they would be federalized under Title 10, while the Active Component deployed to the area. The New York National Guard could mobilize its Army brigades with enablers such as logisticians, military police, and engineers.[14]

Depending on the size of the outbreak, an Army division headquarters (serving as a joint headquarters) consisting of three combat brigades plus all enablers, supported by a Marine brigade would respond to a medium-size outbreak in an urban area as large as NYC to assume command from the National Guard and the NYPD. The ground force's primary mission would be to cordon the area of the outbreak to prevent the further spread of the plague, and as current doctrine dictates, to protect lives and property. The military would accomplish this through the establishment of heavily armed checkpoints at the major entrances to Manhattan.[15] Soldiers would guard the Lincoln Tunnel, George Washington Bridge, Brooklyn Bridge, Holland Tunnel, and the other various access points into the Bronx from across the Bronx River.

While stopping and inspecting the passengers of every vehicle exiting the city for signs of bite or scratch marks, the troops would don Nuclear, Biological and Chemical (NBC) suits to prevent the spread of infection, until the plague's transmission means were better understood. The soldiers and Marines at these frontline checkpoints would turn over to the military police (MPs) anyone suspected of having contact with zombie saliva. From there, the MPs would process the potential victims and turn them over to special teams of doctors enlisted by the Centers for Disease Control and Prevention (CDC). Teams of MPs would transport the infected to quarantine areas established within the city limits at places like Roosevelt Island, in the East River.

Armored cavalry units would patrol the major city thoroughfares of Manhattan and support the checkpoints as needed, patrolling the intervals between them.[16] Aerial drones would patrol along Manhattan's waterfront in support of the Navy and Coast Guard, where troops were not conducting reconnaissance and security duties. In terrain difficult to observe with military assets, such as the base of the George Washington Bridge, soldiers would deploy ground-based early warning sensors that detect the footfalls of approaching individuals.

Troops would keep under constant surveillance critical government and financial centers, such as the United Nations and Wall Street. Great cultural sites like the Metropolitan Art Museum and Yankee Stadium would receive military assets as well, if police were unavailable to guard them. The other New York City boroughs would receive similar assets, which would operate in kind. Only in the most dangerously affected neighborhoods would the military abandon the streets to the zombie horde. Any zombies threatening the cordon that could not be turned back would be eliminated with deadly force. The military would at first avoid the wholesale slaughter of the zombie horde in anticipation of an antidote that might save the victims. Only once medical personnel established that those infected would be lost, would the military resort to slaughtering all zombies. This would be a case similar to what was portrayed in *The Walking Dead*, where the secret of what CDC doctor, Edwin Jenner told the protagonist, Rick Grimes was hidden from the audience during the first season. As long as the ground forces maintained distance and enough ammunition to engage the zombies, they would be largely free from danger. The largest threat in the city would be for zombies to overwhelm or cutoff a checkpoint. If the plague spread throughout the country, however, then the military logistics system would be threatened, putting at risk the Army's capability to sustain itself. A zombie plague infecting a medium-size city simultaneous to a large one like New York would result in less military support for the former urban area.

The Air Force and Navy would play limited roles. Lacking its own ground element other than security and Special Operating Forces personnel, the Air Force would maintain control of the city's airspace and adjoining areas, as well as conduct aerial reconnaissance in the event that a terror group or foreign power attempted to capitalize on the outbreak to its own advantage. The Navy and Coast Guard would perform a similar function on water, patrolling the Hudson, East, Bronx, and Harlem Rivers, as well as the coastal areas adjacent to Long and Staten Islands. These waterborne patrols would also ensure that citizens that had not been properly screened for the zombie-plague did not abscond via water. The Navy would also position hospital ships in New York Harbor to assist the medical teams in the city.

In the New York scenario, one might expect the course of the mission to resemble that of Katrina. The initial response would be slow, confused, and inadequate, but as long as the National Guard held the line until significant Active Component forces materialized, then the plague would be contained and beaten back. If it spread beyond New York City into neighboring areas or states, then the military's mission would be more difficult. If the zombie-virus spread simultaneously around the country, the military would be forced to determine which parts of the country to protect, and Americans would abandon large swaths of the nation, if they did not succumb to the disease. This would operate in principle like a large-scale triage system, where the most critical areas received the limited military resources.

The U.S. military, particularly the Army, can stave off a zombie apocalypse and the destruction of large portions of the United States if the organization can overcome inherent weaknesses. Namely, the lack of sound cooperation in past crises between active duty forces, the Reserve Component, and civilian agencies, has hampered effective responses in a timely manner. The Army would also have to overcome its hesitation to use violence against civilians in the post–Vietnam era to react effectively to the outbreak of a zombie-plague. If the Reserve Component could hold the line efficiently against the zombies until the Active Component deployed into threatened areas, and the cooperation between the two components went more smoothly than in the past, there is a high probability of success in defeating the walking dead. The Reserve Component's availability of modern communication equipment would play a critical role in the effectiveness of its units' cooperation with the Active Component. Any delays in cordoning infected zones and a failure to coordinate efforts from an effective chain of command would risk the spread of the disease.

The military does maintain certain advantages on a zombie-dominated battlefield. Soldiers and Marines could potentially rely on recent combat experience, particularly in an unconventional environment that required adaptation and the implementation of civic skills. The discernment between friend and foe in such a combat zone would make a military response to zombies more effective in protecting innocent lives and property. Reserve

Component soldiers also would rely on domestic law enforcement skills developed in their civilian jobs, which would translate directly into the mission of maintaining civil order against the zombies. Using these advantages, while responding in a timely manner, would stack the odds against the zombie threat in favor of the U.S. military, and lead to a better outcome than the overwhelming of a near-helpless farmstead of survivors in *The Walking Dead*.

NOTES

1. United States Marine Corps, Concepts and Programs 2013, Chapter 2, 14, http://www.hqmc.marines.mil/Portals/136/Docs/Concepts%20and%20Programs/2013/CP13%20CH2_WEBFA_5FEB13.pdf, accessed 1 August 2014.

2. Headquarters, U.S. Department of Army, *Field Manual (FM) 3-94, Theater Army, Corps, and Division Operations* (Washington, D.C., April 2014), 1–20.

3. I used the following sources, which captured the U.S. military experience, especially from the perspective of the Army, in responding to both Hurricane Katrina and the Los Angeles Riots: U.S. Army Combined Arms Center Combat Studies Institute Press, *Army Support During the Hurricane Katrina Disaster*, by James A. Wombwell, The Long War Series Occasional Paper 29 (Fort Leavenworth: 2009); Foreign Military Studies Office, *Combat in Cities: The LA Riots and Operation Rio*, by William W. Mendel (Fort Leavenworth: July 1996). These two recent examples exemplify Army trends in responding to domestic U.S. crises of similar nature to a zombie invasion.

4. U.S. Army War College video teleconference, 2013.

5. Headquarters, U.S. Department of Army, *Army Doctrine Publication (ADP), 3-28, Defense Support of Civil Authorities* (Washington, D.C., 26 July 2012).

6. Pew Research Poll, http://www.pewforum.org/2013/07/11/public-esteem-for-military-still-high/#strong-consensus-about-military, accessed 28 July 2014.

7. Beth Bailey, *America's Army: Making the All-Volunteer Force* (Cambridge: Harvard University Press, 2009).

8. Carl Von Clausewitz, Michael Howard, and Peter Paret, *On War*, eds. (Princeton: Princeton University Press, 1989), 121.

9. John Keegan, *A History of Warfare* (New York: Alfred A. Knopf, 1993) argues that Clausewitz's theory does not account for the irrational nature of war.

10. Thucydides, *The Peloponnesian War*, trans. Steven Lattimore (Indianapolis: Hackett, 1998).

11. For a historical discussion of the laws of war and ethics in warfare see Michael Howard, George J. Andreopoulos, and Mark R. Shulman, eds., *The Laws of War: Constraints on Warfare in the Western World* (New Haven: Yale University Press, 1994).

12. Larry A. Tritle, *A New History of the Peloponnesian War* (West Sussex: Wiley-Blackwell, 2010), describes the combat trauma of veterans from the Greeks to the present day.

13. I used William L. Richter, *The Army in Texas During Reconstruction, 1865–1870* (College Station: Texas A&M University Press, 1987) as my source for Texas Reconstruction for the above discussion.

14. This scenario is based on the above-cited operations and manuals, as well as the author's military experience.

15. Headquarters, U.S. Department of Army, *FM 3-39, Military Police Operations* (Washington, D.C., August 2013).

16. Headquarters, U.S. Department of Army, *FM 3-20.96, Reconnaissance and Cavalry Squadron* (Washington, D.C., 12 March 2010).

Bibliography

Bailey, Beth. *America's Army: Making The All-Volunteer Force.* Cambridge: Harvard University Press, 2009.

Clausewitz, Carl V. *On War.* Eds. Michael Howard and Peter Paret. Princeton: Princeton University Press, 1989.

Howard, Michael, George J. Andreopoulos, Mark R. Shulman, eds. *The Laws of War: Constraints on Warfare in the Western World.* New Haven: Yale University Press, 1994.

Keegan, John. *A History of Warfare.* New York: Alfred A. Knopf, 1993.

Mendel, William W. *Combat in Cities: The LA Riots and Operation Rio.* Fort Leavenworth: Foreign Military Studies Office, July 1996.

Pew Research Poll. http://www.pewforum.org/2013/07/11/public-esteem-for-military-still-high/#strong-consensus-about-military, accessed 28 July 2014.

Richter, William L. *The Army in Texas During Reconstruction, 1865–1870.* College Station: Texas A&M University Press, 1987.

Thucydides. *The Peloponnesian War.* Trans. Steven Lattimore. Indianapolis: Hackett, 1998.

Tritle, Larry A. *A New History of the Peloponnesian War.* West Sussex: Wiley-Blackwell, 2010.

United States. Department of the Army, *Army Doctrine Publication, 3-28, Defense Support of Civil Authorities.* Washington, D.C., 26 July 2012.

_____. *Field Manual 3-39, Military Police Operations.* Washington, D.C., August 2013.

_____. *Field Manual 3-20.96, Reconnaissance and Cavalry Squadron.* Washington, D.C., 12 March 2010.

_____. *Field Manual 3-94, Theater Army, Corps, and Division Operations.* Washington, D.C., April 2014.

United States. Marine Corps. Concepts and Programs 2013. Chapter 2, 14, http://www.hqmc.marines.mil/Portals/136/Docs/Concepts%20and%20Programs/2013/CP13%20CH2_WEBFA_5FEB13.pdf, accessed 1 August 2014.

Wombwell, James A. "The Long War Series Occasional Paper 29." *Army Support During the Hurricane Katrina Disaster.* Fort Leavenworth: U.S. Army Combined Arms Center Combat Studies Institute Press, 2009.

Neurobiology of a Zombie

Steven Schlozman

Gedankenexperiment … German … *"a thought experiment."*

In a thought experiment one literally plays around with an idea. This is a time honored intellectual somersault that allows a person to make sense of hard-to-prove and world-altering concepts.

Physicist Edwin Schrodinger famously engaged in a thought experiment that featured a dead cat with just an ounce or two of radiation. From Schrodinger's celebrated intellectual dexterity … his *Gedankenexperiment* … Schrodinger was able to make clear the perplexing ambiguity of quantum inconsistency.[1]

Schrodinger's experiment, it turns out, is as good a place as any to start talking about zombies. After all, a dead cat, a tincture of radiation, crippling uncertainty—these are all central to the modern zombie trope. I have written about the neuroscience of zombies in both fiction and scientific papers.[2] I have given international lectures on the putative biology of the shambling ghouls from *Dawn of the Dead*, and I have spoken for the National Academy of Sciences about how one might conceptualize what ails the modern flesh-eating Romero-esque movie ghoul.[3]

In other words, I have done my own *Gedankenexperiment*. For the sake of higher learning, this particular thought experiment will almost certainly fail to achieve the heralded notoriety of Schrodinger's cat. Still, it seems worth at least fifteen minutes or so of infamy. One must be remembered for something.

Here is the way the experiment goes, and I will take the liberty of describing this exercise in the present tense for dramatic effect. (Also, who can write about zombies without at least a little cheeky drama?)

It is late at night in 2009. My wife, who is thank goodness utterly healthy now, had just been diagnosed with breast cancer. I am scared, even though I know her prognosis is excellent. I have just gotten my children to bed and I have decided to enjoy a beer, some potato chips, and whatever happens to be on the television. I click on the remote and am treated to black and white images of stumbling blank faces converging on a lone house at nighttime. The windows of the house are clumsily boarded. The situation looks bleak.

Strangely, I feel *better*. As a life-long lover of horror movies, and as an especially rabid fan of zombie tales, I am delighted, though not at all surprised, that *Night of the Living Dead* is once again playing. The movie is in fact imminent domain.[4] Mr. Romero did not initially fancy himself a filmmaker and therefore did not secure copyrights for the movie. Any given night, somewhere in the world, this movie is on the TV or the big screen. It is even entirely on YouTube.

So, I sublimate. I cannot cure my wife's cancer, but I can try to cure these zombies. I do my thought experiment. If Schrodinger can play with the uncertainty of life or death in a cat, then I can play with that same ambiguity with zombies. I am going to make biological and behavioral sense of Romero's monsters.

The Goals

In this essay, I will engage in a particular thought experiment. Let us ponder the medical, neurobiological and psychological mechanisms of the modern zombie. After all, in an honest-to-goodness zombie outbreak, these poor souls would ultimately end up in our emergency rooms. We would not blow their heads off. That is what happens in movies. In the real world, we would examine humans who have contracted the zombie disease, we would study that disease, we might conceptualize some approximate animal models, and we would eventually postulate an etiologic source of the disturbance. If we are lucky, we would cure those who are afflicted.

Therefore, we are going to build a zombie by examining an imagined patient with the zombie disease. Then we are going cure him ... or her ... or it, as best we can. (We should let the philosophers discuss the sticky issues of gender at a later date.)

Let us begin.

The Scenario

This discussion will focus on the zombies that first came to life (so to speak) in George Romero's classic film back in 1968. Romero did not in fact call his monsters zombies. He called them ghouls. Various Hollywood magazines referred to the shuffling brain-dead cannibals in his first film as zombies, and the name therefore stuck. For now it is safe to say that if the word "zombie" is uttered in Western culture, it will most likely conjure some version of Romero's vision. Those zombies are the focus of this scientific inquiry.

This discussion will also focus entirely on the so-called "slow-moving zombies." The "fast-moving zombies" may have gained popularity of late, but one can convincingly argue that the slow zombies are in fact more frightening and also happen to conform more readily to neurobiological explanation. This is a central focus of this essay.

So, a slow moving zombie

- is slow
- lacks coordination
- lacks the ability to solve even simple problems
- is ravenously hungry
- has a propensity for human or at least animal flesh.
- seems always ready to attack
- retains the capacity to lunge with fluid movements (as opposed to the earlier mentioned lack of coordination)
- is often depicted as febrile (having a fever)
- is often depicted as having an elevated respiratory rate and heart rate, either through a formal medical examination or simply by the observation by the viewer of rapid breathing and accelerated pulse.
- can "reanimate" (a pseudo-scientific way of saying can rise from the dead)
- appears to be increasing in numbers, suggesting the early stages of a frightening pandemic

For the sake of the science here, we are also going to define the modern cinematic zombie as suffering from a contagion. Whether that contagion is a biological agent or an inert contaminant is unclear (it is all unclear, since zombies are not real), but we can use this as a jumping off place for our inquiry. Certainly any highly organized biological entity that is breathing rapidly, does not walk well, and has an elevated heart rate and temperature is likely very sick and infected. These are the cardinal signs of sickness and sickness itself is central to this zombie thought experiment.

We are also going to take "rising from the dead" out of the picture. Nothing rises from the dead. So, although the patient might be "as good as dead" (again, an existential quandary best left to the philosophers) the heart beats and the lungs breathe.

Next, we shall do what medicine always does. How many times have you gone to the doctor and been told what you *do not* have? This is the great strength of modern medicine. Tests in modern diagnostics rule things out. One finds that one's leg is not broken with an X-ray. The absence of a fracture, however, fails to explain the pain in the leg that led to the X-ray in the first place. The process of "ruling out" suggests that clinicians and public health officials would approach a zombie pandemic similarly. What can one rule out when examining the modern-day zombie?

If a slow, shambling, confused, rageful and aggressive person comes into the emergency department, that person would most likely be considered intoxicated until proven otherwise. Typically, this patient is brought in by outside officials. This could be the police or perhaps a paramedic. Often the patient arrives in restraints, and this kind of presentation is not at all unusual in a busy emergency ward. Illicit substances are unfortunately common and these substances are among the most common causes of rageful confusion among patients in the emergency settings. Most admitting physicians would first draw blood from the patient to check for the ingestion of particular substances.[5] Methamphetamine, PCP, LSD ... these would comprise the most likely culprits. The patient might also be suffering from a toxic reaction to already prescribed medications. Too much thyroid replacement medications might, for example, share some of the same characteristics that one observes in a zombie tale.[6] These are all diagnoses that can be tested for, and in the case of our imagined patient, the tests come back negative.

Next, the physician would examine the vital signs. These would already have been measured by the paramedics or the admitting nurse, and if the measurements were alarming, the physician would be viewing these even as the blood is being drawn. To a medical professional, an elevated heart rate or a high fever in the setting of confusion and difficulties ambulating would immediately trigger concerns for a serious infection. The blood work would thus also include a white blood cell count, and the results would show that these cells are elevated above normal. White blood cells are created to fight infection. An elevated white blood cell count would trigger the beginning of a search for a contagion.

Still, there is one major problem with the infection hypothesis. The patient is hungry. Being sick and also hungry is indeed very rare. However, one could be infected with something that in fact *causes* hunger. Those infections exist, and we shall get to those later, but let us first take a second to debunk a common zombie misconception.

Contrary to popular belief, infection with rabies is not a good zombie proxy. Although the "rage virus" of movies like *28 Days Later* is often compared to the rabies virus, there are fundamental characteristics among those suffering from rabies that makes this particular infection an unlikely cause for the zombie outbreak. Rabies does in fact make its victims slow and confused and aggressive. Rabies victims even harbor a desire to bite others. However, people with rabies are never hungry. They are terrified to even swallow anything. Show a person with rabies a glass of water and he will cower in terror, a condition called hydrophobia.[7] The rabies virus so inflames the apparatus of swallowing that even the barely consciously recognized cue to swallow—a glass of water—is met with revulsion. The patient in question cannot have rabies or anything like it as long as this aspect of rabies is one of the defining characteristics. Just to be sure, a biopsy the brain could be done (only in very dire circumstances) and there one would, in a patient with rabies, plainly detect the rhabdovirus characteristic of this particular disease. The current patient, however, has neither the requisite hydrophobia nor a version of the rabies virus after brain biopsy. This patient has something else.

Plus, the patient spits and coughs and wheezes. That is a very big clue. Very few people spit and cough and wheeze on purpose. The production of phlegm is a not a desirable activity. A contagion, however, loves it when the host spews it around. This allows the contagion to happily reproduce. The patient is potentially spreading whatever he has just by breathing. There is at least a chance, in a coughing patient, that the infection is airborne.

This is because the current thought experiment also involves breaking news that a potential pandemic is brewing. Emergency wards are notified by federal and local officials whenever there are signs of a rapidly spreading infection. If a pandemic is suspected, and if patients are coughing, then the infection is airborne until proven otherwise. In fact, epidemiologists will argue that it is difficult to imagine a zombie pandemic that spreads through biting alone. This is another reason that rabies is a poor zombie proxy. The rabies virus is spread through saliva … through biting. It would be too easy to isolate those with the affected behavior and therefore stop the logo-rhythmic spread of the disease. For a pandemic scenario, one needs an air-borne germ. It is only in nations with poor medical infrastructure that dis-eases that are not airborne can happily prosper. Western nations are otherwise too quick to isolate and confine.

In this scenario, the pandemic is blossoming in the most sophisticated medical settings on the planet. If the dead could come back to life, then a non-airborne pandemic would remain possible. In other words, it is conceiv-able in the "dead rising" scenario that public health officials would initially fail to isolate potential biters once they had died. Thus, the un-isolated dead would rise and the disease would progress.[8] Since the dead in this thought experiment do not come back to life, we have to turn to an airborne agent to create our quickly growing apocalyptic scenario.

So, in medical terms, this is a *28-year-old Caucasian male who presents to the emergency department after being pulled off of a helpless bystander on whose nose he was nibbling. The patient is restrained by EMTs, but is confused, disoriented, indeed unable to talk but makes instead snarling, grunting sounds. He appears ill, is diaphoretic (sweating), is tachycardic to 135 (heart rate is 135) and he is coughing. His temperature is 104.3°F. He cannot follow even simple commands, and he snaps his mouth at you as you bend over to examine his chest. Toxicology screens are negative. He has an elevated white blood cell count. He attempts to bite the nurse's fingers.*

This is something new. This looks like a new contagion that cannot yet be identified but that is characterized by the symptoms of the patient and countless others like it across the nation. Extreme caution is advised.

An Animal Model

Next we try to replicate the disease in a setting where we can study it. We are not going to do this with humans, of course. Remember, this is not a movie. We are not going to use human subjects to study a new and deadly and so-far incurable contagion. Although this particular discussion does not explore the murky ethical waters of animal experimentation, in the medical world there is no doubt that one would next search for ways to replicate the disease in non-human subjects. Chimps get Ebola. Dogs get liver disease. Mice can be transgenetically designed to mimic all sorts of ailments.

Still, we have not yet even identified the contagion. We just know that dogs that are exposed to sick humans with the zombie bug do not change. The same goes for primates and mice and fish and fruit flies and every other kind of organism we commonly turn to in these frightening circumstances. That does not mean that the other creatures are not infected. It just means that they do not get sick. Since we do not have an identified pathogen, we cannot get much further with this approach. We also cannot utilize the transgenetically engineered mice. We cannot use genes to make something if we do not know what it is we are making.

So we do the next best thing. We look for examples of similar processes in nature that may or may not have been categorized as diseases. For example, scientists have studied the neuropathways of pit bulls as a means of understanding human aggression.[9] Scientists have studied the tendency for fruit flies to fight one another if the affections of a female fruit fly are at stake.[10] Neither the pit-bull nor the jealous fruit fly is considered ill. Thus, comparable behavior in humans might meet the criteria that we have established as defining characteristics for the spate of zombies that are currently increasing.

Pit bulls are too smart to be zombies. They can solve simple problems. And fruit flies, on the other hand, are too limited. Our zombies may be slow and mentally challenged, but that is still a long way from the cognitions of a fruit fly. So, what is a good animal model?

Consider the infected humans. They are hungry, carnivorous, responsive only to the most rudimentary stimuli, and clumsy. If one gets near an infected human, he will snap his jaws in the direction of the stimuli. Nearer still and they may surprise the scientist with an athletic lunge.

Our zombified humans are *reptilian*, in other words. They're human incarnates of crocodiles, or maybe snapping turtles. This is because the reptile brain is wired to either attack or to flee. There exists no higher cortical powers in the reptile. If it is a carnivorous reptile and it is hungry, then it will attack. If it is startled, then it will retreat. These are the only options that the primitive brain of the reptile allows.

Except that crocodiles walk without stumbling. Lizards do not really shamble. So our animal model is most consistent with a sort of drunken crocodile. In this case, however, we need not get our crocodiles drunk. We can instead use our already present understanding of functional neurobiology to infer that certain regions of the crocodile's brain are damaged or dysfunctional. Specifically, if we damage the cerebellum and the basal ganglia of the crocodile, we essentially create a zombie within a crocodile's body. The crocodile cerebellum allows for balance, and the crocodile basal ganglia allows for fluid motion. These are primitive brain regions that are present as well in humans but nevertheless seem lacking in function in the infected patient. For the animal model to be maximally accurate, one needs to damage the cerebellum and the basal ganglia of the crocodile and then starve it. Now it is hungry and clumsy. Now it is much like the human patient.

This is where one starts the neurobiological inquiry. Anyone who thinks about the brain cannot help but to wonder about the pathology of the zombie. If one knows a bit about a crocodile's brain, one can begin to infer what works and what fails to work in a zombie's brain. This is how zombie science moves forward.

Neurobiology and the Zombie

To understand our zombie, we need to understand our crocodile. To understand our crocodile, we need to understand its brain.

A crocodile has barely changed since if first crawled out of the primordial muck. It is efficient at propagating and therefore has had little adaptive pressure to be anything more (or less) than it is. Although ancient crocodiles were much larger, their brains were pretty much the same.

The crocodile brain is almost entirely driven by the amygdala. The amygdala is a structure in the reptile brain that makes up the entire higher brain of the reptile. Here it is important to remember that humans too have amgydalae, but humans enjoy much more sophisticated brain structures that govern the amygdala's instructions. Left alone, the amygdala is devoted only to the basest of emotions. Fight, flight, and maybe rudimentary lust are all situated in the amygdala. If the crocodile is hungry, then the amygdala says "fight." If the crocodile is frightened (a rare event) then the amygdala says "flight." If the crocodile needs to reproduce, then the amygdala gives it the cue to make more of itself. Like all higher organisms (as compared to insects, for example) there are two hemispheres to the crocodile brain. That means that the higher brain of the crocodile has two amygdalae. Experiments have shown that if you introduce damages to these amygdalae, the crocodile just sits still and does not move.[11] The essence of the crocodile is the amygdala. There is no higher brain to solve problems. Show a crocodile a locked window, and short of trying to walk right through it, it will not get very far. Show a crocodile a door, and even if it had fingers and opposable thumbs, it will not be able to open the door. Now starve it and introduce damages as well into its neuro-apparatuses for keeping its balance, and the animal model is complete. The crocodile, like the infected human, is now a shambling, ravenous creature that seems always ready to attack. It will not try to escape (too hungry) and it will not try to procreate (too hungry). In short, make that crocodile a human and one makes a zombie.

There is still, however, that pesky and unexpectedly coordinated lunging that the crocodile and the zombie can do, but there's a biology to that as well. This is a good place to talk about the regions of the human brain relative to the crocodile brain. Then we can overlay the behavior of our zombified humans with our slightly damaged crocodiles, and compare and contrast the corresponding neurobiology. For this, we need a cast of anatomical characters. These include the frontal lobe, the region of the brain involved in solving complex problems and inhibiting impulsivity. There is also the cerebellum, as we've mentioned, that helps us to remain balanced and upright. Add to this the basal ganglia, and we can stay upright and move with impressive coordination. The amygdala is shaped like an almond—amygdala is Greek for "almond." It sits deep in the human brain beneath the frontal lobe but, as has also been mentioned, is almost all there is to the higher brain of the reptile. Finally, there is the ventromedial hypothalamus. This is the bottom, middle part of the region of the brain that sits below—"hypo"—to the thalamus. It is responsible for telling the rest of the body when it is sated with food. This is why it is called the satiety region of the brain. When one has eaten enough, the ventro-medial hypothalamus gives the signal to stop feeding.

Crocodiles cannot solve problems and neither can zombies. Humans have frontal lobes to solve problems but crocodiles do not. Therefore, our zombies have lost, or at least suffered a significant detriment, in frontal lobe function. This is an important clue towards the kind of suspected contagion for the zombie pandemic.

Crocodiles with damaged mechanisms of balance and our zombies also both shamble. That means that the cerebellum of the zombie is impaired. The lack of coordination also implicates the basal ganglia. This is another clue.

Finally, crocodiles are all amygdala. Hungry crocodiles use their amygdala primarily as a call to arms. Our zombies are therefore hyper-amygdala driven. That means that despite the loss of coordination and the seeming cluelessness of the zombie, the amygdala remains intact.

Our starving crocodile and our zombies are both frighteningly hungry. What causes pathological hunger? From a disease standpoint, one might consider a tapeworm. These zombies do not have tapeworms. In the thought experiment, the clinician looks for tapeworms after noting the significant hunger. So, can something make the zombie, who is always eating, feel like the crocodile who is starving? If that something exists, it would have to affect the satiety region—the ventromedial hypothalamus. It turns out that there are some germs one can implicate in the creation of hunger. We shall get to these in the next section.

Consider now the most vexing neurobiological issue of all. Remember, the working theory is that the brain of the zombie suffers dysfunction of the frontal lobe, dysfunction of the cerebellum and basal ganglia, increased amygdalar activity, and increased activity of the ventromedial hypothalamus. Still, the zombie can lunge with impressive dexterity. Recall the elevator scene in *Dawn of the Dead*. Zombies move fluidly when confronted with prey. How does one understand this observation?

People with Parkinson's disease have dysfunctional basal ganglia. They fail to secrete enough dopamine, the neurotransmitter responsible for movement and present in the basal ganglia, to allow for steady, coordinated activity. Toss a person with Parkinson's disease a ball, however, and he will sometimes involuntarily catch it with one fluid and graceful motion. For the infected humans in our thought experiment, we will postulate that just as with the person with Parkinson's disease, an arousing stimuli can produce an involuntary and coordinated motion. Show a zombie food and it will move like a professional athlete, but only for a second or two. That's something we know an injured basal ganglia will allow.[12]

Finally, why the desire for flesh? This gets back to the reptile brain, and here one must stretch the science. Certainly some reptiles are devoted vegetarians, but no self-respecting crocodile would ever stoop to eating a turnip or a pear. Crocodile brains are primed for meat. To that end, we will say that our infected humans are similarly primed. Why they avoid other zombies as food is unclear. It might be because the other zombies smell bad. Primitive animals will avoid the scent of decay as an evolutionary mechanism of evading disease. Still, probably a better answer is that the zombie story works better if zombies eat humans only.

Thus, using the damaged brain of a hungry crocodile, one can start to infer about the brain of newly infected humans. In other words, one can start to make sense of the neurobiology of the zombie.

Infectious Disease Models

By now your city is under martial law. Zombies are popping up every-where, and despite the best efforts of the medical community, people still feel compelled to kill every zombie that they see. A slow, shambling old lady might be somebody's grandmother with a twisted ankle or might instead be a dead-eyed flesh-eating ghoul who cannot possibly recognize anyone. One cannot know the answer until she's close. (Closer ... closer.) By then it is too late. This is why the slow moving zombies are unnerving. The neurobiological deficits prevent them from running, so one is left with time to think ... to hope. Time to think is a dangerous thing if one lacks the information one needs. What one in fact needs most of all is to understand just what exactly is infecting what used to be that grandmother. Then one can go about making a cure.

So start with what is already known. The infection is airborne. The pub-lic health officials have shown everyone the logarithmic curves. This is an infection that spreads like the flu. One case leads to three cases leads to fifteen cases and so on ... every cough, every wheeze, every contaminated doorknob is a zombie waiting to happen.

So, what if this *is* a strain of the flu? Or at least what if the flu virus is the airborne component? This disease is so new, so strange, that doctors would not first think of the obvious. Humans are very susceptible to the flu, and the flu virus is highly airborne. Additionally, one can test for antibodies to flu and these tests come back positive. In other words, the infected humans are making cells that are specifically triggered by a flu infection. Flu is at least part of the answer.

Doctors have seen the flu. People with the flu stay home from work. They take Tylenol. They binge watch Netflix. People with the flu do not sham-ble. People with the flu do not become violent. People with flu do NOT eat other people.

Plus, people with the flu can pick up the phone and call their doctors. They can solve both simple and complex problems. Flu alone will not turn a person into a zombie. It might be part of the picture, but it is certainly not the whole picture.

What about the cognitive deficits? What can take the frontal lobes off-line but maintain functionality of lower brain regions? Many encephalopathic viruses—herpes, for example—can do this, but these viruses progress more quickly towards death. Also, for the sake of this thought experiment, we must note in our imagined scenario that there have been no antibodies to herpes detected so far.

Clinicians and researchers would inevitably perform autopsies on a deceased, infected human. In all cases, the higher brain of the deceased is full of holes. It looks like Swiss cheese, or a sponge. This is called "spongiform encephalopathy," and this is found most commonly with horrible Prion-related illnesses.[13]

Prion is shorthand for "proteinaceous infection." Prions are not even technically alive. They don't have RNA or DNA. They are just proteins, and these proteins exist in all of us all of the time. However, healthy brains degrade the prions before they can fold onto themselves and become, for unclear reasons, deadly. If someone is unlucky enough to ingest a prion, the prion is coded to wander from the gut to the brain where it stops the degradation of the already present prions enough to allow brain damage to begin.

Mad Cow disease … Creutzfeldt-Jakob disease … Scrapie … these are all prion related illnesses. To date, once healthy prions become unhealthy, there's no known cure. These diseases are usually 100 percent fatal.

As if this were not already enough, there is still one more puzzle to solve. The shambling, confused, and febrile patient is also very, very hungry. Tests have already ruled out tapeworms and other parasites that live in the intestines and steal food. What about the ventromedial hypothalamus? Are there any infections that can disrupt the satiety sensors in the brain?

It has recently been noted that certain strains of Adenoviruses and Bornaviruses can make the ventromedial hypothalamus fail to respond to satiety signals from the gut. Other, less common viruses have been implicated as well. This is not part of the thought experiment. These viruses are real.[14] Not everyone becomes dangerously hungry when these infections strike, but when these infections *do* cause hunger, they cause hunger to the point of pathological obesity. Further, these viruses are both zoonotic and primarily human contagions. The Bornavirus is in parrots and sheep. The Adenovirus is passed from human to human through the creation of common cold–like symptoms.

To review then, the new contagion, the source of this frightening new pandemic, is likely a flu virus that carries some kind of prion and may also harbor aspects of something like special strains of the Adenovirus. The chances, outside of our thought experiment, of any virus carrying a prion, are close to zero. Still, one can postulate a genetically engineered virus with an embedded prion. Remember that prions are just proteins. They are even smaller than viruses.

A Cure or a Conclusion

We can treat the Adenovirus and the Bornavirus and we can treat the flu as well. We can vaccinate, provide supportive care, treat coughs, and isolate infections. But prions? We cannot stop those yet. Prions cause incurable diseases like Mad Cow disease and Creutzfeldt-Jakob disease. There are no cures for these illnesses. In a fictional scenario, that happens to be sort of a good thing. What's a good zombie story without a sequel? The infection needs to stick around so that one can see the interesting stuff happen. Zombie stories are, after all, never about the zombies. They're instead cautionary tales about ourselves.

Zombie movies don't, by definition, end well. That leaves us plenty of time to solve the prion problem in the next movie.

NOTES

1. Erwin Schrödinger, "Die gegenwärtige Situation in der Quantenmechanik," *Naturwissenschaften* 23 (1935): 807–812; 823–828; 844–849.

2. Steven Schlozman, *The Zombie Autopsies: Secret Notebooks from the Apocalypse* (New York: Grand Central, 2011.

3. "Programme," Conference Program (2013), http://www.kcl.ac.uk/innovation/groups/chh/Narrative-Medicine-Conference-/NMC-Programme.pdf; and "Science of the Living Dead," *The Exchange*, October 23, 2009.

4. *Night of the Living Dead*, written by George Romero and John A. Russo, directed by George Romero, Image Ten, 1968; see also U.S. Copyright Office, Circular 92, Copyright Law of the United States of America, Chapter 4: Copyright Notice, Deposit, and Registration, Omission of notice on certain copies and phonorecords.

5. E.A. Ross, G.M. Reisfield, M.C. Watson, C.W. Chronister, and B.A. Goldberger, "Psychoactive 'bath salts' intoxication with methylenedioxypyrovalerone," *The American Journal of Medicine* 125, no. 9 (September 2012): 854–858.

6. E.B. Gordon, "Thyroxine causing aggression: Letter," *British Medical Journal* 2, 5961 (April 5, 1975): 36–37.

7. P. Carrara, P. Parola, P. Brouqui, and P. Gautret, "Imported human rabies cases worldwide, 1990–2012," *Tropical Disease* (2013).

8. Philip Munz, Ioan Hudea, Joe Imad and Robert J, Smith, "When Zombies Attack!: Mathematical Modelling of an Outbreak of Zombie Infection," *Infectious Disease Modelling Research Progress*, eds. J.M. Tchuenche and C. Chiyaka (New York: Nova Science, 2009), 133–150.

9. G.S. Berns, A.M. Brooks, and M. Spivak, "Functional MRI in awake unrestrained dogs," *PLoS One* 7, no. 5 (2012).

10. Jill K.M. Penn, Michael F. Zito, and Edward A. Kravitz, "A single social defeat reduces aggression in a highly aggressive strain of Drosophila," *Proc Natl Acad Sci* 107, no. 28 (July 13, 2010): 12682–12686.

11. E.G. Keating, L.A. Kormann, and J.A. Horel, "The behavioral effects of stimulating and ablating the reptilian amygdala (Caiman sklerops)," *Physiol Behav* 5, no. 1 (January 1970): 55–59.

12. B.J. Robottom, W.J. Weiner, F. Asmus, H. Huber, T. Gasser, and L. Schöls, "Kick and rush: paradoxical kinesia in Parkinson disease," *Neurology* 73, no. 4 (July 28, 2009): 328.

13. A.-T. Van Ba, T. Imberdis, and V. Perrier, "From Prion Diseases to Prion-Like Propagation Mechanisms of Neurodegenerative Diseases," *Int J Cell Biol* (2013).

14. A.K. Mitra, and K. Clarke, "Viral obesity: fact or fiction?" *Obes Rev.* 11, no. 4 (April 2010): 289–296.

BIBLIOGRAPHY

Berns, Gregory S., Andrew M. Brooks, and Mark Spivak. "Functional MRI in awake unrestrained dogs." *Plos ONE* 7, no. 5 (2012): 1–7.

Carrara, Philippe, Parola Phillipe, Brouqui Phillipe, and Gautret Philippe. "Imported human rabies cases worldwide, 1990–2012." *Plos Neglected Tropical Diseases* 7, no. 5 (2013): 1–7.

Gordon, E.B. "Thyroxine causing aggression: Letter." *British Medical Journal* 2, no. 5961 (April 5, 1975): 36–37.

Keating, E.G., L.A. Kormann, and J.A. Horel. "The behavioral effects of stimulating and ablating the reptilian amygdala (Caiman sklerops)." *Physiol Behav* 5, no. 1 (January 1970): 55–59.

Mitra, A. K., and K. Clarke. "Viral obesity: fact or fiction?" *Obesity Reviews* 11, no. 4 (2010): 289–296.

Munz, Philip, Ioan Hudea, Joe Imad, and Robert J Smith. "When Zombies Attack! Mathematical Modelling of an Outbreak of Zombie Infection." *Infectious Disease Modelling Research Progress*, edited by J.M. Tchuenche and C. Chiyaka. New York: Nova Science, 2011: 133–150.

Night of the Living Dead. Written by George Romero and John A. Russo and directed by George Romero. Image Ten, 1968.

Penn, Jill K. M., Michael F. Zito, and Edward A. Kravitz. "A single social defeat reduces aggression in a highly aggressive strain of Drosophila." *Proceedings of the National Academy of Sciences of the United States of America* 107, no. 28 (2010): 12682–12686.

"Programme." Conference Program (2013). http://www.kcl.ac.uk/innovation/groups/chh/Narrative-Medicine-Conference-/NMC-Programme.pdf.

Ross, E.A., G.M. Reisfield, M.C. Watson, C.W. Chronister and B.A. Goldberger. "Psychoactive 'bath salts' intoxication with methylenedioxypyrovalerone." *The American Journal of Medicine* 125, no. 9 (September 2012): 854–858.

Schlozman, Steven. *The Zombie Autopsies: Secret Notebooks from the Apocalypse.* New York: Grand Central, 2011.

Schrödinger, Erwin. "Die gegenwärtige Situation in der Quantenmechanik." *Naturwissenschaften* 23 (1935): 807–812; 823–828; 844–849.

"Science of the Living Dead." *The Exchange.* October 23, 2009. http://www.scienceandentertainmentexchange.org/blog/science-living-dead.

U.S. Copyright Office. Circular 92, Copyright Law of the United States of America, Chapter 4: Copyright Notice, Deposit, and Registration, Omission of notice on certain copies and phonorecords. Washington, D.C.: U.S. United States Government Printing Office.

Van Ba, Isabelle Acquatella-Tran, Thibaut Imberdis, and Véronique Perrier. "From prion diseases to prion-like propagation mechanisms of neurodegenerative diseases." *International Journal of Cell Biology* (2013): 1–8.

Communications in a Zombie Apocalypse
Usage, Control and Collapse of Mass Media

DIEM-MY T. BUI

Many zombie texts, television shows, and movies speculate on the role mass media plays leading up to and during a zombie apocalypse. In George Romero's Dead film series beginning with *Night of the Living Dead*, radio and television played a central role in reporting on the unfolding crisis of a zombie outbreak.[1] The survivors turned to the media for information on the animated dead, the cause of the outbreak, expert opinion, where to go for refuge, and on what was being done by institutions, such as the government, police, and military. In the film *28 Days Later,* survivors from a zombie apocalypse in the United Kingdom sought out protection from a group of soldiers, who broadcasted information about their settlement through a radio transmission.[2] While mass media is imagined as playing a key role in the dissemination of information and emergency management, many texts, such as television shows and movies, offer a more cynical view of what might be expected from mass media in the event of a zombie epidemic. *Resident Evil: Apocalypse* suggests that the mass media would be manipulated by powerful corporations, such as in this case, the Umbrella Corporation who was behind the zombie outbreak.[3] Max Brooks' novel *World War Z* presents an interview with a fictional former presidential Chief of Staff who described the media as owned by the largest corporations in the world.[4] It is suggested that these corporations would act in their own best interest, even if that meant covering up an impending disaster. In all of these scenarios, mass media is more than just a bystander, but the key to survival in or acceleration of a modern zombie apocalypse.

In a mathematical modeling where scientists determine a basic model for zombie infection and explore scenarios of a zombie outbreak and its solutions, such as quarantine and a cure, the scenarios point to disastrous outcomes for civilization unless the outbreak is dealt with swiftly and aggressively.[5] Such scenarios in this modeling can be related to how we deal with pandemics in general, but mass media's role has yet to be accounted for in dealing with a possible outbreak. How might mass media facilitate or hinder collective response, utilization of resources, quarantine, or spread of the infection? The role of media in previous real world disasters in the twenty-first century suggests that a variety of media approaches are possible. Yet, these approaches are influenced by the ways in which media is used and how the media industry is structured in the twenty-first century. This essay examines what might be expected from the media in a possible zombie apocalypse based on media practices and systems that were in place during recent disasters and tragedies in the digital media age.

Kyle Bishop, a professor of English at Southern Utah University who works on zombies and popular culture, characterizes current zombie stories as a barometer of society's anxieties rooted in post–9/11 insecurities and fears.[6] Borrowing from his overview of contemporary zombie texts as infection leading to death and desertion, this essay speculates that a zombie apocalypse would be precipitated by a zombie infection, in which the infection spreads rapidly leading to a breakdown in the social order and a survivalist situation.[7] Attention is focused on the period leading up to such an infectious outbreak and at the beginning stages of an outbreak with the assumption that mass media still will be somewhat functional in emergency broadcasting systems, such as Romero imagines in the film *Dawn of the Dead*.[8] In this film, survivors of the zombie apocalypse still were able to receive television and radio broadcasts in which so-called experts on the zombies passed on information about them and engaged in heated debates in chaotic studios about best approaches in handling the zombie infection. Eventually the broadcasts end, and the survivors are left to deal with the zombies on their own. To consider possible media responses in such a doomsday scenario helps to further a conversation on the kind of media that serves a democracy, particularly when the unimaginable, the collapse of civilization, occurs. A media system not restrained by commercial interests allows for a more open media in which ideas and information may be freely distributed and accessed. In the event of a zombie outbreak, any information disseminated through media outlets would be essential for survival.

In the 21st century, traditional mass media, including radio, television, film, and the press, converges with new media, which refers to interactive media and practices in the digital age. Mark Deuze, professor of telecommunications at Indiana University, uses the term digital culture to describe people's lives lived within media: "I consider digital culture … as an emerging value system and set of expectations as particularly expressed in the activities of news and information media makers and users online."[9] The Internet allows for limitless mediated content in which users can participate in consuming and contributing to it. Users are active agents in meaning-making in a digital culture. Further, the dependence on professional storytellers, such as journalists, is no longer as salient to making sense of a complex world as journalism once was in helping people understand global complexities in the twentieth century. Social media, such as Reddit (www.reddit.com), YouTube (www.youtube.com), Facebook (www.facebook.com), and Twitter (www.twitter.com) help to facilitate an interactive user experience where exchange of information is a key function allowing for participants to contribute to a narrative, however, at the risk of disseminating misinformation. Yet, access to the Internet, both in consumption and production, has been controlled and limited. Participation in the digital culture is not as open as was imagined at the advent of the Internet in the late twentieth century. Robert W. McChesney, a professor of communication at the University of Illinois at Urbana-Champaign, observed that policymakers in the U.S. believed the Internet would revolutionize communication and access to information.[10] However, as McChesney argues, the trend towards consolidation of media ownership indicates more corporate control of news, information, and communication. Gatekeeping, the journalistic practice of selecting and crafting the content of news media, continues in the new media age. If mass media is able to survive in the beginning of a zombie outbreak, past media practices suggest a cynical expectation of the role of mainstream corporate media, in which mainstream media responses in previous crises forebode putting commercial interests before public interests. Social media, at least in the outset of a zombie outbreak, may offer alternative information from corporate media and separate from their corporate interests. The next section explores why a cynical perspective might be warranted given mass media's role in past disasters and tragedies but suggests that social media may be a more significant contribution to apocalypse survivors.

Previous Crises, Future Response

The U.S. government funded Centers for Disease Control and Prevention (CDC) posted their survival guide for a zombie outbreak titled "Preparedness 101: Zombie Pandemic."[11] In a graphic narrative similar to *The Walking Dead* comic book series by Robert Kirkman, this guide delineates a story in which a man and his companion watches a television news report of a strange virus spreading across the country.[12] The report advises viewers to go to the CDC emergency website for more information. The man looks up the CDC website on his computer and prints out the emergency preparedness checklist. He returns to watching the news to follow the virus story and eventually switches over to the radio when the television broadcasting is no longer accessible. The man and his companion follow the instructions given on the radio to seek safety and pack the radio for their journey. This story, eventually revealed to be a dream, acknowledges the important role that mass media plays in a disaster, both in the prevention of and during a developing crisis where normal institutional functions of society break down. For prevention, the CDC publishes the "Preparedness 101" guide as well as a blog on what the zombie television series, *The Walking Dead* (AMC, 2011–present) can tell us about how to stay alive.[13] As the virus spreads in a state of emergency, the CDC guide and blog recommends following radio and television and checking the CDC website for updates. The guide and blog promote awareness of the CDC functions, as well as how to prepare for emergencies. Media is one of several sources that people in high-risk disaster areas use to assess their situation and decide whether to flee or stay.[14] Mass media, then, is a source that people turn to in times of crisis for information, communication, and emergency management. It is essential to survival. What happens when communication systems breakdown and people cannot depend on the media? If previous disasters offer any foretelling of how mass media might respond in the event of a zombie apocalypse, citizens have reason to fear for a collapse of society in which structural institutions, such as government, the economy, law enforcement, transportation, and health care would cease to exist.

Hurricane Katrina is a recent example of a communications breakdown during a real life disaster. Emergency communications systems experienced several difficulties and interruptions during this massive hurricane that hit the Gulf Coast in August 2005. The hurricane and its aftermath killed more than 1,000 people and damaged heavily populated areas along the coast, such as Biloxi and New Orleans. Fifty thousand people were displaced from their homes in what became an immense relocation effort. The news media who were on the scene from before the hurricane hit reported on the failure of President George W. Bush's administration and the Federal Emergency Management Agency (FEMA) to organize rescue and supplies for the evacuees. The rest of the world watched in horror as the news broadcasted the poor conditions the survivors lived in, many of whom were without shelter, food, lavatories, and clean water. Radio, television, cell phone service, and satellites were used for emergency communications. There were a number of major problems identified with the communications system during and after the hurricane. First, lessons from the 9/11 emergency communications system breakdown in 2001 were not addressed by the time the hurricane arrived in 2005.[15] During 9/11, the main radio channel used by firefighters was inundated with transmissions rendering them unintelligible, while police officers had no functioning communications system at all. This crisis revealed the need for better preparation with resources and training, more investment in technology, and better cooperation among first responder agencies. Emergency communications during Katrina also were overwhelmed, failed to work properly, were not fully coordinated in disaster preparation, and had a communications infrastructure in disarray. New Orleans experienced a communications blackout in which electronic mass media shut down as the hurricane passed through the area. Senator John McCain at the time criticized commercial broadcasters who held onto their analog spectrum during the disaster, which could have been used by emergency responders to communicate with each other.[16] These commercial television broadcasters were reluctant to return part of the radio spectrum in the 700MHz band that was loaned to them by the government in 1997 to transition from analog to digital. Second, the Emergency Alert System (EAS) that was established originally as the Emergency Broadcast System in 1963 to alert citizens in the event of an emergency was never activated by the White House or by state or local governments.[17] Many residents in the path of the hurricane did not have a sense of the urgency to leave what was almost guaranteed to be a life or death situation. Finally, while broadcast television and radio, cable, and satellite providers are required by law to transmit national emergency alerts, participation in state and local alerts is voluntary. A March 2007 U.S. Government Accountability Office Report to Congressional Committees stated that there were some business concerns by media operators over having to allocate time to alerts because it ate up time for commercials.[18]

Since the events of both 9/11 and Katrina, emergency systems have been updated so that technology can be utilized for alerts through traditional, mobile, and social media in the case of a crisis. Besides investigating a possible zombie outbreak and providing technical assistance, the CDC has set up popular social media options to send messages in the event of a zombie apocalypse or similar disaster. The CDC provides widgets for announcements, Twitter and Facebook accounts, and buttons and badges to post to individuals' own websites, blogs, and social media sites. The CDC uses these social media tools to post information and reach wider audiences. Social media users can share health and safety information and stay informed through mobile devices.

All of this preparation simulates the use of social media in real life crises showing how important social media has become in sharing information. The Boston Marathon bombing in April 2013 was an attack that began with two planted bombs that exploded during the marathon near the end of the course. The threat of the attack spanned several days as investigators hunted down suspects around the Boston area. During this time, investigators and government officials used social media, such as Twitter and Facebook, to update the public on the situation and suspects, as well as to pass along information on how to remain safe with the suspects at large in the city. In an ongoing crisis where fear of the unknown is widespread, mass media can make a precarious situation even more unstable by encouraging speculation and panic. Social media backfired when speculation on Reddit about the identities of the suspects in security footage led to false accusations of individuals.[19] For instance, *The New York Post* published a story that a Saudi national was identified as a suspect and conservative columnist Erik Rush suggested racially profiling Saudis in a personal tweet on Twitter.[20]

This type of speculation situation also was observed in the 2001 anthrax attacks on individuals and was considered the worst bioterrorist attack in U.S. history.[21] The 2001 anthrax attacks in the U.S. demonstrate how mass media frames and contextualizes threats to society. About a month after the 9/11 terrorist attacks on the U.S., the first reported anthrax death in the news was a photo editor for a tabloid publisher based in Florida. Although initially not made public, a week after the 9/11 attacks NBC news anchor Tom Brokaw received a letter in the mail containing anthrax powder. Other contaminated letters were discovered testing positive for anthrax. Twenty-one more people became ill after coming in contact with anthrax spores, and five people died from exposure. Government officials acknowledged the attacks as a possible terrorist connection. Public panic was at a high given the immediately preceding events of the 9/11 attacks. Kristen Alley Swain, professor of journalism at the University of Mississippi, observed that the uncertainty of the anthrax situation bred fear in the populace and that media coverage emphasized speculation and offered conflicting messages.[22] In an age of instant communication, journalists feel pressured to deliver instant information that may not exist yet in a crisis, such as to give information on perpetrators of crimes, to interview experts who may have only limited knowledge on a specialized area, or to offer definitive answers to questions before all facts are known.[23] Despite the media's misinformation, Swain noted that explanations in media coverage that included quasi-scientific stories and definitions of anthrax and infections helped to contextualize the uncertainty and, therefore, mitigated outrage rhetoric, or negative public reactions to crises. She concluded that in times of frightening crises, journalists should not offer vague information but, rather, provide concrete advice about what people should do to minimize their risk and protect themselves. Romero depicts the uncertainty factor, conflicting messages, and vague advice in *Night of the Living Dead* when news reports show scientists and military experts disagreeing with one another on causes. As the media describes the reanimation of the dead, one survivor, Barbra, vacillates from shock to hysteria while recalling her brother who fell victim to a zombie. The media coverage in *Night of the Living Dead*, much like Swain posits in her study, was both functional and frightening. The news reports on the radio and television for the film's survivors sometimes offered concrete, helpful information, such as how to kill the zombie and how one becomes infected. Yet, it also fueled fear and outrage when it showed experts speculating on the infection's origin and uncertainty of what institutional assistance would be given.

During the anthrax attacks, the CDC provided as much information as possible given the uncertainty of the situation and perpetrator. Vicki Freimuth, who was director of communication of the CDC during the attacks, described the tension between the CDC and the press who required constant updates on the situation: "The media had little background to draw on in covering anthrax, a rare disease prior to this time. The information was new and highly technical. Moreover, some of the attacks were on the media themselves, adding to their outrage and anxiety."[24] The CDC and the media had to adapt to each other since each of them depended on the other. The CDC needed the media to communicate accurate information in a public health crisis while the media demanded the CDC provide constant new information and advice. In order to have a more effective and less contentious relationship with each other, they had to reorganize their communication and the system for public inquiries, such as communication teams dedicated to specific tasks like analyzing media information or organizing leading clinicians who could be activated to disseminate information quickly in a crisis.[25] Learning from previous crises and taking advantage of digital technology, the CDC diversified their public relations communications in which the *Preparedness 101: Zombie Pandemic* is a result. This included incorporating social media and popular culture references, organizing media and communications into teams dedicated to specific tasks, and centralizing their website for consistent information. When the zombie apocalypse or other disaster arrives, the CDC has plans to utilize mass media to its maximum potential as long as the communication infrastructure stays intact. By emphasizing being prepared with emergency kits and safety plans through its Zombie Preparedness website and Public Health Matters blog, the CDC hopes everyone will be ready for the zombies.

The Effect of Mass Media

Utilizing and engaging mass media creates an experience for the user. In a critical essay in *Extra! Magazine*, freelance writer Josmar Trujillo contends that during a disaster, the media too frequently turns to either positive perspectives celebrating heroism or the bungled response of emergency organizations and government.[26] Trujillo cites Hurricane Katrina and Hurricane Sandy as examples of mainstream media gushing about or criticizing government officials and agencies. Instead, Trujillo argues for more inclusion of disaster victims' voices and concerns. When mainstream media offers a political reality disconnected from the people experiencing the crisis, people often turn to other media platforms to broadcast their thoughts, pool information, or express their concerns. While mainstream digital media, such as online newspapers, allowed reader input in the comments section, social media had more potential for collective organizing and sharing of information due to the interactive nature of the social media tools. Beginning in 2009 with the Iranian presidential election protests and in 2010 with the Arab Spring, understood as collective political protests against established governments and regimes in the Arab world, several tools of social media, in particular, Facebook, YouTube, and Twitter, were used by protesters to communicate with each other and the outside world. As many foreign journalists were banned from reporting on demonstrations and governments shut down communication systems, protesters turned to proxy services and social media to report on the streets. The rest of the world showed unity and support with protesters through posting of street reports and of protesting colors. A number of researchers believe that social media encouraged a collective feeling of an already brewing anti-government sentiment that also acted as an alternative to government-controlled media.[27] In one 2012 study on Arab students and their relationship to social media, one Arab university student in the United Arab Emirates articulated the effect of social media this way, "Using the social media sites like Facebook, Twitter and YouTube you get to know the information immediately and so fast. Even faster than the TV. You feel you are connected to the whole world no matter where you are. Otherwise you feel so isolated."[28] Journalists, Howard Kurtz and Lauren Ashburn in a PBS interview noted that social media, such as Twitter, was used by people during the Boston Marathon bombing to get in touch with each other immediately and crowdsourced information faster than radio and television.[29] With the rise of social media, crowdsourcing, or the sharing of data and information by users, has been employed in creating a wide variety of projects from building weather databases to organizing knowledge, such as in Wikipedia. Lea Winerman, a writer for *Nature*, documented how people used social media to crowdsource immediately after disasters, such as the 2007 Virginia Polytechnic Institute and State University shooting or during wildfires in Southern California, to exchange and share information.[30] Regardless of whether social media is used in a crisis or not, new communication technologies open up new possibilities for relating to information and other social groups. What might be the possibilities of how social media is used in a zombie apocalypse if communications systems were to stay intact, at least for the start of the outbreak? One such recent disaster as described below, the March 11, 2011, Fukushima nuclear meltdown, demonstrates how social media was used to not only share information but also to express solidarity to those affected by the disaster.

When traditional media access is limited to corporate or government owners, social media may be used to disseminate information and as an emergency management tool for issuing warnings and monitoring user activities, as well as a mass media outlet for individuals and communities who do not have access to mainstream media channels.[31] A 2011 study on the Fukushima nuclear disaster observed that one of the benefits for individuals of using social media, such as Twitter, during times of disasters is that social media can act as a conduit for intimacy offering both commentary and social cohesion for users.[32] Examining the case of the earthquake and the Fukushima nuclear reactor breakdown in Japan, the researchers argue that mobile media extended traditional media practices from the past, such as sending picture postcards of devastated landscapes after disasters, for modern-day emotional connectivity and dealing with grief. Social media users posted their messages of solidarity and concern through media and participated in collective mourning in the virtual space with strangers. During the Boston Marathon bombing, for instance, *#BostonMarathon* was used as a hashtag to organize thoughts and information for users and to build a community of people trying to make sense of the tragedy. During the 2009 Iranian election protests, users uploaded and passed along what became a highly viewed YouTube video of political activist Neda Agha-Soltan being shot dead by a Basij militia member in a peaceful protest. Her death became a symbol of reformists, and the video images helped to give momentum to the reformist movement.[33] Thus, social media and mobile devices can offer new ways to access affective culture, or expressions of emotions and empathy, and becomes an important medium in a crisis or collective action.

In the event of a zombie apocalypse, media users and corporate owners can control media with different results. *Resident Evil: Apocalypse* depicts a television news reporter who initially survives the zombie outbreak. She uses a handheld video recorder to document and record the apocalyptic event. Eventually zombies kill her while the video recorder continues to film. Post-apocalypse, the reporter's footage is released but deemed a hoax as all evidence of the zombie outbreak is erased by the Umbrella Corporation. Thus, survivors may record their story for others, but corporate media owners can censor or alter the material for audiences. Other examples of found video footage include the Spanish horror films *[REC]*, *[REC]2*, and *[REC]3: Genesis*, and Romero's *Diary of the Dead*.[34] These types of films depict fictional narratives of people using their cell phones or video cameras to record the events of a zombie infection unfolding. In the fictional narrative, the footage is eventually found by survivors and edited into a film as a documentation of events. If *RE: Apocalypse* follows contemporary time, then *YouTube* and *Twitter* had yet to be invented. How would social media have changed people's belief in what the carnage was about if the reporter could have uploaded her video in real time as she was documenting the events? How many more lives may have been saved if that information about a zombie outbreak was posted as the outbreak was occurring? Would it have mobilized people into quicker collective action as Agha-Soltan's video did by inspiring outrage and grief? In an era of witnesses using their phones to record traumatic events, such as police brutality or tornadoes, and immediately uploading the videos to social media, media content and sharing may now be in the hands of consumers, unless censored by owners. An apocalyptic situation could be documented for survivors who may want to understand and avoid another apocalypse. By sharing videos through social media or sending them to contacts, the footage is made available to the public and its release is not dependent upon one media owner. Finally, another issue to explore would relate to a post-apocalyptic situation. If communications systems were to remain operational or be rebuilt, social media may be significant in processing and recovery in the event of a zombie apocalypse.

The Corporatization of Mass Media, or Are We All Zombies?

The former White House Chief of Staff in Max Brooks' *World War Z*, a fictional scenario of a post-zombie apocalyptic world, notes mass media's bottom line is profit at the expense of reporting a destructive epidemic: "The 'media'? You mean those networks that are owned by some of the largest corporations in the world, corporations that would have taken a nosedive if another panic hit the stock market? That media?"[35] His embittered statement points to the author's understanding of the mass media industry trends. For the past couple of decades, the mass media industry has consolidated into large media conglomerates that control most of global media today.[36] What does this mean for mass media in a zombie apocalypse? As a zombie outbreak unfolds, can survivors afford to trust mainstream media to give accurate information or will media be manipulated like the Umbrella Corporation? William Lafi Youmans, a professor of media and public affairs at George Washington University, and Jillian C. York, a freelance writer, remind us that social media platforms are commercial products and services for commercial interests.[37] While social media might provide us with alternative channels to broadcast and exchange information during an apocalypse, media platform owners may control content and access if usage is deemed incompatible with their commercial interests. Youmans and York also note that social media has been used to gather intelligence and spread propaganda. Thus, the concern of the corporatization of mass media is one of the controlling of access and content of the media. Freedom of information, both in its production and its distribution, is imperative for a functioning democracy in which citizens have all the knowledge and resources necessary to make informed decisions in their best interest. For many media scholars, the trend towards the consolidation of the media industry is troubling because of the power to control that information. In the case of the Internet service providers (ISPs) and cable company mergers, this concern of the accessibility of information is articulated in the principle of net neutrality that advocates all data on the Internet is equal and that ISPs cannot restrict access. This control has been demonstrated in past events, in popular culture, and speculated for the future. As the censorship by the Umbrella Corporation in *Resident Evil* shows, if an infection is imminent, information on it may be withheld due to corporate or governmental interests. In considering the coming zombie apocalypse, people must ask what might the media withhold from them, why might the media do that, and what might people be able to do about it? As discussed earlier, media operators admitted that they withheld information broadcasting during Hurricane Katrina due to having to sacrifice commercial air time. *World War Z* describes one situation in which alternative broadcasting took place as the zombie infection spread across the world. In the book's story, "Ulithi Atoll, Federated States of Micronesia,"[38] the storyteller describes how one civilian ship quarantined off the coast of Micronesia became a radio broadcasting center, Radio Free Earth. It was a radio hub and an international network to exchange information on the Zombie War. The storyteller states that part of Radio Free Earth's mission was to combat misinformation about the zombie infection so that more people would have a chance at survival, Radio Free Earth appears to be based off its real life counterpart, Radio Free Europe/Radio Liberty, whose mission is to make information available to all, particularly in regions where free press is limited. Max Brooks' imagining of mass media in a zombie crisis tells us that even a fictional story predicts the untrustworthiness of corporate media and the importance of citizen journalism and media in spreading accurate survival information.

What about the Internet in a disaster? How useful would this mass medium be in exchanging information? Given that the Internet was originally conceived by the U.S. Department of Defense as a decentralized network that would continue to exchange information even if one part of the network was destroyed in a war scenario, the challenge of a functional Internet would be access to satellite and cellular communications, neither of which are guaranteed to be standing in an apocalypse. In the event of an apocalypse, electricity is likely to fail, and mass media, such as television, the Internet, and landline phones cannot function. The likeliest scenario is radio broadcasting on battery or solar power, as depicted in the film *28 Days Later* where the military broadcasted with HAM radio, or point-to-point communications between radio devices, such as two-way radios. Joshua Kopstein, in an article for *The New Yorker*, investigated the various agencies, ranging from non-profits to the U.S. Department of Homeland Security, who imagine apocalyptic scenarios in which the Internet is shutdown.[39] Previous state-mandated Internet shutdowns, such as seen in the Arab Spring uprisings beginning in 2010, motivated many people to come up with solutions to a possible Internet blackout by practicing communication setups in an apocalypse, or Practo-calypse. Kopstein described the most likely communication would be a "wireless mesh network," in which mobile devices would be connected between nodes, rather than relying on centralized servers. He concluded that the world is ill-prepared for an Internet shutdown given the wireless mesh network was rudimentary and not guaranteed to work for everyone, nor was it far-reaching. The Zombie Research Society, an organization with a team of experts studying the living dead, on the other hand, puts its faith in functioning satellite phones, as long as one has battery power.[40]

Conclusion

Romero reflects upon the message of his iconic film as the fear of isolation and loneliness in the breakdown of communication: "*Night of the Living Dead* is specifically saying that you're not talking to each other, electronic media doesn't work, people don't communicate."[41] Given that we depend so much on mass media for our identities and social relationships, the loss of communication technology may add to our feeling of isolation. When the world collapses, information and relationships via media may be a survivor's saving grace.

Robert W. McChesney argues that since the twentieth century our media system has been trending towards a consolidation to a few large conglomerates who control the majority of the world's media. As described above in both popular culture zombie texts and in real life disasters, people depended upon the media for information, affective culture, and communication. If a few corporate media owners decided to withhold information from its audiences as seen in *Resident Evil: Apocalypse*, people have no real perspective in which to assess the situation and make the necessary preparation for a disaster. Further, if journalists do not do their job or are restricted by the commercial interests of their news organizations and access to media becomes limited in the increasing digital divide so that citizen journalists cannot participate widely, who will help provide the context to make sense of an apocalypse? The experience of mainstream news reporting immediately after 9/11 showed that alternative voices, especially those who offer unpopular perspectives, can be silenced and shut out of media. One such instance was when former talk show host Phil Donahue in 2002 allowed antiwar perspectives on his MSNBC program, which the network soon canceled despite being the highest-rated show on the network at the time.[42] Finally, if media owners refuse to turn over communications systems to emergency broadcasters and instead hoard the broadband for their commercial interests, communications in an emergency situation such as a zombie outbreak will break down almost guaranteeing more loss of life, as demonstrated after Hurricane Katrina.

NOTES

1. *Night of the Living Dead*, directed by George A. Romero, 1968, Elite Entertainment, 2002, DVD.

2. *28 Days Later*, directed by Danny Boyle, 2002, Fox Searchlight Pictures, 2003, DVD.

3. *Resident Evil: Apocalypse*, directed by Alexander Witt, 2004, Sony Pictures Home Entertainment, 2004, DVD.

4. Max Brooks, *World War Z* (New York: Broadway, 2006).

5. Philip Munz et al., "When Zombies Attack! Mathematical Modelling of an Outbreak of Zombie Infection" in *Infectious Disease Modelling Research Progress*, eds. Jean Michel Tchuenche and C. Chiyaka (New York: Nova Science, 2009), 133–150; and Robert Smith, *Mathematical Modelling of Zombies* (Ottawa: University of Ottawa Press, 2014).

6. Kyle Bishop, "Dead Man *Still* Walking: Explaining the Zombie Renaissance," *Journal of Popular Film & Television* 37, no. 1 (2009): 16–25.

7. Ibid.

8. *Dawn of the Dead*, directed by George A. Romero, 1978; United Film Distribution Company, 2004, DVD.

9. Mark Deuze, "Participation, Remediation, Bricolage: Considering Principal Components of a Digital Culture," *The Information Society* 22 (2006): 63.

10. Robert McChesney, *Rich Media, Poor Democracy: Communication Politics in Dubious Times* (Champaign: University of Illinois Press, 1999).

11. Maggie Silver, "Preparedness 101: Zombie Pandemic," Centers for Disease Control, accessed October 4, 2013, http://www.cdc.gov/phpr/_media/cymkPrint/11_225700_A_Zombie_Final.pdf.

12. Robert Kirkman, *The Walking Dead* (Berkeley: Image Comics, 2003–present).

13. Craig Fugate and Ali S. Khan, "First there were Zombies; then came Hurricanes!" Centers for Disease Control: Public Health Matters (blog), May 26, 2011 (1:31 p.m.), http://blogs.cdc.gov/publichealthmatters/2011/05/first-there-were-zombies-then-came-hurricanes/.

14. Randolph Burnside, DeMond Shondell Miller, and Jason Rivera, "The Impact of Information and Risk Perception on the Hurricane Evacuation Decision-Making of Greater New Orleans Residents," *Sociological Spectrum* 27, no. 6 (2006): 727–740.

15. Mark Benjamin, "Communications Breakdown," Salon.com, September 9, 2005, accessed October 14, 2013, http://www.salon.com/2005/09/10/comm_meltdown/.

16. McCain argued for a faster transition to digital broadcasting and reserving some spectrum to first responders: "Let us remember that Congress provided additional spectrum for the first responders in the Telecommunications Act of 1996...."

So after spending millions of dollars in funding and additional spectrum for our nation's first responders, why aren't we better off than we were on 9/11 when it comes to interoperable communications? Because the spectrum Congress provided to first responders in 1996 is being held hostage by television broadcasters even though broadcasters have been given new spectrum." In John Eggerton, "McCain Blasts Broadcasters," *Broadcast & Cable*, September 19, 2005, 28.

17. Ken Kershbaumer, "Broadcasters Seek Better Emergency Alert," *Broadcast & Cable* 135, no. 7 (2005): 8.

18. United States Government Accountability Office, Report to Congressional Committees, "Emergency Preparedness: Current Emergency Alert System Has Limitations, and Development of a New Integrated System Will Be Challenging," GAO-07–411, Washington, D.C., 2007.

19. Leslie Kaufman, "Bombings Trip Up Reddit in Its Turn in Spotlight," *New York Times*, April 28, 2013.

20. Post Staff Report, "Authorities ID Person of Interest as Saudi National in Marathon Bombings, Under Guard at Boston Hospital," *New York Post*, April 15, 2013, http://nypost.com/2013/04/15/authorities-id-person-of-interest-as-saudi-national-in-marathon-bombings-under-guard-at-boston-hospital/.

21. Joby Warrick, "FBI Investigation of 2001 Anthrax Attacks Concluded; U.S. Releases Details," *Washington Post*, February 20, 2010.

22. Kristen Alley Swain, "Outrage Factors and Explanations in News Coverage of the Anthrax Attacks," *Journalism & Mass Communication Quarterly* 84, no. 2 (2007): 335–352.

23. Ibid., 338.

24. Vicki S. Freimuth, "Order Out of Chaos: The Self-Organization of Communication Following the Anthrax Attacks," *Health Communication* 2, no. 2 (2006): 144.

25. Vicki S. Freimuth, "Order Out of Chaos: The Self-Organization of Communication Following the Anthrax Attacks," *Health Communication* 2, no. 2 (2006): 141–148.

26. Josmar Trujillo, "Covering Sandy Relief in Superhero Mode," *Extra! Magazine*, January 2013.

27. Simon Cottle, "Media and the Arab Uprisings of 2011: Research Notes," *Journalism* 12, no. 5 (2011): 647–659; Christos A. Frangonikolopoulos and Ioannis Chapsos, "Explaining the Role and the Impact of the Social Media in the Arab Spring," *Global Media Journal: Mediterranean Edition* 7, no. 2 (2012): 10–20; and Tapas Ray, "The 'Story' of Digital Excess in the Revolutions of the Arab Spring," *Journal of Media Practice* 12, no. 2 (2011): 189–196.

28. Julie B. Wiest and Nahed Eltantawy, "Social Media Use Among UAE College Students One Year After the Arab Spring," *Journal of Arab & Muslim Media Research* 5, no. 3 (2012): 219.

29. Howard Kurtz and Lauren Ashburn, interviewed by Gwen Ifill, *PBS Newshour*, April 16, 2013, http://www.pbs.org/newshour/bb/media-jan-june13-dd_04-16/.

30. Lea Winerman, "Social Networking: Crisis Communication," *Nature*, January 21, 2009, 376–378.

31. Bruce R. Lindsay, "Social Media and Disasters: Current Uses, Future Options and Policy Considerations," *Journal of Current Issues in Media and Telecommunications* 2, no. 4 (2011): 287–297.

32. Larissa Hjorth and Kyoung-hwa Yonnie Kim, "The Mourning After: A Case Study of Social Media in the 3.11 Earthquake Disaster in Japan," *Television & New Media* 12, no. 6 (2011): 552–559.

33. Andrea Duranti, "The Green Screen: Neda and the Lost Voices," *International Journal of Communication* 7 (2013): 1344–1370.

34. *[REC]*, directed by Jaume Balagueró and Paco Plaza, 2007, Sony Pictures Home Entertainment, 2009, DVD; *[REC]2*, directed by Jaume Balagueró and Paco Plaza, 2009, Magnolia Pictures, 2009, DVD; *[REC]3: Genesis*, directed by Paco Plaza, 2012, Sony Pictures Home Entertainment, 2012, DVD; and *Diary of the Dead*, directed by George A. Romero, 2008, Weinstein, 2008, DVD.

35. Brooks, *World War Z*, 62.

36. Robert McChesney, *Digital Disconnect: How Capitalism Is Turning the Internet Against Democracy* (New York: The New Press, 2013), and Robert McChesney, *Rich Media, Poor Democracy: Communication Politics in Dubious Times* (New York: The New Press, 1999).

37. William Lafi Youmans and Jillian C. York, "Social Media and the Activist Toolkit: User Agreements, Corporate Interests, and the Information Infrastructure of Modern Social Movements," *Journal of Communication* 62, no. 2 (2012): 315–329.

38. Brooks, "Ulithi Atoll, Federated States of Micronesia," 194–199.

39. Joshua Kopstein, "How to Survive an Internet Apocalypse," *The New Yorker*, April 11, 2014, http://www.newyorker.com/tech/elements/how-to-survive-an-Internet-apocalypse.

40. ZRS Contributor, "Hi-Tech Zombie Survival Gear," *Zombie Research Society*, undated, http://zombieresearchsociety.com/archives/23848.

41. Stephen Jones, comp., *Clive Barker's A–Z of Horror* (New York: Harper-Collins, 1996), 243.

42. Dennis J. Bernstein, "Silencing Donahue and Anti-War Voices," Consortiumnews.com, January 15, 2012, http://consortiumnews.com/2012/01/15/silencing-donahue-and-anti-war-voices/.

Bibliography

Benjamin, Mark. "Communications Breakdown." Salon.com. September 9, 2005. Accessed October 14, 2013. http://www.salon.com/2005/09/10/comm_melt down/.

Bernstein, Dennis J. "Silencing Donahue and Anti-War Voices." Consortiumnews. com. January 15, 2012. http://consortiumnews.com/2012/01/15/silencing-dona hue-and-anti-war-voices/.

Bishop, Kyle. "Dead Man *Still* Walking: Explaining the Zombie Renaissance." *Journal of Popular Film & Television* 37, no. 1 (2009): 16–25.

Brooks, Max. *World War Z*. New York: Broadway, 2006.

Burnside, Randolph, DeMond Shondell Miller, and Jason Rivera. "The Impact of Information and Risk Perception on the Hurricane Evacuation Decision-Making of Greater New Orleans Residents." *Sociological Spectrum* 27, no. 6 (2006): 727–740.

Cottle, Simon. "Media and the Arab Uprisings of 2011: Research Notes." *Journalism* 12, no. 5 (2011): 647–659.

Dawn of the Dead. Directed by George A. Romero. 1978. United Film Distribution Company, 2004. DVD.

Deuze, Mark. "Participation, Remediation, Bricolage: Considering Principal Components of a Digital Culture." *The Information Society* 22, no. 2 (2006): 63–75.

Diary of the Dead. Directed by George A. Romero. 2008. Weinstein, 2008. DVD.

Duranti, Andrea. "The Green Screen: Neda and the Lost Voices." *International Journal of Communication* 7 (2013): 1344–1370.

Eggerton, John. "McCain Blasts Broadcasters." *Broadcast & Cable.* September 19, 2005, 28.

Frangonikolopoulos, Christos A., and Ioannis Chapsos. "Explaining the Role and the Impact of the Social Media in the Arab Spring." *Global Media Journal: Mediterranean Edition* 7, no. 2 (2012): 10–20.

Freimuth, Vicki S. "Order Out of Chaos: The Self-Organization of Communication Following the Anthrax Attacks." *Health Communication* 2, no. 2 (2006): 141–148.

Fugate, Craig, and Ali S. Khan. "First there were Zombies; then came Hurricanes!" Centers for Disease Control: Public Health Matters (blog). May 26, 2011 (1:31 p.m.), http://blogs.cdc.gov/publichealthmatters/2011/05/first-there-were-zom bies-then-came-hurricanes/.

"Hi-Tech Zombie Survival Gear." *Zombie Research Society.* Undated. http://zombiere searchsociety.com/archives/23848.

Hjorth, Larissa, and Kyoung-hwa Yonnie Kim. "The Mourning After: A Case Study of Social Media in the 3.11 Earthquake Disaster in Japan." *Television & New Media* 12, no. 6 (2011): 552–559.

Jones, Stephen, comp. *Clive Barker's A–Z of Horror.* New York: HarperCollins, 1996.

Kaufman, Leslie. "Bombings Trip Up Reddit in Its Turn in Spotlight." *New York Times,* April 28, 2013.

Kershbaumer, Ken. "Broadcasters Seek Better Emergency Alert." *Broadcast & Cable* 135, no. 7 (2005): 8.

Kirkman, Robert. *The Walking Dead.* Berkeley: Image Comics, 2003–present.

Kopstein, Joshua. "How to Survive an Internet Apocalypse." *The New Yorker.* April 11, 2014, http://www.newyorker.com/tech/elements/how-to-survive-an-Internet-apocalypse.

Kurtz, Howard, and Lauren Ashburn. *PBS Newshour.* By Gwen Ifill. April 16, 2013, http://www.pbs.org/newshour/bb/media-jan-june13-dd_04-16/.

Lindsay, Bruce R. "Social Media and Disasters: Current Uses, Future Options and Policy Considerations." *Journal of Current Issues in Media and Telecommunications* 2, no. 4 (2011): 287–297.

McChesney, Robert. *Digital Disconnect: How Capitalism Is Turning the Internet Against Democracy.* New York: New Press, 2013.

_____. *Rich Media, Poor Democracy: Communication Politics in Dubious Times.* New York: New Press, 1999.

Munz, Philip, et al. "When Zombies Attack! Mathematical Modelling of an Outbreak of Zombie Infection." In *Infectious Disease Modelling Research Progress,* edited by Jean Michel Tchuenche and C. Chiyaka, 133–150. New York: Nova Science, 2009.

New York Post Staff Report. "Authorities ID Person of Interest as Saudi National in Marathon Bombings, Under Guard at Boston Hospital." *New York Post.* April 15, 2013. http://nypost.com/2013/04/15/authorities-id-person-of-interest-as-saudi-national-in-marathon-bombings-under-guard-at-boston-hospital/.

Night of the Living Dead. Directed by George A. Romero. 1968. Elite Entertainment, 2002. DVD.

Ray, Tapas. "The 'Story' of Digital Excess in the Revolutions of the Arab Spring." *Journal of Media Practice* 12, no. 2 (2011): 189–196.

[REC]. Directed by Jaume Balagueró and Paco Plaza. 2007. Sony Pictures Home Entertainment, 2009. DVD.

[REC]2. Directed by Jaume Balagueró and Paco Plaza. 2009. Magnolia Pictures, 2009. DVD.

[REC]3: Genesis. Directed by Paco Plaza. 2012. Sony Pictures Home Entertainment, 2012. DVD.

Resident Evil: Apocalypse. Directed by Alexander Witt. 2004. Sony Pictures Home Entertainment, 2004. DVD.

Silver, Maggie. "Preparedness 101: Zombie Pandemic." Centers for Disease Control. Accessed October 4, 2013. http://www.cdc.gov/phpr/_media/cymkPrint/11_2257 00_A_Zombie_Final.pdf.

Smith, Robert. *Mathematical Modelling of Zombies*. Ottawa: University of Ottawa Press, 2014.

Swain, Kristen Alley. "Outrage Factors and Explanations in News Coverage of the Anthrax Attacks." *Journalism & Mass Communication Quarterly* 84, no. 2 (2007): 335–352.

Trujillo, Josmar. "Covering Sandy Relief in Superhero Mode." *Extra! Magazine*. January 2013.

28 Days Later. Directed by Danny Boyle. 2002. Fox Searchlight Pictures, 2003. DVD.

United States. Government Accountability Office, Report to Congressional Committees. "Emergency Preparedness: Current Emergency Alert System Has Limitations, and Development of a New Integrated System Will Be Challenging." GAO-07-411. Washington, D.C., 2007.

Warrick, Joby. "FBI Investigation of 2001 Anthrax Attacks Concluded; U.S. Releases Details." *Washington Post*. February 20, 2010.

Wiest, Julie B., and Nahed Eltantawy. "Social Media Use Among UAE College Students One Year After the Arab Spring." *Journal of Arab & Muslim Media Research* 5, no. 3 (2012): 219.

Winerman, Lea. "Social Networking: Crisis Communication." *Nature*. January 21, 2009, 376–378.

Youmans, William Lafi, and Jillian C. York. "Social Media and the Activist Toolkit: User Agreements, Corporate Interests, and the Information Infrastructure of Modern Social Movements." *Journal of Communication* 62, no. 2 (2012): 315–329.

Reaction of Health Care Providers to the Zombie Apocalypse

LINDA W. THOMPSON

What will be the reaction of health care providers (HCP) to a zombie apocalypse? Assuming that the zombie "infection" has a gradual onset and then progresses to a pandemic level, health care provider reactions will progress thorough several stages. The hypothetical causative agent for this paper will be a mild to moderate infection to which most of the world's population was exposed, but which is only activated to "zombism" upon death. This type of exposure would allow for a somewhat gradual expression of the zombie pandemic. In this case, there would also be multiple ground zero incidents. Frontline providers including emergency medical service (EMS) personnel, nurses, and physicians working in hospitals will be the providers most directly affected by an infection that creates zombies. This essay will explore possible scenarios that might occur in patient care settings in each of the proposed reaction stages. Reactions to the zombie apocalypse by health care providers would progress through five stages from providing care, to confinement, to contagion, to chaos and, finally, to collapse of the health care system.

Caring Stage

In the first stage of the epidemic, many people will be infected and the symptoms could be like those associated with the common cold, such as fever, coughing and sneezing for 24 to 48 hours, and then recovering from the acute stage. Thus the illness would gradually spread throughout the general population. Many people would treat it symptomatically with over the counter medications and recover from the initial stage of the illness. Others would see their primary care providers who would treat the symptoms and some infected individuals would recover from the illness. The elderly and those with suppressed immune systems would have more severe symptoms and might be hospitalized. The issues of treatment and care will be similar to those with any unknown infectious illness.

As patients present themselves to primary care providers and at emergency rooms with this unknown illness, some would be admitted to the hospital and placed on protective precautions or isolation. In this initial stage there would be concerns and risks associated with caring for and treating infected individuals that would intensify once the first case of "zombism" occurs.

Despite the risk, all heath care providers have a professional duty to care for those who are ill, based on ethical standards. The American Medical Association's (AMA) *Principles of Medical Ethics* indirectly indicates that physicians have a duty to treat. Statement VIII provides a clear obligation to the patient: "A physician shall, while caring for a patient, regard responsibility to the patient as paramount."[1] No principle directly addresses care in an epidemic when one's own life is threatened through possible infection, but the AMA has addressed this in an opinion statement (non-binding professional behavioral recommendation) in their *Code of Medical Ethics*. This statement indicates that physicians do have a duty during epidemics to provide care to infected individuals, even in the light of extreme personal risks, but that this duty must be balanced by being able to continue providing care to others.[2] Samuel J. Huber and Matthew K. Wynia, two experts in medical ethics, indicated that this position statement was the result of issues arising from the refusal to treat HIV patients and the response to the 9/11 World Trade Center attacks. Thus, the AMA issued the opinion statement to clarify the role that physicians have during life threatening epidemics. This would mean that physicians have a non-binding professional obligation to continue providing care during disasters, even during a zombie epidemic.[3]

The American Nurses Association's *Code of Ethics* also indicates that nurses have a duty to their patients. Provision 2 states that "the nurse's primary commitment is to the patient whether an individual, family, group or community."[4] Unfortunately, in Provision 5 of the code, this obligation is contradicted by an obligation to self: "The nurse owes the same duties to self as to others, including the responsibility to preserve integrity and safety, to maintain competence and to continue personal and professional growth."[5] Two nurses who have written about the ethical duties of nurses in epidemics, Mary Elizabeth Grimaldi and Diane Twedell, point out that the ambiguity in the ANA Code of Ethics would leave the individual nurse in conflict over his or her obligations to patients and self.[6,7] Additionally, Twedell indicates the nursing profession needs to clarify the nurse's responsibilities in a pandemic.[8] This would present a nurse with an ethical dilemma as the epidemic progresses.

The other major group of front line health care providers who would be involved in caring for patients during a zombie apocalypse would be emergency medical service technicians or paramedics. The *Code of Ethics for Emergency Medical Services* states that they have a professional duty "to conserve life, alleviate suffering, promote health, do no harm, and encourage the quality and equal availability of emergency medical care."[9] This statement clearly indicates that there is an ethical duty to provide care. Paramedics would be the ones called to scenes where death was imminent and might be the first care providers to encounter a reanimated person or zombie.

In real life many health care professionals make the decision to stay on duty and provide care under extreme circumstances. Paramedics were among those who risked their lives in the 9/11 World Trade Towers terrorism attack and nurses and physicians stayed on duty at hospitals or volunteered during and after Hurricane Katrina and during the 2013 Ebola outbreak.[10] Physicians and nurses have provided care throughout history to people with life threatening diseases including cholera, tuberculosis, pandemic influenza, and AIDS.[11] One zombie movie that directly depicted physicians and nurses staying on duty to care for patients is *Planet Terror*.[12] This movie contains several scenes where victims of zombie attacks receive care and the physicians attempt to determine the cause of the illness. So there is every reason to think that many health care professionals would do the same during the initial stage of a zombie epidemic.

Legally, there are no absolute mandates related to the duty to treat or care for patients in disasters or pandemics. Professional practice is regulated by state law and thus varies. Most state statutes address agency preparedness including having disaster and fire plans, and require staff training for emergencies.[13] Each professional is responsible for knowing and abiding by the laws within their state of practice and have legal limits to their scope of practice. For example, physicians can perform surgery, but nurses cannot. No professional can legally perform duties outside those legal mandates. This would become an issue during a zombie apocalypse as the number of healthcare professionals available would shrink. At some point in time there could be blurring of roles in order to provide care. Media portrayals have already addressed this issue. For example, in *The Walking Dead* TV series, Hershel Greene who is a veterinarian performs emergency surgery on Carl Grimes and later Maggie Greene who has no medical training does a cesarean section on Lori Grimes.[14]

Health professionals are covered to some degree under Good Samaritan

Laws when providing care during emergency situations.[15] Good Samaritan Laws protect health care providers from malpractice suits when care is provided in non-work situations as long as the care provided meets the standards that any reasonable and prudent provider would give in a similar situation. In the initial stage of a zombie epidemic, most health care providers would be providing care in an employment situation and thus contract law would apply. Health care professionals can be drafted in the event of an emergency via Public Law 100-180: Health Care Personnel Delivery System 50 U.S. Code Appendix, Section 460(h), but this code has never been activated for use.[16] In a basic discussion of legal issues in nursing, Laura R. Mahlmeister, a nurse, indicates that some states including Vermont, Minnesota and Wisconsin, have passed laws requiring that nurses have a legal mandate to provide care in any emergency situation, whether or not they are on duty.[17]

Contract law may require employees to report to work during an emergency depending upon agency policy. Health care providers would also be expected to report to work in case of infectious diseases since there is both a societal expectation of that and the argument could be made that exposure to disease is "part of the job," thus refusal to work could be treated as patient abandonment.[18] Any health care provider who refused to provide care or who left their post could be sued for gross negligence or malpractice.[19] All hospitals have disaster policies and procedures that require essential staff report for or to remain on duty during any event with mass casualties, such as tornados, airplane crashes, or terrorist attacks. Thus, in the initial stage of a zombie epidemic, there is a legal mandate that staff would stay on duty until relieved by the next shift of workers. During the caring stage and early contagion stage of a zombie epidemic, health care providers would be expected to report to work. Once the zombie epidemic reached the chaos and collapse stages, due to the rapid disintegration of societal structures, this would no longer be an issue and more and more healthcare providers would refuse to provide care. Should such an epidemic occur, each provider will be making their own individual decisions related to if, and for how long, they would stay on duty. During Hurricane Katrina, the medical and nursing staff at Memorial Medical Center in New Orleans which was surrounded by floodwaters and then loss of electricity, made the decision that patients with terminal and irreversible conditions would be evacuated last.[20] This decision then led to some patients being "euthanized" instead of being abandoned.[21] The abandonment of patients has been depicted in the television series, *The Walking Dead*, where Rick Grimes was abandoned at Harrison Memorial and in the movie *28 Days Later* were Jim was abandoned by the staff at the London hospital.[22]

At the hospital the physician would be primarily concerned with diagnosing the patient(s). This would require a battery of laboratory tests in an attempt to identify the causative agent, reporting the infection to public health authorities, and keeping abreast of health care alerts. The infection would be treated symptomatically by way of antibiotics, fluid replacement via intravenous infusion, and prescribing medications for fever, pain and nausea. The physicians involved in caring for the infected individuals would have limited contact with the infected patients after the initial assessment at the time of admission to the hospital. They would encounter risks by way of their daily contacts during rounds to continue assessing and treating the patients. Physicians would be treating an unknown illness and would be reporting the illness to public health officials and contacting the Centers for Disease Control and Prevention.

For nurses, the care of the infected individual would be more direct and more prolonged but still considered routine care. All hospital personnel and visitors entering the room would need to use personal protective equipment (PPE). Typically that means wearing gowns, gloves, masks, eye protectors and even respirators plus following Universal Precautions, like hand washing before and after contact and when leaving the room. PPE is used at different levels for different types of infections. For example, gloves are used for any direct patient care contact involving body fluids. Gowns would be added for infectious diseases like hepatitis, cholera and methicillin-resistant *Staphylococcus aureus* infections. Respirators are used with any respiratory infections such as SARS, antibiotic resistant tuberculosis, and influenzas.

Unfortunately, healthcare providers' compliance with PPE use is a major problem. An integrative review of research on compliance indicated that observed use of protective eye equipment ranged from only 4 percent up to 47 percent, gloves from 45 percent to 67 percent and masks from 32 percent to 59 percent.[23] It stated that nurses report they would use PPE with high risk patients 69 percent of the time and with expected blood contact 75 percent, however, only 50 percent or less may actually use PPE.[24] In the same review, the frequent reason for lack of use was limited access, followed by time to don, lack of peer use, that it affects patient relationships, and degree of perceived risks.[25] This research indicates that nurses and other healthcare professionals place themselves at risks due to noncompliance and this would also increase the spread of the zombie infection to other patients, and even their own families. In relation to supplies, the United States Department of Labor has estimated that in a high risk situation which one can assume the zombie infection would be considered, there would be a need for four respiratory protective devices per day for each hospital employee and that 33 percent of the staff would be considered at high risk for becoming infected.[26] For example, if a hospital had a nursing staff of 100 nurses, the nursing staff at high risk (33 percent of 100 or 33 nurses × 4 per nurse) would need 132 advanced respirators per day, plus those needed for physicians, other staff, and family members. While this information is relevant to infections spread by air droplets and by contact with body fluids as seen with the Rage virus in *28 Days Later*[27] and *28 Weeks Later*,[28] PPE of this nature would not provide any protection against bites as depicted in *The Walking Dead* TV series.[29] The data indicates that as time passed, it would be more and more difficult for an agency to have the necessary supplies, equipment, and medicines to continue provision of care.

Assuming the infection became life threatening, medical and nursing care would intensify, but despite all efforts eventually patients will die. When any patient has a cardiac or respiratory arrest, an attempt is made to revive them through cardio-pulmonary resuscitation (CPR). The nurse finding the patient would activate a "code" to bring others to help and then begin CPR by way of chest compressions and month-to mouth breathing. If the deceased individual immediately reanimated as a zombie, this would lead to the nurse being attacked by the zombie and becoming severely injured or killed before others arrived on the scene. Assuming a more gradual reanimation, the patient would be intubated and CPR continued for 10 to 15 minutes until there was no response to defibrillation (electro-shocks to the heart) and other medical treatments during the resuscitation attempt. The deceased person could reanimate at any time during this procedure attacking and injuring staff members.

If the reanimation was more gradual, the patient would be pronounced dead by the physician, and the nurse would prepare the body for transport to the morgue or to a funeral home. This care involves removing tubing and monitoring equipment, washing the body and placing it into a plastic shroud. It is very possible that a nurse would be providing this care alone when the first patient turns into a zombie. Another possibility would be that a family member would be present at the time of reanimation. If the deceased person was left alone due to demands of other patients, the newly arisen zombie might wander out of their room and attack another patient, visitor, or a staff member. The first reaction of the nurse would be shock and disbelief and for the family member surprise and even relief. Those reactions would be short lived once the dead person, now a zombie, attacked them. If a family member is the first contact, then very likely they would be attacked, killed or badly injured. The nurse, visitor, family member, or other staff member would find themselves confronting a seemingly out of control, violent "patient." Hopefully, they would have time to scream for help and someone would activate the code system for a violent incident before anyone actually was bitten.

Most media depictions of a zombie apocalypse depict initial attacks being on family members or strangers, but not within a hospital setting. For example, in the original *Night of the Living Dead* the daughter of the family in the farmhouse first attacks her mother when she dies and turns into a zombie.[30] Again a movie that depicts attacks within a hospital setting is *Planet*

Terror. In this movie there are scenes where infected people are brought to the emergency room. After arrival and initial assessment, the infected individuals die and turn into zombies attacking first responders, staff and other patients.[31] There is little or no time in this movie for actual care since the situation rapidly deteriorates, skipping the confinement and contagion stages, to chaos and collapse. This essay assumes a more gradual onset and progression, and thus the need to confine people who have turned into zombies.

Confinement Stage

Once the violent patient code is activated the agency has moved into the confinement stage. All health care agencies have a procedure in place for controlling violent individuals. In general that policy includes staff coming from all points to restrain the violent person. The staff reporting to the violent incident would include psychiatric nursing staff, security personnel, and male nursing staff. In this case the first action taken is an attempt to talk to the out of control individual, which of course would be to no avail with a zombie. The next action would be for four to five staff members to physically restrain the zombie by one staff member seizing each limb and another the trunk of the person and bringing them to the ground. It is very likely that at least one person would be bitten in this process. The now somewhat controlled zombie would be placed in four point restraints and then moved to a psychiatric ward and placed in seclusion. Once there, the patient has to be monitored every 10 to 15 minutes and offered food and water and an opportunity for toileting after a 2- to 4-hour time period. Vital signs have to be checked and there would be a realization that the person has no heart rate or respiratory rate, and in fact they are indeed dead.

Assuming that at first there was only one zombie, there might be actual attempts to provide care for the person in a humane way. Caring for the zombie would be a major challenge since they would not only refuse care, but continually fight the care provider. Assuming that physical restraint was a success, the physician would still be at risk in attempting any physical assessment to determine the cause of this new condition, the medical technician challenged to get blood for lab tests (there would be none), and the nurse to provide basic care. The four point restraints and even an added body restraint would not protect anyone from bites. It can be imagined that the physician would take tissue samples, electrocardiogram (EKG) readings to confirm lack of cardiac function, and electroencephalogram (EEG) readings to rule out any presence of brain activity. It is even possible that the zombie might be intubated and placed on a respirator. Psychiatrists called in to evaluate the zombie would be at a loss since there would be no coherent communication.

At that time, the psychiatrist would confirm the restraint and seclusion orders and add psychotropic medications in an attempt to calm the out of control person. Nurses would be asked to give psychotropic medications every hour until calm. It is possible that a zombie could lapse into a calm state until a staff member entered the room and then the violent behavior would resume as the zombie recognizes the presence of prey. Offers of food and water would be met with refusal and any attempts to encourage this would provoke violence. Staff would soon only monitor and record behavior but not put themselves at risk by entering the seclusion room.

As the number of zombies increases, the agency would have to consider where these "violent dead" individuals could be confined. Since most psychiatric units only have one or two seclusion rooms and the supply of four point leather restraints are limited to a less than a half a dozen sets, the ability to confine individuals in that manner would cease. Psychiatric units could become a holding place for zombies since they can be locked off from the rest of the hospital. The "live" psychiatric unit patients would have to be moved or discharged to allow for this action. It can be assumed that efforts would be made to discharge all these patients to family members whenever possible. Thus, psychiatric units would provide at least one holding area for zombies. Then the issue would be that of placing new zombies into this holding area without letting those already confined there out. Staff would have to be inventive in placing newly dead as yet unanimated bodies there while protecting themselves. It might be possible to use distraction techniques like noise or a person at window to draw the already animated bodies away from the doorway long enough to push new bodies into the area. Soon this area would be filled to capacity and the agency would have to consider other places of confinement.

Another area that could be used as a confinement area for zombies would be the morgue. With luck this would occur before individuals turn into zombies. Some zombies could be confined to locked cubicles in the hospital morgue. After the first patient turned and reanimated, it can be hypothesized that an order would exist to place all patients who die into a cubicle immediately. Unfortunately, the morgue as a place of confinement would only work until all the cubicles had been filled with deceased individuals. It might be possible to continue to place a few there until those not in a cubicle turn into zombies. At that point time staff would not be able to place other bodies there without an attempt on their life. Hospital administration might ask law officials to take zombies to jails. It is even possible that the police or military would be asked to "kill" the zombies, if, as in fictional portrayals, someone had realized the zombies could be killed by a shot to the head. If so, this would give hospitals more time and resources to care for living persons.

By this time the hospital will have activated emergency management plans and be under full agency lockdown. In an article, René Steinhaur, a nurse and paramedic, and Jeff Bauer, editor, outline the four phases of emergency management that all Joint Commission on Accreditation of Healthcare Organizations (JCAHO) must have in place: mitigation, preparedness, response and recovery. In a "zombie" created infection, which no agency would be prepared for, the plan might resemble that for other kinds of infectious diseases, like SARS, or violent outbreaks, like multiple victim shootings. Mitigation is a hazard vulnerability analysis for the types of emergency situations likely to occur. Preparedness includes staff training, practice drills, inventory of needed supplies and agreements to obtain additional supplies, and a backup communication system. The response phase is the actions in the actual event, in this case the zombie apocalypse. The final phase is recovery or return to normal operations which is unlikely to occur in the zombie apocalypse. In the response phase, there should be a Hospital Emergency Incident Command System which would provide an organized internal and external communication system for notification of other agencies like police, the military and the Centers for Disease Control of the epidemic. This command center would also notify staff's families of the situation and the need for staff to remain on duty, provide a plan for the use and level of PPE, and order a 24-hour backup supply of medications (not adequate for a zombie apocalypse).[32]

At this time, all admissions would cease except emergency cases, like heart attacks and motor vehicle accidents. All patients who could be discharged would be sent home with family or friends to reduce the risk of exposure. In a survey of family physicians in Ottawa, Canada, 75 percent of the physician's responding indicated a willingness to be contacted during a public health emergency, but 50 percent indicated their offices were not prepared to respond to a serious respiratory emergency like SARS and 81 percent indicated that they were not prepared to respond to an emergency like an earthquake.[33] This survey data indicates a concern for the ability of the outpatient system to provide care during a zombie apocalypse as care shifts away from an inpatient setting. In this same survey, areas in which physicians were willing to assist and that might be needed in a zombie apocalypse included 58 percent in diagnosis centers, 52 percent in treatment centers, 41 percent in declaring deaths, 26 percent in home treatment and 23 percent in phone counseling.[34] Only 55 percent of the practices reported having nursing staff that could assist with care.[35] So while patients are being discharged home, the resources to care for people may not be present. This presents an added ethical and legal issue since arrangements for follow-up care must be made before a patient can be discharged home. As the number of zombies present in neighborhoods and in the community as a whole increases, both numbers of health care providers and family members present outside the hospital to provide any care would become severely limited.

Agency lockdown will require additional security staff. This means calling in the police, the National Guard, or the national military in that order and depending on the extent of the apocalypse. In the best case scenario, the hospital would have an electronic lock system in place that would lock off each unit from each and every other unit. One can envision that the agency might use upper floors as quarantine floors as the number of zombies present in the agency grows. This would mean locking off elevators and stairwells and moving living patients to lower floors. Either the military or staff would be required to erect barricades and invent ways to lock floors off if there was no electric lockdown system in place. Time, staff and material resources would make these efforts difficult to impossible. This would mean that each agency would progress to the chaos and collapse stages relatively quickly.

Contagion Stage

Assuming that the numbers of zombies grew somewhat slowly, health care agencies and providers would either move into or concurrently be in a stage of contagion. Individuals who were attacked and bitten by reanimated patients would need care and attention. Bites would have to be cleaned, debrided and sutured in the emergency room. Any bitten individual would be placed on antibiotics as they developed fever and became sicker. This would begin the cycle of care, death, and violence once again. Staff would be more concerned with personal protection via observing isolation and Universal Precautions, increasing the demand for that equipment. As the awareness that one can be infected by bites, concern for safety among staff members would start to escalate. At this time, health care providers would have to make a decision of whether or not they would stay on duty to provide care. Since there is no research on zombie epidemics, health care provider reactions and patient care issues that would arise at this time can be compared to those found in research related to Ebola and SARS outbreaks.

Ryan C.W. Hall, Richard C.W. Hall and Maria J. Chapman, physicians in the United States, discussed some of the issues that arose during the 1995 Ebola outbreak in the Congo. Effects on health care providers included lack of adequate history on exposure, problems with performing general examinations, even checking blood pressure due to issues in disinfecting equipment, lack of medical records for patients in isolation, and lack of supplies including medications and intravenous fluids. In the Ebola outbreak, there was a need to medicate agitated patients which further decreased the availability of certain medications, like Haloperidol (Haldol) and Diazepam (Valium). Due to the demands on staffing, family members were providing personal care needs to the patients which increased the need for PPE equipment. Staff caring for the infected experienced fear of contracting the illness themselves and fear that they might infect others. These fears were lessened because the staff was too busy to reflect on them, but increased as staff contracted the illness and died. In addition, staff found themselves isolated and shunned by others due to fear of infection which led to many staying at the hospital and not returning home at end of shifts. Those who were quarantined had the highest negative psychological reactions, as well as reduced capacity to function. The nurses and other staff who stayed on duty to provide care demanded more autonomy, better infection control measures, and increased personal safeguards.[36] One can see that these reactions which occurred during the Ebola outbreak might also occur in a zombie apocalypse.

Other studies have addressed nurses' reactions during epidemics. A qualitative study was done of nurses in several countries in central Africa after the Ebola outbreaks.[37] In this study, themes expressed that might relate to care during a "zombism" outbreak included lack of resources, especially protective equipment and stigmatization due to fear that all health care workers could be infected; but despite these issues, a strong commitment to continue to provide care.[38] A qualitative study of nurses in China who cared for patients during the SARS outbreaks indicated that the nurses interviewed had three major reactions: "personal challenge," "essence of care," and "self-growth."[39] Challenges included providing care wearing bulky protective equipment, dealing with the death of patients, and not being able to do enough to save patients.[40] At the same time, they coped with the pressure by total involvement in their work, doing things to care for each other, and finding ways to stay healthy.[41] The major issue for caring was a lack of knowledge about the infection, trying to organize their work, and concerns for the patients and their fears.[42] Self-growth involved the nurses seeing themselves as heroes going into battle, and developing a collaborative work environment.[43] Again, all these reactions might be the same during a zombie apocalypse.

While the studies above indicated that there were some positive responses to providing care in a disaster, other studies provide information about both the percentage of staff who would not respond and the reasons for that lack of response. A survey of hospital personnel on their willingness to respond to various disasters[44] ranged from 74.6 percent for a SARS outbreak, 79.4 percent for a smallpox epidemic and 85.1 percent for an influenza epidemic, but dropped to 69.1 percent for a radiological event.[45] The two most common reasons given for being unwilling to report were concern or fear for family at 45.8 percent and concern or fear for self at 30 percent.[46] The same study also found that those in clinical or direct care positions would be significantly less willing to report than nonclinical positions.[47] A survey of Taiwanese nurses in four community hospitals after a SARS outbreak found that 45.7

percent of the nurses no longer wanted to provide care or were considering resigning due to the risks.[48] The main factors predicting that nurses might leave their positions included a short time in the nursing position, increased workload and stress, perceived fatality of SARS, and effect on social relationships.[49] Finally, EMS personnel surveyed in one Midwestern ambulance service, when asked if they would stay on duty if there was a smallpox epidemic with no vaccine and no protective gear, 83 percent indicated that they would definitely or probably not do so.[50] Thus the research indicates that as risks increase, fewer providers would report to work.

Like other epidemics, one can envision that in a zombie apocalypse staff might convince themselves to stay on duty by seeing themselves as heroes. Just the fact that the providers would be very busy providing care and thus not stopping to think about the risks involved, would keep some providers on duty for at least some period of time. But fear, not only of infection, but of death itself, would increase as the number of zombies present at the agency increased. Work conditions would decline as supplies and equipment became increasingly scarce. It would be impossible to replace supplies as the infrastructure of society collapsed. Since there is no way to prevent or treat "zombism," as health care providers recognized the advanced risks to self, more and more staff would not report for duty or choose to leave the work setting. One can speculate that the stages of caring, confinement and contagion might be overlapping and only last a few days.

Chaos and Collapse Stages

So as the zombie contagion spreads, the sheer numbers of infected individuals would eventually lead to chaos and collapse of the health care system and the surrounding infrastructure. Eventually, the ability of the agency to confine zombies to specified areas would no longer be possible. The areas of the hospital with the living staff and patients would shrink day by day and then hour by hour. As staff recognized the increasing risks, many would abandon their patients after doing all they could to move them to safe places. While horrible to imagine, many health care providers would themselves die and then turn into zombies. Thus staff, who had been providing care, would as zombies attack and kill the patients they had cared for and tried to save. The few remaining staff would do their best to protect themselves and patients by barricading themselves into a few remaining safe patient care areas. In some cases, staff would have no weapons with which to defend themselves and would depend on calls to officials for rescue. In other cases, staff might have made weapons out of surgical instruments and medical equipment. One can imagine an intravenous pole being cut and sharpened to use as a weapon.

There might be some security personnel, police, or military present with weapons to fight off zombies. The military might be sent in to try to rescue and evacuate survivors and move then to smaller Mobile Army Surgical Hospital (M.A.S.H.) like units in secure settings like military bases. The M.A.S.H units would then serve as the only available health care facilities (at least for some period of time). Actual hospitals and medical centers would be overrun by zombies and would likely become the last place a person would want to be at or enter for any reason.

The health care system would collapse along with the rest of the infrastructure of our society. At that point, health care would be limited or nonexistent for most survivors. There would be individual health care providers out there doing their best to survive along with others. A lucky group of survivors would have a physician, nurse, or EMS as a member of their group who could provide care and treat injuries. These individuals would have limited resources and have to make do with what supplies they could scavenge or find to use. The survivor groups might see them as individuals to be protected by other group members. At this point, the scope of practice would be expanded and blurred and any living provider would use whatever skills they have to save the lives of those still living.

Notes

1. American Medical Association, *Principles of Medical Ethics* (Chicago: American Medical Association, 2001), para. 9.
2. American Medical Association, *AMA's Code of Ethics; Opinions on Professional Rights and Responsibilities, Physician Obligation in Disaster Preparedness and Emergency Response, Opinion 9.067* (Chicago: American Medical Association, 2001), para. 1.
3. Samuel J. Huber and Matthew K. Wynia, "When pestilence prevails ... physician responsibilities in epidemics," *American Journal of Bioethics* 4, no. 1 (Winter 2004), W5.
4. American Nurses Association, *Code of ethics for nurses with interpretive statement* (Washington, D.C.: American Nurses Association, 2001), 5.
5. Ibid., 9.
6. Mary Elizabeth Grimaldi, "Ethical decisions in times of disaster: Choices healthcare workers must make," *Journal of Trauma Nursing* 14, no. 3 (July 2007), 164.
7. Diane Twedell, "Duty to care," *Journal of Continuing Education in Nursing* 40, no. 2 (February 2009), 54.
8. Ibid.
9. National Association of Emergency Medical Technicians, *Code of Ethics for EMS Practitioners,* provision 1, para. 2.
10. John P. Proyor. "The 2001 World Trade Center disaster: summary and evaluation of experiences," *European Journal of Trauma & Emergency Surgery* 35, no. 3 (June 2009): 212–224; Wendy Bonifazi and Lisette Hilton, "Nurse heroes 2007: ten nurses honored for extraordinary feats under extraordinary circumstances." *Nurseweek (15475131)* 14, no. 26 (December 17, 2007): 10–12; M.J. Miller and Thérèse Dowd, "FEMA nurses face 'mission impossible': remembering Katrina," *Journal of Holistic*

Nursing 26, no. 1 (March 2008): 56–62; and Larry Copeland, "Ebola now taking its toll on doctors," *USA Today*, July 18, 2014, retrieved from http://www.usatoday.com/story/news/world/2014/07/27/ebola-africa-disease-epidemic/132 36743/.

11. Samuel J. Huber and Matthew K. Wynia, "When pestilence prevails ... physician responsibilities in epidemics," *American Journal of Bioethics* 4, no. 1 (Winter 2004), W8.

12. *Planet Terror*, directed by Robert Rodriguez, 2007, Dimension Films, Genius Productions, 2007, DVD.

13. University of Minnesota, School of Public Health, *State Regulations Pertaining to Disaster/Emergency Preparedness*, 2011.

14. *The Walking Dead*, produced by Gale Ann Hurd and David Alpert, 2010– , American Movie Classics, TV series.

15. Matthew K. Wynia, "Ethics and public health emergencies: Encouraging responsibility," *American Journal of Bioethics* 7, no. 4 (April 2007), 3.

16. Ibid.

17. Laura R. Mahlmeister, "Legal issues in health care," in *Contemporary Nursing: Issue, Trends & Management*, 6th ed), edited by Barbara Cherry and Susan R. Jacob (St. Louis: Elsevier, 2014), 120–164.

18. "Could EDs be sued for requiring staff to work during disasters?" *ED Legal Letter* 19, no. 3 (March 2008), 32.

19. Laura R. Mahlmeister, "Legal issues in health care," in *Contemporary Nursing: Issue, Trends & Management*, 6th ed), edited by Barbara Cherry and Susan R. Jacob (St. Louis: Elsevier, 2014), 120–164.

20. Sherri Fink, "The deadly choices at Memorial," *New York Times*, August 25, 2009, retrieved from http://www.nytimes.com/2009/08/30/magazine/30doctors.html?pagewanted=all&_r=0.

21. Ibid.

22. *28 Days Later*, directed by Danny Boyle, 2002, DNA Films and British Film Council.

23. Felicia Neo, Karen-Leigh Edward, and Cally Mills, "Current evidence regarding non-compliance with personal protective equipment—an integrative review to illuminate implications for nursing practice," *ACORN: The Journal of Perioperative Nursing in Australia* 25, no. 4 (December 2012), 26.

24. Ibid.

25. Ibid., 26, 28.

26. United States Department of Labor, "Proposed Guidelines on Workplace Stockpiling of Respirators and Facemasks for Pandemic Influenza," accessed November 30, 2013, from https://www.osha.gov/dsg/ guidance/proposed GuidanceStockpilingRespirator.pdf., 10.

27. *28 Days Later*, directed by Danny Boyle, 2002, DNA Films and British Film Council.

28. *28 Weeks Later*, directed by Juan Carlos Fresnadillo, 2007, Fox Atomic.

29. *The Walking Dead*, produced by Gale Ann Hurd and David Alpert, 2001– , American Movie Classics.

30. *Night of the Living Dead*, directed by George Romero, 1969, Image Ten, 1994, movie/DVD.

31. *Planet Terror*, directed by Robert Rodriguez, 2007, Dimension Films, Genius Productions, 2007, movie/DVD.

32. René Steinhauer and Jeff Bauer, "A readied response. The emergency management plan," *RN* 65, no. 6 (June 2002), 41–45.

33. William Hogg, Patricia Huston, Carmel Martin, and Enrique Soto, "Enhancing public health response to respiratory epidemics: Are family physicians ready and willing to help?" *Canadian Family Physician* 52 (2006), 1257.

34. Ibid., 1258.

35. Ibid., 1259.

36. Ryan C.W. Hall, Richard C.W. Hall, and Marcia J. Chapman, "The 1995 Kikwit Ebola outbreak: Lessons hospitals and physicians can apply to future viral epidemics," *General Hospital Psychiatry* 30, no. 5 (September 2008), 448, 449.

37. Bonnie L. Hewlett and Barry S. Hewlett, "Providing care and facing death: Nursing during Ebola outbreaks in Central Africa," *Journal of Transcultural Nursing* 16, no. 4 (October 2005), 293–296.

38. Ibid.

39. Huaping Liu and Patricia Liehr, "Instructive messages from Chinese nurses' stories of caring for SARS patients," *Journal of Clinical Nursing* 18, no. 20 (October 15, 2009), 2883, 2885, 2886. 2880–2887.

40. Ibid., 2883–2884.

41. Ibid., 2884.

42. Ibid., 2885.

43. Ibid., 2886.

44. Lavonne M. Adams and Devon Berry, "Who will show Up? Estimating ability and willingness of essential hospital personnel to report to work in response to a disaster," *Online Journal of Issues in Nursing* 17, no. 2 (2012), 3.

45. Ibid., 18.

46. Ibid., 20.

47. Ibid., 6.

48. Judith Shu-Chu Shiao, David Koh, Li-Hua Lo, Meng-Kin Lim, and Yueliang Leon Guo, "Factors predicting nurses' consideration of leaving their job during the SARS outbreak," *Nursing Ethics* 14, no. 1 (2007), 8.

49. Ibid., 13.

50. Niklas Mackler, William Wilkerson, and Sandro Cinti, "Will first-responders show up for work during a pandemic? Lessons from a smallpox vaccination survey of paramedics," *Disaster Management & Response* 5, no. 2 (2007), 46.

Bibliography

Adams, Lavonne M., and Devon Berry. "Who will show up? Estimating ability and willingness of essential hospital personnel to report to work in response to a disaster." *Online Journal of Issues in Nursing* 17, no. 2 (May 2012): 1–2, doi:10.3912/OJIN.Vol17No02PPT02.

American Medical Association. *AMA's Code of Ethic; Opinions on Professional Rights and Responsibilities, Physician Obligation in Disaster Preparedness and Emergency Response, Opinion 9.067.* Chicago: American Medical Association, 2001.

_____. *Principles of Medical Ethics.* Chicago: American Medical Association, 2001.

American Nurses Association. *Code of ethics for nurses with interpretive statement.* Washington, D.C.: American Nurses Association, 2001.

Bonifazi, Wendy, and Lisette Hilton. "Nurse heroes: ten nurses honored for extraordinary feats under extraordinary circumstances." *Nurseweek (15475131)*14, no. 26 (December 17, 2007): 10–12.

Copeland, Larry. "Could EDs be sued for requiring staff to work during disasters?" *ED Legal Letter* 19, no. 3 (March 2008): 31–33.

_____. "Ebola now taking its toll on doctors," *USA Today*, July 18, 2014. http://www.

usatoday. com/story/news/world/2014/07/27/ebola-africa-disease-epidemic/132 36743/.

Fink, Sherri. "The deadly choices at Memorial." *The New York Times*, August 25, 2009. http://www.nytimes. com/2009/08/30/magazine/30doctors.html?pagewanted=all &_r=0.

Grimaldi, Mary Elizabeth. "Ethical decisions in times of disaster: Choices healthcare workers must make." *Journal of Trauma Nursing* 14, no. 3 (July 2007): 163–164.

Hall, Ryan C.W., Richard C.W. Hall, and Marcia J. Chapman. "The 1995 Kikwit Ebola outbreak: Lessons hospitals and physicians can apply to future viral epidemics." *General Hospital Psychiatry* 30, no. 5 (September 2008): 446–452.

Hewlett, Bonnie L., and Barry S. Hewlett. "Providing care and facing death: Nursing during Ebola outbreaks in Central Africa." *Journal of Transcultural Nursing* 16, no. 4 (October 2005), 289–297.

Hogg, William, Patricia Huston, Carmel Martin, and Enrique Soto. "Enhancing public health response to respiratory epidemics: Are family physicians ready and willing to help?" *Canadian Family Physician* 52 (October 2006): 1255–1260.

Huber, Samuel J., and Matthew K. Wynia. "When pestilence prevails ... physician responsibilities in epidemics." *American Journal of Bioethics* 4, no. 1 (Winter 2004): W5–W11.

Lui, Huaping, and Patricia Liehr. "Instructive messages from Chinese nurses' stories of caring for SARS patients." *Journal of Clinical Nursing*, 18, no. 20 (October 15, 2009): 2880–2887.

Mackler, Niklas, William Wilkerson, and Sandro Cinti. "Will first-responders show up for work during a pandemic? Lessons from a smallpox vaccination survey of paramedics." *Disaster Management & Response* 5, no. 2 (April–June 2007): 45–48.

Mahlmeister, Laura R. "Legal issues in health care." In *Contemporary Nursing: Issue, Trends & Management*, 6th ed., edited by Barbara Cherry and Susan R. Jacob, 120–164. St. Louis: Elsevier, 2014.

Miller, M. J., and Thérèse Dowd. "FEMA nurses face 'mission impossible': remembering Katrina." *Journal of Holistic Nursing* 26, no. 1 (March 2008): 56–62.

National Association of Emergency Medical Technicians. 2013. *Code of Ethics for EMS Practitioners,* provision 1, para. 2. http://www.naemt.org/about_us/emtoath.aspx.

Neo, Felicia, Karen-Leigh Edward, and Cally Mills. "Current evidence regarding non-compliance with personal protective equipment—an integrative review to illuminate implications for nursing practice." *ACORN: The Journal of Perioperative Nursing in Australia* 25, no. 4 (December 2012), 22–30.

Night of the Living Dead. Directed by George Romero. Image Ten, 1969. Movie. Elite Entertainment, 1994. DVD.

Planet Terror. Directed by Robert Rodriguez. Dimension Films, Genius Productions, 2007. DVD.

Pryor, John P. "The 2001 World Trade Center disaster: summary and evaluation of experiences." *European Journal of Trauma & Emergency Surgery* 35, no. 3 (June 2009): 212–224.

The Walking Dead. Produced by Gale Ann Hurd and David Alpert. American Movie Classics, 2012– . TV series.

Twedell, Diane. "Duty to care." *Journal of Continuing Education in Nursing* 40, no. 2 (February 2009).

28 Days Later. Directed by Danny Boyle. DNA Films and British Film Council, 2002. Movie.

28 Weeks Later. Directed by Juan Carlos Fresnadillo. Fox Atomic, 2007. Movie.

Shu-Chu Shiao, Judith, David Koh, Li-Hua Lo, Meng-Kin Lim, and Yueliang Leon Guo. "Factors predicting nurses' consideration of leaving their job during the SARS outbreak." *Nursing Ethics* 14, no. 1 (January 2007): 5–17.

Steinhauer, René, and Jeff Bauer. "A readied response. The emergency management plan." *RN* 65, no. 6 (June 2002): 41–45.

Twedell, Diane. "Duty to care." *Journal of Continuing Education in Nursing* 40, no. 2 (February 2009): 53–54.

United States Department of Labor. "Proposed Guidelines on Workplace Stockpiling of Respirators and Facemasks for Pandemic Influenza." Accessed November 30, 2013, from https://www.osha.gov/dsg/guidance/proposedGuidanceStockpiling Respirator.pdf., 10.

University of Minnesota, School of Public Health. *State Regulations Pertaining to Disaster/Emergency Preparedness*, 2011. http://www.hpm.umn.edu/nhregsplus/NH% 20Regs%20by%20Topic/NH%20Regs%20 Topic%20Pdfs/Disaster%20Prepared ness/category_administration_disaster_emergency_preparedness.pdf

Wynia, Matthew K. "Ethics and public health emergencies: Encouraging responsibility." *American Journal of Bioethics* 7, no. 4 (April 2007): 1–3.

Day of the Engineer
Engineering and the
Zombie Apocalypse

JEFF MOEHLIS

The word "engineer" means "one who practices ingenuity."[1] Over the years, engineers have practiced ingenuity in many different and important ways, using their training in theoretical and mathematical analysis, computational approaches, experimental methods, and design techniques to invent and improve useful devices and solve practical problems. Engineers have been crucial to the development of planes, trains, and automobiles, plus computers, phones, robots, and refrigeration. With this in mind, it should not be surprising that engineers will make significant contributions during a zombie apocalypse.

Clearly, "Plan A" is to survive a zombie apocalypse. As a human, that is, not as a zombie. This essay will show that this will be much more likely thanks to the efforts of some well-trained engineers. For example, they will devise novel strategies for killing zombies, fortify buildings and cities to protect humans, design robots to carry out dangerous tasks, and maintain modern conveniences such as power sources, communication channels, and safe food and water. Engineers will also play a huge role in reestablishing civilization after a zombie apocalypse subsides, and, from a technological standpoint at least, might actually make things better than they were beforehand. This will include establishing more robust power and communication networks, and designing energy efficient buildings and transportation systems. Indeed, with their technical knowledge, engineers will be key players in preserving humanity and civilization during and after a zombie apocalypse.

One of the most important ways that engineers will be useful will be in the development of better ways to kill zombies, which at a fundamental level is a problem in structural mechanics, a subfield of mechanical engineering.

Recall that the only way to kill a zombie is to destroy its brain,[2] which is housed in a skull that is roughly as strong as that for a healthy human. There have been studies using skulls from human cadavers which consider the geometrical, material, and structural properties of human skulls, with particular interest in determining when a skull fails due to impact. One such study showed that impact loading from a hemispherical anvil with a radius of 48 millimeters (2 inches) moving at 8 meters per second (18 miles per hour), which delivers a force on the order of 15 kilonewtons (370 pounds), will typically lead to complex fracture patterns in the skull, and thus severe brain damage.[3] Stated more bluntly, such an impact can destroy a zombie's brain. This force is within the range that a panicked individual could achieve, for example, by wildly swinging a frying pan, with multiple blows giving a greater chance of success. Of course, a good engineer who is not in imminent danger from a group of hungry zombies in deadly pursuit will set up a finite element model on a computer to validate these numbers and to investigate various crushing strategies, including finding impact locations which are most likely to lead to severe skull fractures. If functioning computers aren't available, they can rely on back-of-the-envelope calculations, intuition, and trial-and-error experimentation.

The current state of the art is to kill zombies using conventional weapons such as firearms, swords, and machetes, or tools which can be used as weapons such as chainsaws, hammers, crowbars, and hatchets. Each of these can do the job, but has disadvantages. For example, firearms and chainsaws are quite loud, and the noise they produce might attract other zombies. Moreover, despite what one might see in movies and in video games, it can be difficult for an untrained marksman to shoot a zombie from a safe distance with enough accuracy to destroy its brain. Swords, machetes, hammers, crowbars, hatchets, and the like present their own challenges, most notably the fact that one has to be dangerously close to a zombie to use them to crush its brain. Remember from *28 Days Later* that a single drop of infected blood in the eye can be problematic. Moreover, as early episodes of The Walking Dead have illustrated, it may require repeated blows from a weapon such as a rock or a baseball bat to kill a zombie. The 2004 remake of *Dawn of the Dead* also shows that one must be careful about chainsaw mishaps.

Fortunately, engineers are trained to design devices which meet the necessary specifications to accomplish challenging tasks, such as crushing zombies' skulls and brains. Here the device must supply the requisite amount of force and the impact must be of minimum speed to cause severe fracture of the skull, the device must operate within allowable power constraints, the user should be able to use the device while maintaining a minimum safe distance from zombies, the device must be ready for re-use within a short time interval in case other zombies are closing in, and the device should be quiet enough that its use will not alert other zombies to the user's presence. There are at least two ways to produce the necessary force to crush a zombie's head which, until now, have been underutilized in zombie fiction. One is pneumatics, in which pressurized gas is used to produce mechanical motion. Examples of pneumatic devices are nail guns, some types of jackhammers, and bolt pistols as used by the psychopathic hitman in *No Country for Old Men*. Another is hydraulics, in which pressurized liquids in pipes and tubes are used to do mechanical work. Examples of hydraulic devices include car crushers and some types of construction equipment.

The author has already proposed killing zombies with a portable, pneumatic, telescopic head crusher,[4] which was not patented in order to keep the idea in the public domain. (Who has time for intellectual property disputes when under relentless attack from a group of ravenous zombies?) The basic idea is to have a lightweight telescopic beam which extends at least ten feet, with a trigger which activates pneumatics to cause spiked jaws mounted on the far end to very rapidly close on the zombie's head. Such head crushers can be designed with a mechanism to quickly reopen the jaws so that they can be used on the next zombie. Because the device is telescopic, it is portable in its retracted form but can be extended if needed so that the user can destroy the zombie's brain from a safe distance. It can also be designed to be quiet enough that its use should not alert other zombies to the user's location. This device is expected to work best when the density of zombies is low and/or they are moving slowly.

If using a telescopic head crusher is still too close for comfort for a user (a common complaint: "I wouldn't touch a zombie with a ten foot pole"), an alternative strategy is to lead the zombies into large hydraulic devices which crush their entire body, including their head. These could be similar in concept to car crushers or trash compactors, but with their design adapted by engineers to meet the particular specifications for crushing zombies. Fortunately it's easier to crush a zombie than a car or the typical waste on a Death Star in a galaxy far, far away, so sophisticated hydraulics should not be necessary. These devices will be somewhat loud, but in this case the noise is a bonus as one actually wants the zombies to be attracted to their imminent doom; recall that in *The Walking Dead*, noise is sometimes used to lure zombies into the walker pit near Woodbury.

Of course, it is best to avoid contact with zombies altogether, which leads to another reason why it is desirable to have engineers around. Engineers can design and build structures and cities that keep zombies out. Since zombies like to break through windows to feast on humans, strong windows are a must. It's a no-brainer! A good rule of thumb is that windows should be able to survive a hurricane or an explosive blast, which is roughly the level of abuse that they will receive from a typical horde of zombies. It is common to attempt to protect windows by covering them with plywood, but this blocks one's ability to visually monitor potential threats, and cuts off natural light, leading to the need to waste limited power on lighting. Fortunately, engineers have recently been developing a new class of super-strong window systems with innovations like ductile energy-absorbing mechanisms, composite layers, and laminates.[5] It has been demonstrated, for example, that laminated glass with a polyvinyl butyryl interlayer can be very effective at preventing penetration of a window from an impact.[6] An engineer can do the calculations to determine which windows are right for zombie-protection needs, plus what reinforcements are necessary when mounting them. Of course in a pinch, plywood might still be the best option.

Strong doors, door jambs, and locks are also necessary. It's best to have an engineer compute the stresses on the doors and door jambs due to the repeated beating on them by zombies in pursuit of fresh flesh. This will provide crucial information for choices of materials and design to keep the zombies from breaking down the doors. A fence is also a great idea, especially a tall one because zombies typically cannot climb due to muscle deterioration. As is true for doors, the fence must be able to withstand the pushing and clawing of insatiably hungry zombies. This will require sensible choices of materials, design, and construction, all of which can be done by a well-trained engineer.

Another way to limit potential contact with zombies is to design and build robots to do necessary but dangerous tasks such as carrying water back from a lake or a well, or for the surveillance of the perimeter of a compound. Naturally, engineers are at the forefront of robotics research. It should be noted that robots are already being used for tasks such as bomb disposal,[7] shipwreck exploration,[8] and surgery[9]; a robot even aided in the capture of the alleged Boston Marathon bomber.[10] While many of today's robots are controlled remotely by a human, it is becoming more and more common for them to be autonomous, meaning that they can perform tasks in complex and changing environments without continuous guidance or oversight from humans. Such robots have cameras and sensors which collect data about their own state and the state of their environment, which is continually processed by onboard computers to determine what the robot should do next.

An example of an autonomous robot/vehicle is a driverless car, which uses cameras, radar, Global Positioning System (GPS), and other sensors to characterize its surroundings, including obstacles and signs. This data is processed by onboard computers so that the car can determine, on its own, how to move in order to safely reach its destination. The development of driverless cars has been rapid. In 2004, the Defense Advanced Research Projects Agency (DARPA) issued a "Grand Challenge" to develop cars that could navigate a course in the desert; none finished, and the one that went furthest didn't even make it eight miles. For the next year's challenge, five vehicles were able to complete a 132-mile course. Things got more sophisticated in 2007 with the "Urban Challenge" in which driverless cars had to navigate a 60-mile urban course while obeying traffic laws and dealing with other vehicles.[11] Six teams finished the course, with the fastest averaging 14 miles per hour. Progress on driverless cars continues; for example, Google has a fleet of driverless cars that have collectively driven hundreds of thousands of miles without major accidents. It must be noted that during a zombie apocalypse it will not be as crucial to obey traffic laws or to avoid hitting pedestrians who happen to be zombies, which in the best of times would simplify the design and operation of driverless cars; however, this means that other vehicles on the road are less predictable, which presents new challenges.

Clearly, driverless cars could be very useful during a zombie invasion, since they could transport food, water, and supplies without the danger of a human driver being attacked. Indeed, one can imagine setting up secure stations which distribute needed supplies via driverless cars to people who are living in their own compounds. Various logistical issues would need to be sorted out, and care would have to be used to make sure that the driverless vehicles don't bring back unwanted zombie passengers, and that zombies don't enter through open gates or doors as has happened in the 2004 *Dawn of the Dead* remake and episodes of *The Walking Dead*. This could be guarded against through the use of airlock-like double door entry systems. Overall, driverless cars open up exciting survival possibilities.

An indication of the future of robotics is captured by another DARPA challenge, the 2012 Robotics Challenge to program humanoid robots to carry out "complex tasks in dangerous, degraded, human-engineered environments,"[12] a competition which was inspired by the recognition that such robots would have been useful for doing dangerous tasks at the crippled Fukushima nuclear power plant. At the December 2013 trials for this competition, the robots were able to do tasks such as climbing ladders, opening doors, and walking on uneven terrain.[13] With further expected rapid progress on this challenge, which had its final competition in June 2015, and an extension to a greater variety of environments, engineers will soon be creating autonomous robots which can execute tasks such as retrieving water from a lake or a well in an area infested with zombies. Notably, being fleshless entities, such robots are expected to attract only minor interest from the zombies.

Robots can also be used to do important security tasks including surveillance.[14] For example, engineers can program a group of mobile robots to follow paths around the perimeter of a compound, with cameras and computer vision being used to detect intruders and sound an alarm. Moreover, engineers have developed algorithms which allow a group of robots to work together to systematically (and correctly) determine if intruders are present in a building, a task which could be highly dangerous for humans in a zombie-infested world.

Robots could also be used to kill zombies. Robots have the advantage that they can get very close to a zombie, opening up options such as pneumatically or hydraulically driven hammers and pinchers to crush a zombie's head. Combat drones and other high-tech military weapons[15] designed by engineers could also easily be adapted to carry out zombie elimination. The robots could be controlled remotely by a human who uses images from cameras to decide what the robot should do, or they could be autonomous agents, using computer vision and onboard computer processing to identify, track, and eradicate zombies. Image identification algorithms can be designed to be accurate enough to distinguish zombies from humans, even in the low light conditions of common zombie hangouts like cemeteries and abandoned nuclear power plants; however, it would be prudent to program autonomous robots to spare an individual that yells, "Stop!" in any language, something that is beyond a zombie's capability.

While holed up in a secure compound, perhaps with robots in zombie-infested areas doing the dirty work, people will want to maintain a lifestyle which is as close as possible to the one they were accustomed to before the zombies came. This is another area to which engineers can make significant contributions. One of the most important requirements for modern comfort is access to a reliable power source; in fact, the robot applications above assume that such a power source is present. As a zombie invasion progresses, it is expected that power from public utilities will become unavailable fairly rapidly.[16] Also, fuel and batteries will rapidly be exhausted or hoarded, which could force many people to adopt very primitive lifestyles.

Fortunately, engineers have made significant progress developing alternative energy sources.[17] One of the most promising is the use of solar panels made up of photovoltaic cells that convert light energy from the sun into electricity, through a semiconductor material that absorbs photons and releases electrons. Thanks to the hard work of engineers, the efficiency of solar panels is approaching 20 percent, meaning that 20 percent of the incident radiation from the sun can be converted into useful energy. Proposed designs, such as special nanosized antennas to capture solar energy, could eventually push the efficiency above 50 percent.[18] Outfitting a home with solar panels and an appropriate solar energy storage system can help to keep the power on during a zombie apocalypse.

Of course, solar energy will not always be available, for example at nighttime, when it is cloudy, or in a nuclear winter scenario. Another promising alternative energy source is wind energy, which can be converted into mechanical and electrical energy with a wind turbine. Wind turbines can also pump needed fresh water out of the ground, as windmills have done for hundreds of years. Engineers are at the forefront of wind energy technology and can help to design and build turbines to suit the needs of those during a zombie apocalypse. Other alternative energy sources that might be useful are thermoelectrics, which convert temperature gradients into electric voltage, and fuel cells which convert energy from oxidation chemical reactions into electricity.

It is also crucial to maintain lines of communication, including the abilities to receive and transmit information. Rapid communication requires electrical devices, such as telephones (cellular or otherwise), radios, television, and computers, plus infrastructure such as land lines, telephone exchange systems, cellphone towers and antennas, television and radio transmission stations, internet routers, and the like. All of these were designed by engineers, who can help to maintain them or repair them if they break down. But it is not advisable to rely on the internet or cellphones during a zombie apocalypse, even with engineers around. One of the most robust forms of communication during a crisis is amateur radio ("ham radio"), which has little reliance on manmade infrastructure. It also has great flexibility, with the possibility of using improvised antennas made of wire or metal poles (such as pneumatic telescopic head crusher), and power sources such as car batteries. Note that amateur radio set-ups are more powerful and have more frequency flexibility than walkie-talkies, which still have shown their usefulness in *The Walking Dead* and elsewhere. Amateur radio has been very helpful in providing emergency communication for crises such as the September 11 terrorist attacks,[19] the 2003 Northeast blackout,[20] and Hurricane Katrina,[21] when other forms of communication were non-functioning or overloaded. Engineers should have the necessary training in electrical circuitry to keep these radios operational, and thus the lines of communication open even in extreme conditions.

It is also of utmost importance to keep food and water supplies safe, a definite challenge when people might not have access to adequate refrigeration. Having to deal with zombies is bad enough, just imagine trying to do it while suffering from food poisoning or an intestinal infection! Fortunately, engineers will be able to help preserve food and water supplies.[22]

There are various relatively low-tech methods that engineers have developed for ensuring a supply of safe water. Good examples can be found in the efforts of different chapters of the humanitarian engineering organization Engineers Without Borders,[23] which have had success with projects such as low-cost solar powered water purification systems, slow sand filter water purification systems, and rainwater capture systems for rural communities. Alternatively, if appropriate resources are available, chemical engineers will have the expertise to purify water using chlorination or alternative oxidation treatments such as ultraviolet light or the use of hydrogen peroxide or ozone.

To ensure safe food, consider pasteurization, a technique used since the pioneering work of Louis Pasteur in the 1800s which kills disease-causing pathogens in liquids such as milk. Here the liquid is heated for a short time without altering its chemical properties, and then it is rapidly cooled. Done properly, this destroys harmful bacteria including salmonella and *E. coli*, and can help prevent diseases such as tuberculosis, diphtheria, and scarlet fever. Successful pasteurization can be accomplished by trial and error, but it is far safer to draw upon the expertise of a chemical engineer to plan and carry out the procedure. Another method for keeping food and liquids safe is irradiation, in which radiation such as X-rays is used to kill pathogens. Irradiation is particularly useful for treating meat and poultry. It should be noted that it must be done before the food spoils and it cannot be used to make spoiled food safe again. Unfortunately this process may not be feasible during a zombie apocalypse as it requires a well-trained individual (engineer), special equipment, a suitable power supply, and most of all, the food.

There are other methods for lengthening the shelf-life of food, most notably freeze-drying; it is noted that having a stockpile of freeze-dried food is recommended by Doomsday Preppers.[24] Here the food is deeply frozen, then dried by the chemical process called sublimation, for which the ice in the frozen food undergoes a phase transition directly to water vapor. This requires the frozen food to be in very low air pressure, which is accomplished by having it in a vacuum. Successful execution of the freeze-drying process will require engineers to design and build the appropriate temperature control and vacuum apparatus, and to carry out the procedure. Obviously this is not practical for every household, but secure freeze-drying stations could be established, with driverless cars making deliveries. An easier alternative to preserve food, although it will not keep as long, is food dehydration.

Another important role for engineers is in the production of alcohol, which when consumed in moderation could boost the morale of individuals who otherwise spend every moment worrying about imminent death through zombie attacks. Of course, moderation is key here, because one doesn't want to be so incapacitated that he or she cannot fight or run away from zombies. Why engineers? Well, the production of alcohol involves chemistry and, particularly when distillation is involved, thermal science; these are topics in which chemical and mechanical engineers have extensive training. (One could argue that home beer brewers and wine and spirit makers are effectively amateur engineers.) Engineers are also typically game for a good-spirited drinking session to relieve stress, something they are particularly familiar with if they went to graduate school.

But there's more. Engineers could contribute to one of the most amazing scientific feats that can even be imagined: curing zombies through targeted treatment. This might sound like a pipe dream, and many would argue that it is impossible. But engineers are known for thinking big. They helped to put a man on the moon, right? This would be superior to treatments such as Zombrex from the Dead Rising games which only slow down zombification.[25]

People become zombies when they are infected with the zombie virus, which is spread through contact with the zombie's bodily fluids, typically saliva from a zombie bite. It is argued that the virus attacks the victim's brain, but the details about how this transforms the victim into a zombie are not clear. Right now, the only known way to kill a zombie is to destroy the infected brain.

An intriguing idea is to develop nanobots to attack and destroy the zombie virus in a targeted manner in an infected individual. Nanobots are robots whose size is on the order of a nanometer, which is just one billionth of a meter and is 100,000 times smaller than the width of a human hair. It has been proposed that nanobots can be used for identifying and destroying cancerous cells in human patients[26]; here, nanobots would be injected into the bloodstream, and when they come into contact with an undesirable cell they correct the genetic mutations that lead to cancer, or deliver targeted medications. It is worth emphasizing that nanobots are mechanical devices, unlike the Krippin virus in the movie *I Am Legend* which was a genetically re-engineered measles virus that was supposed to kill cancer cells but ended up mutating and wiping out most of humanity. A thus far unanswered question is whether or not such a nanobot could somehow fix brain cells which are infected with the zombie virus. It is not clear if this is possible, because the virus completely highjacks the cell's genetic machinery. Answering this question would require experiments to be done on captured zombies, which can only be safely be performed by robots controlled remotely by humans. If fixing the cells is not possible, the nanobots could still be used to deliver targeted medications to destroy virus infected cells.

There are many challenges associated with developing nanobots. Because they are so small, it is necessary to have special facilities to fabricate them. An advantage of their small size is that they should be able to cross the blood-brain barrier, which normally functions to keep harmful chemicals out of brain cells. Once they are in the bloodstream, the principles of fluid mechanics imply that because of their small size, they will be in the regime known as Stokes flow, in which viscous forces dominate inertial forces; this prevents certain types of locomotion from working, although bio-inspired mechanisms such as using rotating helical flagella for propulsion hold promise.[27] One cannot rely on blood flow to bring the nanobots to the zombie's brain, because they do not have a heartbeat.

With the above considerations in mind, here is an idea of how a zombie cure might work in practice. Engineers will develop mobile robots to safely administer injections into zombies, perhaps even directly into their brains. The injections will contain nanobots, which are able to detect gradients of zombie virus, and propel themselves using their flagella toward the areas of highest concentration. The nanobots would then use genetic technology or targeted medicine delivery to disrupt the infected cells, either healing them or killing them and thus preventing them from infecting other cells. The nanobots then move on to other infected cells until the infection is eradicated. It's almost too easy! Provided the damage done to the victim's brain is not too severe, the brain can then heal and the victim can recover. If they have lost limbs while they were a zombie, engineers can help to develop artificial limbs so that they can regain more functionality. If it is too late for the victim to recover, or killing the infected cells ends up killing the zombie/person, then at least there is one less infectious zombie wandering the streets. One hopes that the development and implementation of this treatment methodology will not be cost prohibitive. Unfortunately the research funding climate will likely be even worse than it is now,[28] so it may be necessary to solicit funding contributions from filthy rich overlords and evil masterminds who have exploited the zombie apocalypse for their own selfish ends and who also happen to want to save loved ones infected with the zombie virus. If this is the case, it may be necessary for the scientists and engineers who develop this technology to leak it so that it can be used more widely.

If all else fails and the zombies are unstoppable and will soon take over the world, there will likely be a final push to launch a group of healthy humans on rockets in search of another planet on which to reestablish civilization. Engineers will play a crucial role in making such a plan work. In particular, a team of engineers will use their knowledge of science and technology to design the rockets.[29] In addition to aerospace engineering, this will involve combustion science to determine how best to propel the rocket, materials science and structural mechanics to make the rocket lightweight but strong, fluid mechanics to design the rocket to be aerodynamic enough to be able to escape the Earth's atmosphere, and control engineering to keep its flight stable under extreme conditions. It will be possible to use insight from previous rockets that have been designed over the decades, but this is more ambitious than those cases: the rocket will be traveling much further, and will need to carry enough supplies to make it to a new, suitable home.

By now, it should be obvious that it would be prudent to include some engineers as passengers on the rocket. Not only will they help to operate, maintain, and repair the rocket, but they will also be major contributors to the reestablishment of civilization, as described in the following.

Reestablishing Civilization After an Apocalypse

Whether civilization needs to be reestablished on Earth because the zombie apocalypse has left much of the infrastructure destroyed, or it is going to be established on a different planet, engineers will play a key role. Supposing that humans are able to or forced to remain on Earth, the following will describe some of the many ways that engineers will be useful after the post-apocalyptic chaos dies down.

A top priority will be to reestablish reliable power generation and delivery. In the short term, the local capture of solar and wind energy can provide power to individual households. In the longer term, a large power network will need to be restored. The enormity of this task should not be minimized. Keep in mind that the United States power grid has been referred to as "the largest and most complex machine devised by man."[30]

To understand the scope of this problem, it is instructive to consider some facts about the present United States power grid. It consists of three parts: one that serves the Eastern United States, one that serves the Western United States, and one that serves Texas. The components of the power grid are power generation plants (for example, coal plants, nuclear plants, and hydroelectric plants), transmission lines (at high voltage to minimize losses), and distribution stations (which transform from high to low voltage for usage). The U.S. power grid is approximately one hundred years old and operates near its limits, which can lead to problems such as the Northeast blackout of 2003 that affected more than 50 million people.[31] In reestablishing a national-scale power network, it would be desirable to change the overall design in several significant ways. First of all, it should be designed to readily incorporate renewable energy sources, which are not as consistent as conventional power generation but will allow for long-term sustainability. Second, it should be a smart grid, in which information about power generation and usage is continuously collected and used to automatically improve efficiency and reliability of the network. Finally, it should be designed to be robust against failures, for example situations in which a single downed power line or power generation plant due to stray zombies leads to a cascade of shutdowns, concluding with a blackout affecting a large number of people. All three of these improvements are possible, but will require the efforts of many engineers. Once power is available, refrigeration will become a reliable way to keep food safe. It is also expected that pasteurization and irradiation will become more common.

It is likely that many buildings and houses will need to be rebuilt when the zombie apocalypse is under control, due to damage from the zombies themselves and efforts to kill them. These should be rebuilt with the goal of making them as energy efficient as possible, ideally as zero-energy buildings which have zero annual net energy consumption and carbon emissions. This will reduce the need for fixing damaged power plants or constructing new ones, and would be a vast improvement over present buildings which are responsible for approximately 40 percent of our nation's energy consumption and carbon dioxide emissions.

The energy efficiency of buildings can be greatly improved in many ways. Easy-to-implement energy-saving technologies include energy efficient windows, lighting, and heating/ventilation/air conditioning. To achieve zero-energy status, the building must have energy production capabilities from solar and wind energy, and building management that takes into account the complex and dynamic interactions between the building and its environment and occupants. For example, one can use data from multiple wireless sensors positioned throughout the building to determine heating, cooling, and ventilation strategies, including natural ventilation. The construction of new energy efficient buildings, or the adaptation of existing buildings to be more energy efficient, will require the efforts of engineers trained in areas including thermal sciences, control engineering, fluid mechanics, materials, advanced lighting systems, and sensors and actuators. Communication systems will also need to be reestablished through the efforts of engineers. This will involve the repair of old hardware and design and construction of new hardware to transmit and receive information, and the development and implementation of new communication protocols and software. This will be an opportunity to improve upon present communication systems, which have developed in a rather ad hoc manner, leading to inefficiencies, areas of low coverage, mismatched supply and demand of bandwidth, and frequent network failures. For example, consider the allocation of the radio spectrum by the Federal Communications Commission, which is a partition of the electromagnetic spectrum into bands used for different purposes such as broadcast television and audio, cellphones, radar, WiFi, GPS navigation, military applications, and much more. The current allocation scheme is a patchwork which has been driven as much by economics as by engineering considerations, and could be improved by more rational allocation schemes and innovations such as cognitive radio, in which wireless devices adaptively change their operation to best take advantage of user needs and the locally available radio spectrum.[32] Communication networks can also be improved through the development of fault management strategies and the implementation of results from network science on the relationship between node connectivity and network robustness.

There will also be opportunities for engineers to improve transportation systems. With the likely destruction during the zombie apocalypse of much of the infrastructure associated with fuel processing and delivery, it will be very important for transportation approaches to emphasize energy efficiency. This will include further development of hybrid automobiles and the use of biofuels. Perhaps engineers can even figure out how to turn all those dead zombie bodies left over from the apocalypse into fuel. As long distance travel between cities becomes more common, there will also be a push for high speed rail based on technology such as Maglev, which uses magnets to levitate and propel trains rather than relying on wheels. Such trains will have less reliance on hard-to-obtain fossil fuels, and, as a bonus, they produce less pollution. For travel within a city, Maglev subways would provide efficient transportation, and because there are limited points of entry it will be easier to guard the system from incursion by any remaining zombies.

As described in this essay, engineers have much to contribute both during and after a zombie apocalypse. During the zombie apocalypse, they will develop new ways to kill zombies, including innovations such as telescopic, pneumatic head crushers; hydraulic zombie crushers; and robots which can safely get close enough to the zombies to use hammers and pinchers to crush their brains. Autonomous robots, including driverless cars and humanoid robots, will also be developed to do dangerous tasks such as transporting supplies, recovering water from a lake or a well, and surveillance tasks. Nanobots might even provide a way to cure zombies. Engineers will also help to fortify buildings through stronger windows and doors chosen based on engineering analysis, and keep quality of life of the survivors as high as possible by maintaining power sources, communication channels, and safe food and water.

After the zombie apocalypse is under control, engineers will use their know-how to help to reestablish civilization. In the short term, on a timescale of, say, 28 days, localized and relatively primitive solutions to everyday challenges such as securing safe food and water will benefit from the expertise of engineers. On a timescale of, say, 28 weeks, roads, buildings, and bridges which were damaged during the zombie apocalypse will start to be rebuilt, with the help of engineers. On a timescale of, say, 28 months or more, engineers will design power networks that incorporate renewable energy sources; design buildings that achieve high energy efficiency; build communication networks which have improved hardware, software, and protocols; and design and build transportation systems with energy efficient cars and high speed rail. Such efforts will be constrained by cost and will need to be designed to withstand future zombie onslaughts.

Engineers are trained to practice ingenuity in many contexts, and will clearly play many crucial roles during and after a zombie apocalypse, including many that will only be dreamt up when a zombie apocalypse or other similarly disastrous situation occurs.

NOTES

    1. Celeste Baine, *Engineers Make a Difference* (Calhoun, LA: Bonamy, 2008).
    2. Max Brooks. *The Zombie Survival Guide: Complete Protection from the Living Dead* (New York: Three Rivers Press, 2003).
    3. N. Yoganandan, F.A. Pintar, A. Sunces, Jr., P.R. Walsh, C.L. Ewing, D.J. Thomas, and R.G. Snyder, "Biomechanics of skull fracture," *Journal of Neurotrauma* 12 (1995): 659–668.

4. Jeff Moehlis, "Prof J-Mo at 2010 UCSB Zombie Debate," http://www.youtube.com/watch?v=ZV4usz3Srt0.

5. L.H. Lin, E. Hinman, H.F. Stone, and A.M. Roberts, "Survey of window retrofit solutions for blast mitigation." *Journal of Performance of Constructed Facilities* 18 (2004): 88–94.

6. X. Zhang, H. Hao, and G. Ma, "Laboratory test and numerical simulation of laminated glass window vulnerability to debris impact," *International Journal of Impact Engineering* 55 (2013): 49–62.

7. Noah Shachtman, "The Baghdad Bomb Squad," Wired, November 2005.

8. B. Bingham, B. Foley, H. Singh, R. Camilli, K. Delaporta, R. Eustice, A. Mallios, D. Mindell, C. Roman, and D. Sakellariou, "Robotic tools for deep water archaeology: surveying an ancient shipwreck with an autonomous underwater vehicle" *Journal of Field Robotics.* 27 (2010): 702–717.

9. K.C. Kim, ed., *Robotics in General Surgery* (New York: Springer, 2014).

10. Ryan Villarreal, "Robotics helped capture Boston bombing suspect," *International Business Times*, http://www.ibtimes.com/robots-helped-capture-boston-bombing-suspect-video-1206075.

11. M. Buehler, K. Iagnemma, and S. Singh, eds., *The DARPA Urban Challenge: Autonomous Vehicles in City Traffic* (Berlin: Springer, 2009).

12. "DARPA Robotics Challenge." http://www.darpa.mil/Our_Work/TTO/Programs/DARPA_Robotics_Challenge.aspx.

13. "DARPA Robotics Challenge 2013: A Woodstock for Robots," *New York Times.* http://www.youtube.com/watch?v=w222KFAiMQc.

14. F. Bullo, J. Cortes, and S. Martinez, *Distributed Control of Robotic Networks* (Princeton: Princeton University Press, 2009); T. Theodoridis and H. Hu, "Toward intelligent security robots: a survey," *IEEE Transactions on Systems, Man, and Cybernetics—Part C: Applications and Reviews* 42 (2012): 1219–1230.

15. P.W. Singer, *Wired for War: The Robotics Revolution and Conflict in the 21st Century* (New York: Penguin, 2009).

16. "When the zombies take over, how long till the electricity fails?" *The Straight Dope*, http://www.straightdope.com/columns/read/2165/when-the-zombies-take-over-how-long-till-the-electricity-fails.

17. F.M. Vanek, L.D. Albright, and L.T. Angenent, *Energy Systems Engineering,* 2d ed. (New York: McGraw-Hill Professional, 2012).

18. R. Corkish, M.A. Green, and T. Puzzer, "Solar energy collection by antennas." *Solar Energy* 73 (2002): 395–401.

19. "New York City ARECS Members and the Attacks of September 11, 2001," http://www.nyc-arecs.org/911.html.

20. Stephen Singer, "Ham radios came to rescue in blackout," http://www.govtech.com/public-safety/Ham-Radios-Came-to-Rescue-in.html.

21. Gary Krakow, "Ham radio operators to the rescue after Katrina," http://www.nbcnews.com/id/9228945/ns/technology_and_science-wireless/t/ham-radio-operators-rescue-after-katrina/.

22. N.G. Marriott, *Essentials of Food Sanitation* (New York: Chapman & Hall, 1995).

23. *Engineers Without Borders*, http://www.ewb-usa.org.

24. "Get Prepped: Food," http://channel.nationalgeographic.com/channel/doomsday-preppers/articles/get-prepped-food/.

25. "Zombrex," http://deadrising.wikia.com/wiki/Zombrex.

26. R.A. Freitis Jr., "Current status of nanomedicine and medical nanorobotics," *Journal of Computational and Theoretical Nanoscience* 2 (2005): 1–25.

27. T.E. Mallouk and A. Sen, "Powering nanorobots," *Scientific American* 300 (2009): 72–77.

28. Beryl Lieff Benderly, "On American Scientist's Dismal Morale," http://sci encecareers.sciencemag.org/career_magazine/previous_issues/articles/2014_03_03/ caredit.a1400056.

29. Travis S. Taylor, *Introduction to Rocket Science and Engineering* (New York: CRC Press, 2009).

30. P. Kundur, *Power System Stability and Control* (New York: McGraw-Hill, 1994).

31. J.R. Minkel, "The 2003 Northeast blackout—five years later," *Scientific American*, August 13, 2008.

32. S. Haykin, "Cognitive radio: brain-empowered wireless communications." *IEEE Journal on Selected Areas in Communications* 23 (2005): 201–220.

BIBLIOGRAPHY

Baine, Celeste. *Engineers Make a Difference.* Calhoun, LA: Bonamy, 2008.

Benderly, Beryl Lieff. "On American Scientist's Dismal Morale." http://sciencecareers. sciencemag.org/career_magazine/previous_issues/articles/2014_03_03/caredit. a1400056.

Bingham, B., B. Foley, H. Singh, R. Camilli, K. Delaporta, R. Eustice, A. Mallios, D. Mindell, C. Roman, and D. Sakellariou. "Robotic tools for deep water archaeology: surveying an ancient shipwreck with an autonomous underwater vehicle." *Journal of Field Robotics* 27 (2010): 702–717.

Brooks, Max. *The Zombie Survival Guide: Complete Protection from the Living Dead.* New York: Three Rivers Press, 2003.

Buehler, M., K. Iagnemma, and S. Singh, eds. *The DARPA Urban Challenge: Autonomous Vehicles in City Traffic.* Berlin: Springer, 2009.

Bullo, F., J. Cortes, and S. Martinez. *Distributed Control of Robotic Networks.* Princeton: Princeton University Press, 2009.

Corkish, R., M.A. Green, and T. Puzzer. "Solar energy collection by antennas." *Solar Energy* 73 (2002): 395–401.

"DARPA Robotics Challenge." http://www.darpa.mil/Our_Work/TTO/Programs/ DARPA_Robotics_Challenge.aspx.

"DARPA Robotics Challenge 2013: A Woodstock for Robots." *New York Times* http:// www.youtube.com/watch?v=w222KFAiMQc.

Engineers Without Borders. http://www.ewb-usa.org.

Freitis, R.A., Jr. "Current status of nanomedicine and medical nanorobotics." *Journal of Computational and Theoretical Nanoscience* 2 (2005): 1–25.

"Get Prepped: Food." http://channel.nationalgeographic.com/channel/doomsday-preppers/articles/get-prepped-food/.

Haykin, S. "Cognitive radio: brain-empowered wireless communications." *IEEE Journal on Selected Areas in Communications* 23 (2005): 201–220.

Kim, K.C., ed. *Robotics in General Surgery.* New York: Springer, 2014.

Krakow, Gary. "Ham radio operators to the rescue after Katrina." http://www.nbcnews. com/id/9228945/ns/technology_and_science-wireless/t/ham-radio-operators-rescue-after-katrina/.

Kundur, P. *Power System Stability and Control.* New York: McGraw-Hill, 1994.

Lin, L.H., E. Hinman, H.F. Stone, and A.M. Roberts. 2004. "Survey of window retrofit solutions for blast mitigation." *Journal of Performance of Constructed Facilities* 18 (2004): 88–94.

Mallouk, T.E., and A. Senn. "Powering nanorobots." *Scientific American* 300 (2009): 72–77.

Marriott, N.G. *Essentials of Food Sanitation.* New York: Chapman & Hall, 1995.

Minkel, J.R. "The 2003 Northeast blackout—five years later." *Scientific American.* August 13, 2008.

Moehlis, Jeff. "Prof J-Mo at 2010 UCSB Zombie Debate." http://www.youtube.com/watch?v=ZV4usz3Srt0.

"New York City ARECS Members and the Attacks of September 11, 2001." http://www.nyc-arecs.org/911.html.

Shachtman, Noah. "The Baghdad Bomb Squad." *Wired,* November 2005.

Singer, P.W. *Wired for War: The Robotics Revolution and Conflict in the 21st Century.* New York: Penguin, 2009.

Singer, Stephen. "Ham radios came to rescue in blackout." http://www.govtech.com/public-safety/Ham-Radios-Came-to-Rescue-in.html.

Taylor, Travis S. *Introduction to Rocket Science and Engineering.* New York: CRC Press, 2009.

Theodoridis, T., and H. Hu. "Toward intelligent security robots: a survey." *IEEE Transactions on Systems, Man, and Cybernetics—Part C: Applications and Reviews* 42 (2012): 1219–1230.

Vanek, F.M., L.D. Albright, and L.T. Angenent. *Energy Systems Engineering,* 2d ed. New York: McGraw-Hill Professional, 2012.

Villarreal, Ryan. "Robotics helped capture Boston bombing suspect." *International Business Times,* http://www.ibtimes.com/robots-helped-capture-boston-bombing-suspect-video-1206075.

"When the zombies take over, how long till the electricity fails?" *The Straight Dope.* http://www.straightdope.com/columns/read/2165/when-the-zombies-take-over-how-long-till-the-electricity-fails.

Yoganandan, N., F.A. Pintar, A. Sunces, Jr., P.R. Walsh, C.L. Ewing, D.J. Thomas, and R.G. Snyder. "Biomechanics of skull fracture." *Journal of Neurotrauma* 12 (1995): 659–668.

Zhang, X., H. Hao, and G. Ma. "Laboratory test and numerical simulation of laminated glass window vulnerability to debris impact." *International Journal of Impact Engineering* 55 (2013): 49–62.

"Zombrex." http://deadrising.wikia.com/wiki/Zombrex.

Homeland Security, FEMA and Securing the Masses Against the Zombie Horde

Jeremy Youde

In what is surely one of the first instances of a United States senator directly addressing the prospects of an outbreak of zombieism, Tom Coburn (R–Oklahoma) decried, "Paying first responders to attend a HALO Counter-terrorism Summit at a California island spa resort featuring a simulated zombie apocalypse does little to discourage potential terrorists."[1] Coburn lambasted the Department of Homeland Security (DHS), Federal Emergency Management Agency (FEMA), and other disaster response organizations for wasting money on such a frivolous activity. (Officials from HALO Corporation rebutted Coburn's charge, saying that an event sponsor, not the federal government, paid the $1,000 fee for each participant.[2])

Coburn may have scored political points by laughing at the idea of a zombie outbreak, but he missed the bigger picture. If a zombie outbreak were to occur, DHS and FEMA would have leading roles to play in the federal government's response. DHS and FEMA already have structures in place for responding to a hurricane, massive flood, or other natural disasters, and they may be able to employ these to respond to zombies. Zombies are much like a natural disaster, albeit one that shambles around looking for tasty flesh to eat.

This chapter will explore how the federal government's lead disaster response agencies, DHS and FEMA, would respond to a zombie apocalypse and examine how the Stafford Act and other laws apply. While these structures offer important opportunities for the federal government to marshal an effective response, there are significant issues that may limit their ability to act in a prompt, efficient, and appropriate manner.

Before the Zombies Arrive

The best defense is a good offense, and DHS and FEMA encourage individuals and governments to be ready in case a natural disaster strikes. Before the undead arrive, both offices encourage people to make emergency preparations for themselves and their families. FEMA notes that emergency responders will eventually be on the scene after a disaster, but that they cannot reach everyone immediately. Both FEMA and the Centers for Disease Control and Prevention's (CDC) comic book on zombie preparedness recommend that households create an emergency supply kit with food, water, batteries, a radio, proper clothing, and other essential supplies for at least 72 hours.[3] For zombies, one might assume that a gun would be essential for ensuring safety, but that does not appear on either list of emergency supplies.

FEMA and DHS counsel vigilance and preparation for state and local governments, too. Well before any sort of natural disaster occurs, all the relevant parties should know who will take on what role and they should conduct simulations to reinforce these plans. At the local level, the municipality's chief elected or appointed official, such as the mayor or city manager, is tasked with ensuring public safety and welfare, while the emergency manager, such as the police chief or director of public works, has day-to-day responsibility for implementing emergency management programs in collaboration with local departments and agencies.[4] The private sector and nongovernmental organizations also play a role, providing direct support to victims and helping local areas restore their economies as quickly as possible.[5] At the state or tribal level, the governor or tribal leader, the homeland security advisor, and director of emergency management agency should coordinate to support each other, act as liaisons between the local and federal government, and deploy resources like money and the National Guard as appropriate.[6] The President, Secretary of Homeland Security, and FEMA Administrator work to coordinate actions and support on-the-ground activities in affected regions, and they collaborate with other agencies, offices, and departments as appropriate, such as the Department of Health and Human Services, the Environmental Protection Agency, and the Small Business Administration.[7] These offices involved in emergency preparation and response are already in place and should establish relationships with each other so that they can function smoothly and efficiently if and when their services are called upon.

While these preparations should facilitate timely responses to any type of natural disaster, the University of Florida has emerged as a leader in explicitly incorporating a zombie apocalypse into its preparations. Its plan provides information on how to properly identify a zombie, establishes reporting procedures, and identifies the relevant resources for responding.[8]

Government Disaster Response Structures

While personal and institutional vigilance is useful, it will be necessary for the federal government to step in when the zombie apocalypse finally arrives. Under current U.S. law, three key elements guide federal responses to natural disasters: the Stafford Act, the Federal Emergency Management Agency (FEMA), and the Department of Homeland Security (DHS). These three pieces work together, overlapping and intersecting in various ways. A disaster declaration under the terms of the Stafford Act engages FEMA, which is part of DHS, to mobilize its resources and involve other DHS agencies like the Coast Guard, Customs and Border Protection, and the Federal Protective Service to respond in a timely, efficient manner. They form a web that guides the maintenance and distribution of resources to prepare for and respond to hurricanes, tornadoes, wildfires, and swarms of the undead walking the streets.

Having a single agency to coordinate the federal government's activities in natural disasters is a relatively new innovation. Until the 1930s, the government responded to disasters in piecemeal fashion, merely passing appropriations in response to specific crises. During the Great Depression, new legislation empowered certain government agencies to provide funding and support in response to disasters, but these offices frequently failed to coordinate with each other and would at times work at cross purposes with each other.[9] In 1973, Presidential Reorganization Plan #1 created the Federal Disaster Assistance Administration within the Department of Housing and Urban Development. This organization sought to coordinate the myriad of federal offices and agencies that had some hand in disaster response, but it found itself stymied by political infighting. To better manage federal responses to natural disasters, President Jimmy Carter signed Executive Order 12127 in 1979 to create the Federal Emergency Management Agency as an independent agency.[10]

In 1988, the Robert T. Stafford Disaster Relief and Emergency Assistance Act (Stafford Act) created the current system by which the president can declare a disaster. A disaster declaration triggers financial assistance and resource mobilization under FEMA's guidance. Under this Act, a state's governor must assess whether his or her state has the necessary resources to respond effectively. If not, he or she may contact the federal government to seek a presidential disaster declaration. The president will then decide whether the event in question is of sufficient scale and requiring the resources of the federal government based on reports from local, state, and/or federal officials of the affected areas.[11]

If the president makes a disaster declaration under the Stafford Act, FEMA swings into action. It can provide assistance to state and local officials, helping with food, clothing, shelter, and repairing damaged and destroyed structures. FEMA can also provide loans to state and local governments, as well as individual businesses, to reimburse them for lost tax revenues and sales. It can mobilize other relevant agencies to assist with recovery efforts, such as the Department of Housing and Urban Development, the Small Business Administration, the Department of Transportation, and the Department of Agriculture.[12] The disaster declaration spells out how the federal government and state and local governments will share the costs for various assistance programs. Typically, the federal government covers 75 percent, while state and local governments cover 25 percent; these percentages may be adjusted based on need and the nature of the disaster's effects.[13] It also allows for non-governmental organizations, such as the American Red Cross, to take specific actions to support federal disaster recovery efforts.

While the Stafford Act triggers federal involvement in disaster recovery, it does not mean that the federal government takes over response and recovery activities. Stafford Act declarations follow the "bottom-up" principle, meaning that state and local officials are in charge of any response. FEMA and other agencies step in only when state and local resources are exhausted, and FEMA's activities focus on coordinating state and local involvement with federal programs and aiding state and local officials with accessing resources.[14]

In the aftermath of the September 11 attacks, Congress decided that the government needed a new Cabinet-level department to coordinate and oversee preparations for and responses to events that threaten the United States: the Department of Homeland Security (DHS). By bringing a diverse array of offices and agencies together under a single department, policymakers believed that DHS would promote communication and coordination, increase efficiency and decrease response time. While many of these efforts were geared toward responding to and preventing terrorist attacks, the reorganization efforts also incorporated the federal government's response mechanisms to natural disasters. DHS absorbed twenty-two different existing agencies and offices from throughout the government, ranging from the Immigration and Naturalization Service (previously housed in the Department of Justice) to the U.S. Coast Guard to the Environmental Measurements Laboratory (previously housed in the Department of Energy) to FEMA. By drawing on its various constituent agencies, particularly FEMA, and partnering with state and local offices, DHS' creation was intended to provide a better-coordinated response to emergencies while eliminating confusion over lines of responsibility and poor communications.[15]

The combined energies of FEMA, the Stafford Act, and DHS got their first major test in 2005 responding to the aftermath of Hurricanes Katrina and Rita along the Gulf Coast. They failed spectacularly. The federal response was, at best, a comedy of errors. Policymakers at all levels and of all political persuasions lambasted FEMA and DHS for their failures to provide supplies, communicate with officials on the ground, and exercise flexibility in implementing regulations. Powerful media images of survivors trapped on the roofs of their houses, people huddled into a vastly-undersupplied New Orleans Superdome, and fields of school buses sitting idle instead of evacuating residents seared themselves into the public's consciousness. Bureaucratic wrangling and political turf fights within DHS hampered recovery efforts, and FEMA officials alleged that DHS had removed so many resources from FEMA that the agency could not be effective.[16] DHS' Office of the Inspector General described the response efforts as quickly overwhelmed, with officials unclear about responsibilities and protocols and poorly distributed resources.[17] The White House's assessment of the federal government's disaster response and recovery efforts along the Gulf Coast bluntly stated, "There is no question that the Nation's current incident management plans and procedures fell short of what was needed and that improved operational plans could have better mitigated the Hurricane's tragic effects."[18]

The collective failure of the federal government's disaster response and recovery efforts in the face of this natural disaster spurred significant changes in the structure and framework for federal responses to natural disasters. Congress passed six pieces of legislation that directly affected federal disaster response and recovery activities. The most significant was the Post-Katrina Emergency Reform Act of 2006. This act kept FEMA as a part of DHS, but elevated the agency's status within the department. It also consolidated more of DHS' response functions within FEMA and gave it more autonomy in how it responds to emergencies.[19] These changes organizationally placed FEMA roughly on par with the U.S. Coast Guard and the Secret Service, but with substantial freedom to act.

Responding to the Zombie Outbreak

Although plans and preparations may be in place, it is highly unlikely that the public will have any warning about a zombie outbreak before it happens. Unlike a hurricane, there exists no forecasting models that can provide a sense of when and where the undead will roam. This means that any response by the federal government would rely on the authority of the Stafford Act, Department of Homeland Security, and the Federal Emergency Management Agency. It is likely that the governor of the affected state(s) would approach the federal government for a disaster declaration to combat the zombie hordes. Based on what we have learned from *World War Z*, *Zombieland*, and *28 Days Later*, state and local officials will likely be overwhelmed.[20] Further, zombies are unlikely to respect jurisdictional boundaries, meaning that the outbreak's effects will spread to multiple localities and states. (As a side note, this assumes that the federal government remains intact. If the government collapses because of a zombie apocalypse, as in *Zombieland* and *The Walking Dead*, all bets are off.)

Zombieism's means of transmission may complicate the government's ability to declare a natural disaster. Most descriptions of zombieism posit that it acts essentially like a virus. In films like *28 Days Later* and *Shaun of the Dead* and books like *World War Z*, a person becomes infected when bitten by a zombie, which then turns that person into a zombie.[21] There is no agreement about the origin of the zombie virus itself; suspected origins include mutated animal viruses (*Dead Rising*), mutated human viruses (*Feed*), nuclear fallout (*Night of the Living Dead*), and Haitian Vodou (*The Serpent and the Rainbow*), but some films openly acknowledge they have limited understanding of the virus' genesis (*Dawn of the Dead*).[22] It is unclear, though, whether the federal government can declare a natural disaster under the Stafford Act for "biological disasters." The government did it in response to the discovery of West Nile virus in 2000, but FEMA declared in 2005 that disease outbreaks did not qualify under the Stafford Act because they were not explicitly mentioned in the authorizing legislation.[23] Subsequent government declarations included ambiguous guidance on this issue.[24] When President Obama declared a national emergency in response to the 2009 H1N1 influenza outbreak, he sidestepped the issue by relying on the National Emergencies Act, which gives the president authority to act in response to emergencies other than natural disasters and wars, instead of using the Stafford Act.[25] This approach is consistent with the National Strategy for Biosurveillance, which locates its authority for collecting and disseminating information about infectious disease outbreaks within an explicitly terrorism-oriented framework, but does not clarify whether disease outbreaks constitute a natural disaster that can activate the Stafford Act, FEMA, and DHS.[26]

The ambiguity over federal responses to biological disasters need not prevent an effective response to zombieism. While zombieism may spread biologically, its effects are not merely biological. *Dawn of the Dead*, to give but one example, illustrates that zombies not only attack people, but they also wreak havoc on infrastructure and leave a path of destruction in their wake.[27] Additionally, Jonathan Coulton's "Re: Your Brains" references zombies' negative effects on buildings and public spaces.[28] The combination of a rapidly spreading infectious agent and damaged buildings and infrastructure would justify the president declaring zombieism a major disaster. If the president did declare a major emergency, the Stafford Act would be in effect, triggering various resources, operational plans, and coordination among officials.[29]

Once the Stafford Act is enacted, FEMA and DHS can step in to coordinate and guide disaster response and recovery efforts to combat the zombie plague. This does not put either DHS or FEMA in charge; it merely activates their resources to support state and local actors in a variety of ways. FEMA relies in part on a mathematical formula before it will release financial resources in response to a disaster. The damage assessment must reach at least $1.30 per capita in order to grant Public Assistance funds.[30] As a result of this requirement, a zombie outbreak in California would need to cause nearly $50 million in damage to trigger these funds, while Wyoming could get these funds with less than $1 million in zombie-related damage. Once made available, state and local governments could use these funds to repair infrastructure, remove debris, and pay for additional protective services from police or military forces.

The issue of police and military personnel deserves greater attention. It is reasonable to assume that local officials would require additional police or military support in the face of a zombie outbreak. Rick Grimes, the sheriff's deputy from *The Walking Dead*, demonstrates that some sort of law enforcement presence can be useful in promoting group coherence and providing at least some small degree of protection, though law enforcement may make missteps as they adjust their operations to deal with zombies in addition to criminals.[31] FEMA and DHS designate the Department of Defense as a full partner in responding to domestic disasters and can be incorporated into response activities. When deployed under these frameworks, though, military personnel remain a part of the official military chain of command. Requests for federal military personnel must go through the Secretary of Defense, and any deployed personnel would still ultimately be accountable to their commanders.[32] Further, the 2006 revisions to the Insurrection Act of 1807 allow the president to deploy military troops to "restore public order and enforce the laws of the United States" in response to natural disasters—including public health emergencies.[33] National defense authorities may, in the course of responding to a natural disaster, "respond to save lives, protect property and the environment, and mitigate human suffering under imminently serious conditions."[34]

In this instance, calling on the services of federal military personnel may be incredibly logical and appropriate. To neutralize a zombie, one must shoot it in the head. Specifically, *The Zombie Survival Guide* specifies that a person must destroy a zombie's frontal lobe in order to kill it, and doing that without a firearm is incredibly difficult (and would require getting so close to a zombie that a person would put him/herself in danger of being bitten).[35] Military personnel would be uniquely qualified and equipped for providing this service. Private individuals could also contribute to these efforts, but the military's existing infrastructure and training would make it an invaluable resource in these efforts. The military's transportation equipment and armored vehicles could also be highly useful for moving through infested cities, navigating around debris, and transporting evacuees to safe locations.

Evacuating people from zombie hotspots and providing suitable housing would also be of crucial importance during a zombie outbreak, and coordinating those efforts would fall into FEMA's hands. This could include short-term emergency housing needs and longer-term remediation and rebuilding efforts as part of the general attempts to restore normalcy as rapidly as possible. Indeed, providing housing services is one of FEMA's specific Emergency Support Functions detailed in the *National Response Framework*.[36] None of the official government response and recovery frameworks necessarily dictate the form that emergency housing would need to take. In the past, FEMA has utilized large public spaces like schools and sports arenas to meet immediate needs and obtained trailers for people to use for longer-term housing solutions. It can also provide grants and loans for long-term rental options, home repairs and modifications, or home replacements if repairs are not possible. Both the original and remake of *Dawn of the Dead* conclusively prove that shopping malls are ill-equipped at providing shelter from the shuffling hordes, so FEMA could instead direct its emergency housing resources toward zombieproofing homes and providing long-term rentals in non-infested areas during a zombie outbreak.

Addressing the health needs of those in the midst of a zombie outbreak would rely most heavily on local officials, nongovernmental organizations, and private charities. Both the *National Response Framework* and the *National Disaster Recovery Framework* emphasize their importance in providing such individual level and highly personalized interventions. In the immediate aftermath of an outbreak, FEMA could mobilize its National Disaster Medical System to provide personnel for basic triage services.[37] FEMA could also cooperate with the Department of Health and Human Services (HHS) and the Centers for Disease Control and Prevention (CDC), though both are independent of DHS. Under HHS' authority, it could offer a wide variety of health-related functional aspects, such as providing necessary pharmaceuticals, ensuring food and water safety, assisting with victim identification and removal, and offering mental health support services.[38] Nongovernmental organization like the Red Cross may also assist with providing these vital services in the aftermath of a disaster.

Limitations on DHS, FEMA and Zombies

FEMA and DHS may have the lead responsibilities for coordinating federal responses to natural disasters, but their operating frameworks evince two distinct problems that could impair effective response and recovery activities in the event of a zombie outbreak: an unwieldy structure, and serious operational limitations. From a structural perspective, FEMA and DHS face serious complications. The creation of DHS, with its seemingly haphazard merger of twenty-two distinct agencies and offices, complicated lines of shared governance with state and local officials. These subnational units are expected to carry out most emergency management operations, but their ability to do so is highly constrained by DHS' orders and funding.[39] State and local agencies rely on DHS and FEMA for training and resources to do their jobs, but the federal agencies have provided little of either. As a result, the state and local bodies struggle to maintain the necessary resources, personnel, and capabilities to appropriately respond to disaster situations.[40] At the same time, Congress has demonstrated little will to make the financial and political investments necessary to ensure that FEMA and DHS can mount a credible and robust response to emergencies.[41]

DHS became the largest bureaucracy in the U.S. government when it was formed, but the department has struggled to bring its various offices and agencies together to work toward a common goal with a shared culture. DHS remains an amalgamation of offices, tasked with doing too much with too few resources.[42] The department made substantial changes to bolster communications among agencies and establish clear lines of authority in the wake of Hurricane Katrina, but these adjustments largely exist on organizational flow charts at this point. Until these are tested by a natural disaster, such as shuffling hordes of the undead crowding the streets of American cities, there is little evidence that these changes have trickled down to the operational level.[43] FEMA's self-assessment after Hurricane Sandy hit the East Coast of the United States in 2012 found deficiencies in addressing survivors' needs quickly, coordinating among various governmental levels, and deploying a qualified disaster workforce.[44]

This leads to a second area of weakness in FEMA and DHS' response to zombies—the actual operational plans themselves. One problem is that both FEMA and DHS have poor reputations in the government and the public. Martha Derthick of the University of Virginia has studied the efficacy and public perceptions of the federal government's responses to natural disasters. She notes, "The job of interagency coordination is a hard one at best, and FEMA's performance has never been judged very favorably."[45] She finds that government officials blasted the agency for its response to tornadoes in Pennsylvania in 1988 and Hurricane Hugo's effects on South Carolina in 1989. She even cites FEMA's lackluster response to Hurricane Andrew in Florida in 1992 for contributing to President George H.W. Bush's defeat later that year. FEMA's reputation improved during the Clinton administration, but Derthick reminds readers that the agency faced only one major domestic natural disaster—a major East Coast blizzard in 1993—during that period.[46] DHS' poor performance during Katrina still dogs the department years later. A 2012 survey by the American Customer Satisfaction Index found that DHS had the second-lowest satisfaction score among Cabinet departments. (Only the Treasury Department, which houses the Internal Revenue Service, scored lower.[47]) The lack of faith in DHS' abilities makes it difficult to actuate its operational plans, though the department has made improving its stature a part of its Quadrennial Homeland Security Review.[48]

DHS as an operational body fundamentally orients its activities, resources, and personnel toward preventing terrorism. Public service professor Paul Light argues that 65 percent of the department's budget focuses solely on terrorism prevention and response activities, leaving only a small amount for the department's other responsibilities like responding to natural disasters. As a result, FEMA's budget has been gutted over the years since it became part of DHS. Further compounding its response capabilities, FEMA has lost many of its top personnel due to political infighting and frustrations with the agency's ever-decreasing status.[49] When these skilled personnel leave, they take with them experience, know-how, and institutional memory. This makes it ever more difficult for the agency to learn from its past mistakes or build on its successes.[50]

State and local governments have found it incredibly difficult to implement FEMA's response and recovery frameworks on the ground. A study by public health researchers at the University of Alabama Birmingham examined the guidelines for ensuring the continuity of operations in the event of an emergency distributed to subnational governments by FEMA and DHS. This study found that these guidelines lack the specificity and detail necessary for these local agencies to build an effective disaster response and recovery infrastructure.[51] If the best defense is a good offense, then FEMA and DHS are failing to provide state and local officials with what they would need to create that offense—or what they would need to mobilize after a natural disaster occurs.

The operational plans themselves do not necessarily lend themselves to effective strategies to fight back against zombies or any other infectious disease threat. First, the lines of communication between FEMA and CDC remain weak. During Hurricane Katrina, both agencies deployed medical teams without consulting each other.[52] This was both inefficient and counterproductive, but there is little evidence from the current frameworks that such confusion would be avoided in the future.

Second, while military personnel can indeed serve an important role in maintaining public order during a major disaster, they face a serious public relations hurdle. In short, many people do not trust the military's involvement with FEMA. In some corners, this results in paranoid conspiracy theories about the government manufacturing a national disaster to justify declaring martial law under FEMA's watch or creating concentration camps. One of the most prominent proponents of these theories is former professional wrestler and Minnesota governor Jesse Ventura. Ventura's television program *Conspiracy Theory* (TruTV, 2009–2012) dedicated an entire episode to the idea. Entitled "The Police State Conspiracy," it originally aired on November 12, 2010. Turner Broadcasting System, the owner of TruTV, never re-aired the episode, citing "scheduling concerns," but Ventura alleges that someone pressured the network not to air the show again because it was too close to the truth.[53] In more mainstream outlets, though, the military's response in New Orleans during Hurricane Katrina undermined its ability to do good or make people feel secure. Military officials described their involvement in New Orleans as "combat operations" in response to an "insurgency." The commander of the Louisiana National Guard told *Army Times*, "This place is going to look like Little Somalia."[54] These are not the images or reports that engender a sense of security among the populace—especially in a panicked situation where the reanimated dead are haunting the local population. FEMA and DHS have done a poor job in explaining how and why the military would get involved in disaster response and recovery operations, and they have not established the sorts of accountability measures that would reassure the public that using military personnel does not automatically equate with martial law or the end of democratic rule.

Third, evacuation and emergency housing are of vital importance, but the specter of Katrina looms over these operations as well. From the images of people stranded on their rooftops to the utter chaos and rumors of murder in the Louisiana Superdome to FEMA-provided trailers whose faulty construction led to health problems due to elevated levels of formaldehyde, FEMA's track record is not great. Such incredibly and well-publicized failures have created serious doubts about whether the government would be able to respond effectively to get people out of harm's way and provide them with a place to live until they could return to their homes. The disaster response and recovery frameworks highlight the need to get people to safety, but they remain so vague that it would be incredibly difficult to put this idea into action.

Finally, the operational plans emphasize the importance of devolving more and more responsibilities to state and local governments and giving private actors an increasingly prominent role in overseeing response operations and providing services to the affected. Kevin Fox Gotham, a sociologist at Tulane University in New Orleans who studies post-disaster rebuilding and resilience, objects to this strategy. He sees it as evidence of ever-creeping neoliberalism within the American government, taking what should fundamentally be a government-provided service and making its provision contingent upon market forces. These private actors, he alleges, operate without transparency or cost controls, making corruption and cost overruns more likely.[55] These are problems because the incentives for private businesses are not the same as the incentives for the government. Poor and vulnerable populations may find themselves with fewer services because companies deem them unprofitable, and the government may not have time to properly vet private companies when emergencies strike. Private businesses may take advantage of the panic caused by zombies invading another state to inflate their costs and overcharge the government. Recent history has demonstrated how the government's failure to provide oversight and implement cost controls can lead to substandard outcomes and rampant corruption. Economic professors Peter Leeson and Russell Sobel argue that natural disasters breed corruption because of the massive influx of funds from FEMA with too little oversight. They find that every $100 per capita of FEMA aid increases the average state's corruption by 102 percent.[56]

Conclusion

An outbreak of zombieism in the United States would present a significant challenge to the government. Under the authority vested in the Department of Homeland Security and the Federal Emergency Management Agency, and through the Stafford Act, the federal government may be able to mobilize significant resources to coordinate a response and support state and local officials in implementing effective response and recovery plans. FEMA and DHS have established frameworks that offer a number of possible options for implementing programs, but there remain important questions about whether these frameworks are appropriate or useful in fighting off the undead.

It is worth highlighting that there may be some upsides to a zombie apocalypse. *Shaun of the Dead* suggests that zombies may actually prove useful, providing manual labor and serving as game show contestants, and Colson Whitehead's *Zone One* points out that it may be easier to get a table in New York's most prestigious restaurants.[57] While these possibilities can offer some solace, the fact remains that an outbreak of zombieism remains utterly unpredictable. As such, it behooves the federal government to ensure that its natural disaster plans can effectively counter the effects of masses of the undead.

NOTES

1. Tom Coburn, "Dr. Coburn releases 'Safety at any Price' report on questionable grant spending at Department of Homeland Security," 5 December 2012, http://www.coburn.senate.gov/public/index.cfm/pressreleases?ContentRecord_id=bcd61d71-5454-4a81-9695-e949e0253faa (accessed 22 February 2013).

2. Lauren Fox, "Report: DHS spent money on zombie simulation," *U.S. News &*

World Report, 5 December 2012, http://www.usnews.com/news/articles/2012/12/05/coburn-zombies-napolitano-homeland-security (accessed 21 February 2013).

3. FEMA, "Build a Kit," http://www.ready.gov/build-a-kit (accessed 28 February 2013); Centers for Disease Control and Prevention, *Preparedness 101: Zombie Pandemic* (Atlanta: Centers for Disease Control and Prevention, n.d.), http://www.cdc.gov/phpr/_media/cymkPrint/11_225700_A_Zombie_Final.pdf (accessed 28 February 2013).

4. FEMA, *National Recovery Framework* (Washington, D.C.: FEMA, 2008): 15–17.

5. FEMA, *National Recovery Framework*, 18–21.

6. FEMA, *National Recovery Framework*, 21–23.

7. FEMA, *National Recovery Framework*, 24–26.

8. Nathan Crabbe, "Thank goodness! UF has a plan for zombie invasions," *The Gainesville Sun*, 1 October 2009, http://www.gainesville.com/article/20091001/ARTICLES/910019935/1002?Title=Thank-goodness-UF-has-a-plan-for-zombie-invasions (accessed 28 February 2013). The plan itself is available at http://chalkboard.blogs.gainesville.com/files/2009/10/zombieplan.pdf.

9. Federal Emergency Management Agency, "About the Agency," 15 October 2012. http://www.fema.gov/about (accessed 25 February 2013).

10. "Executive Order 12127—Federal Emergency Management Agency," signed 31 March 1979, http://www.archives.gov/federal-register/codification/executive-order/12127.html (accessed 25 February 2013).

11. Francis X. McCarthy and Jared T. Brown, *Congressional Primer on Major Disasters and Emergencies* (Washington, D.C.: Congressional Research Service, 2012).

12. McCarthy and Brown, *Congressional Primer*, pp. 6–7.

13. McCarthy and Brown, *Congressional Primer*, p. 3.

14. McCarthy and Brown, *Congressional Primer*, p. 4.

15. Department of Homeland Security, "Creation of the Department of Homeland Security," http://www.dhs.gov/creation-department-homeland-security (accessed 23 July 2014).

16. Michael Grunwald and Susan B. Glasser, "Brown's turf wars sapped FEMA's strength," *Washington Post*, 23 December 2005, http://www.washingtonpost.com/wp-dyn/content/article/2005/12/22/AR2005122202213_pf.html (accessed 18 March 2013).

17. Department of Homeland Security Office of the Inspector General, *A Performance Review of FEMA's Disaster Management Activities in Response to Hurricane Katrina* (Washington, D.C.: Department of Homeland Security, 2006).

18. White House, *The Federal Response to Hurricane Katrina: Lessons Learned* (Washington, D.C.: White House, 2006), p. 19.

19. Keith Bea, *Federal Emergency Management Policy Changes After Hurricane Katrina: A Summary of Statutory Provisions* (Washington, D.C.: Congressional Research Service, 2007), p. 5.

20. Max Brooks, *World War Z: An Oral History of the Zombie War* (New York: Three Rivers Press, 2006); *28 Days Later*, directed by Danny Boyle, Fox Searchlight Pictures, 2002; *Zombieland*, directed by Ruben Fleischer, Columbia Pictures, 2009.

21. Boyle, *28 Days Later*; Brooks, *World War Z*; and *Shaun of the Dead*, directed by Edgar Wright, StudioCanal, 2004. See also Philip Munz, Ioan Hudea, Joe Imad, and Robert J. Smith, "When zombies attack! Mathematical modelling of an outbreak of a zombie infection," *Infectious Disease Modelling Research Progress*, J.M. Tchuenche and C. Chikaya, eds. (New York: Nova Science, 2009), 136; and Jen Webb and Sam

Byrnand, "Some kind of virus: the zombie as body and as trope," *Body and Society* 14, no. 2 (2008), 84.

22. *Dead Rising*, produced by Keiji Inafune, developed by Capcom, 2006; Mira Grant, *Feed* (New York: Orbit, 2010); *Night of the Living Dead*, directed by George A. Romero, Image Ten, 1968; *The Serpent and the Rainbow*, directed by Wes Craven, Universal Pictures, 1988; *Dawn of the Dead*, directed by George A. Romero, United Film, 1978 and directed by Zack Snyder, Universal Pictures, 2004.

23. Edward C. Liu, *Would an Influenza Pandemic Qualify as a Major Disaster Under the Stafford Act?* (Washington, D.C.: Congressional Research Service, 2008), 3.

24. Liu, *Would an Influenza Pandemic*, 4–5.

25. White House Office of the Press Secretary, *Declaration of a National Emergency with Respect to the 2009 H1N1 Influenza Pandemic*, 24 October 2009, http://www.whitehouse.gov/the-press-office/declaration-a-national-emergency-with-respect-2009-h1n1-influenza-pandemic-0 (accessed 28 February 2013).

26. White House, *National Strategy for Biosurveillance* (Washington, D.C.: White House, 2012), http://www.whitehouse.gov/sites/default/files/National_Strategy_for_Biosurveillance_July_2012.pdf (accessed 28 February 2013).

27. Original version, directed by George Romero, United Film, 1978; remake, directed by Zack Snyder, Universal Pictures, 2004.

28. Jonathan Coulton, "Re: Your Brains," in *Thing a Week Two*, released by Jonathan Coulton, 2006).

29. McCarthy, *FEMA's Disaster Declaration Process*, 1.

30. McCarthy, *FEMA's Disaster Declaration Process*, 12.

31. Robert Kirkman and Tony Moore, *The Walking Dead* (Berkeley: Image Comics, 2006–present); AMC, *The Walking Dead* (TV series, 2010–present)

32. FEMA, *National Disaster Recovery Framework* (Washington, D.C.: FEMA, 2009): 29; *National Response Framework*, 11.

33. Lydia R. Wilson and Robert McCreight, "Public emergency laws and regulations: understanding constraints and opportunities," *Journal of Homeland Security and Emergency Management* 9 (2012): 3.

34. *National Response Framework*, 26.

35. Max Brooks, *The Zombie Survival Guide: Complete Protection from the Living Dead* (New York: Broadway, 2003), 41–50.

36. *National Response Framework*, 58.

37. *National Response Framework*, 30.

38. Department of Health and Human Services, "Emergency Support Functions—PHE," http://www.phe.gov/preparedness/support/esf8/Pages/default.aspx (accessed 4 March 2013).

39. Patrick S. Roberts, "Dispersed federalism as new regional governance for homeland security," *Publius: The Journal of Federalism* 38 (2008): 416–443.

40. Sivan Kohn, Daniel J. Barnett, Costanza Galastri, Natalie L. Semon, and Jonathan M. Links, "Public health-specific national incident management system trainings: building a system for preparedness," *Public Health Reports (1974–)* 125 (2010): 43–50.

41. Donald F. Kettl, "Is the worst yet to come?" *Annals of the American Academy of Political and Social Science* 604 (2006): 273–287.

42. Eli Lehrer, "The homeland security bureaucracy," *The Public Interest* 156 (2004): 71–85.

43. Beverly A. Cigler, "The 'big questions' of Katrina and the 2005 Great Flood

of New Orleans," *Public Administration Review* 67 (2007): 64–76; Beverly A. Cigler, "Emergency management challenges for the Obama Administration," *International Journal of Public Administration* 32 (2009): 759–766.

44. Federal Emergency Management Agency, *Hurricane Sandy FEMA After-Action Report* (Washington, D.C.: Department of Homeland Security, 2013).

45. Martha Derthick, "Where federalism didn't fail," *Public Administration Review* 67 (2007): 36.

46. Derthick, "Where federalism didn't fail," 37.

47. American Customer Satisfaction Index, "Citizen satisfaction by federal department," http://www.theacsi.org/acsi-results/citizen-satisfaction-by-federal-department (accessed 15 March 2013).

48. Department of Homeland Security, "Quadrennial Homeland Security Review (QHSR)," http://www.dhs.gov/quadrennial-homeland-security-review-qhsr (accessed 23 July 2014).

49. Paul C. Light, "The homeland security hash," *Wilson Quarterly* 31 (2007): 36–44.

50. Cigler, "The 'big questions' of Katrina."

51. Andrew C. Rucks, Peter M. Ginter, W. Jack Duncan, and Cindy Leisinger, "A continuity of operation planning template: translating public policy into an effective plan," *Journal of Homeland Security and Emergency Management* 8 (2011): 4–15.

52. Jason D. Mycoff, "Congress and Katrina: a failure of oversight," *State and Local Government Review* 39 (2007): 25–26.

53. James McGurk, "Jesse Ventura suspects a conspiracy about his show about conspiracies," *The Atlantic*, 17 December 2012, http://www.theatlantic.com/entertainment/archive/2012/12/jesse-ventura-suspects-a-conspiracy-about-his-show-about-conspiracies/266361/ (accessed 28 August 2013). *Popular Mechanics* rebuts Ventura's allegations in an article on its website: "The evidence: debunking FEMA camp myths," http://www.popularmechanics.com/technology/military/news/4312850 (accessed 28 August 2013).

54. "Troops begin combat operations in New Orleans," *Army Times* (3 September 2005). Cited in James Joyner, "Katrina: Army begins combat operations in New Orleans," *Outside the Beltway*, http://www.outsidethebeltway.com/katrina_army_begins_combat_operations_in_new_orleans/ (accessed 6 March 2013).

55. Kevin Fox Gotham, "Disaster, Inc.: privatization and post–Katrina rebuilding in New Orleans," *Perspectives on Politics* 10 (2012): 633–646; Kevin Fox Gotham and Miriam Greenburg, "From 9/11 to 8/29: post-disaster recovery in New York and New Orleans," *Social Forces* 87 (2008): 1039–1062.

56. Peter T. Leeson and Russell S. Sobel, "Weathering corruption," *Journal of Law and Economics* 51 (2008): 667–681.

57. *Shaun of the Dead*; Colson Whitehead, *Zone One* (New York: Doubleday, 2011).

Bibliography

American Customer Satisfaction Index. "Citizen Satisfaction by Federal Department." http://www.theacsi.org/acsi-results/citizen-satisfaction-by-federal-department (accessed 15 March 2013).

Bea, Keith. *Federal Emergency Management Policy Changes After Hurricane Katrina: A Summary of Statutory Provisions.* Washington, D.C.: Congressional Research Service, 2007.

Brooks, Max. *World War Z: An Oral History of the Zombie War.* New York: Three Rivers Press, 2006.

_____. *The Zombie Survival Guide: Complete Protection from the Living Dead*. New York: Broadway, 2003.

Centers for Disease Control and Prevention. *Preparedness 101: Zombie Pandemic*. Atlanta: Centers for Disease Control and Prevention, n.d. http://www.cdc.gov/phpr/_media/cymkPrint/11_225700_A_Zombie_Final.pdf (accessed 28 February 2013).

Cigler, Beverly A. "The 'Big Questions' of Katrina and the 2005 Great Flood of New Orleans." *Public Administration Review* 67, Suppl. 1 (2007): 64–76.

_____. "Emergency Management Challenges for the Obama Administration." *International Journal of Public Administration* 32, no. 9 (2009): 759–766.

Coburn, Tom. "Dr. Coburn Releases 'Safety at any Price' Report on Questionable Grant Spending at Department of Homeland Security." 5 December 2012. http://www.coburn.senate.gov/public/index.cfm/pressreleases?ContentRecord_id=bcd61d71-5454-4a81-9695-e949e0253faa (accessed 22 February 2013).

Coulton, Jonathan. "Re: Your Brains." *Thing a Week Two*, released by Jonathan Coulton, 2006.

Crabbe, Nathan. "Thank Goodness! UF Has a Plan for Zombie Invasions." *The Gainesville Sun*, 1 October 2009. http://www.gainesville.com/article/20091001/ARTICLES/910019935/1002?Title=Thank-goodness-UF-has-a-plan-for-zombie-invasions (accessed 28 February 2013).

Dawn of the Dead. Directed by George A. Romero. United Film, 1978.

Dawn of the Dead. Directed by Zack Snyder. Universal Pictures, 2004.

Dead Rising. Produced by Keiji Inafune. Capcom, 2006.

Derthick, Martha. "Where Federalism Didn't Fail." *Public Administration Review* 67, Suppl. 1 (2007): 36–47.

Fox, Lauren. "Report: DHS spent money on zombie simulation," *US News and World Report*, 5 December 2012. http://www.usnews.com/news/articles/2012/12/05/coburn-zombies-napolitano-homeland-security (accessed 21 February 2013).

Gotham, Kevin Fox. "Disaster, Inc.: Privatization and Post-Katrina Rebuilding in New Orleans." *Perspectives on Politics* 10, no. 3 (2012): 633–646.

_____, and Miriam Greenburg. "From 9/11 to 8/29: Post-Disaster Recovery in New York and New Orleans." *Social Forces* 87, no. 2 (2008): 1039–1062.

Grant, Mira. *Feed*. New York: Orbit, 2010.

Grunwald, Michael, and Susan B. Glasser. "Brown's turf wars sapped FEMA's strength." *Washington Post*, 23 December 2005. http://www.washingtonpost.com/wp-dyn/content/article/2005/12/22/AR2005122202213_pf.html (accessed 18 March 2013).

Joyner, James. "Katrina: Army Begins Combat Operations in New Orleans." *Outside the Beltway*, 3 September 2005. http://www.outsidethebeltway.com/katrina_army_begins_combat_operations_in_new_orleans/ (accessed 6 March 2013).

Kettl, Donald F. "Is the Worst Yet to Come?" *Annals of the American Academy of Political and Social Science* 604 (2006): 273–287.

Kirkman, Robert, and Tony Moore. *The Walking Dead*. Berkeley: Image Comics, 2006–present.

Kohn, Sivan, Daniel J. Barnett, Costanza Galastri, Natalie L. Semon, and Jonathan M. Links. "Public Health-Specific National Incident Management System Trainings: Building a System for Preparedness." *Public Health Reports (1974–)* 125, Suppl. 5 (2010): 43–50

Leeson, Peter T., and Russell S. Sobel. "Weathering Corruption." *Journal of Law and Economics* 51, no. 4 (2008): 667–681.

Lehrer, Eli. "The Homeland Security Bureaucracy." *The Public Interest* 156 (2004): 71–85.

Light, Paul C. "The Homeland Security Hash." *Wilson Quarterly* 31, no. 2 (2007): 36–44.

Liu, Edward C. *Would an Influenza Pandemic Qualify as a Major Disaster Under the Stafford Act?* Washington, D.C.: Congressional Research Service, 2008.

McCarthy, Francis X., and Jared T. Brown. *Congressional Primer on Major Disasters and Emergencies.* Washington, D.C.: Congressional Research Service, 2012.

McGurk, James. "Jesse Ventura Suspects a Conspiracy about His Show about Conspiracies." *The Atlantic*, 17 December 2012. http://www.theatlantic.com/entertain ment/archive/2012/12/jesse-ventura-suspects-a-conspiracy-about-his-show-about-conspiracies/266361/ (accessed 28 August 2013).

Munz, Philip, Ioan Hudea, Joe Imad, and Robert J. Smith. "When Zombies Attack! Mathematical Modelling of an Outbreak of a Zombie Infection." In *Infectious Disease Modelling Research Progress*, edited by Jean Michel Tchuenche and C. Chikaya, 133–150. New York: Nova Science, 2009.

Mycoff, Jason D. "Congress and Katrina: A Failure of Oversight." *State and Local Government Review* 39, no. 1 (2007): 16–30.

Night of the Living Dead. Directed by George A. Romero. Image Ten, 1968.

Popular Mechanics. "The Evidence: Debunking FEMA Camp Myths." http://www.popularmechanics.com/technology/military/news/4312850 (accessed 28 August 2013).

Roberts, Patrick S. "Dispersed Federalism as New Regional Governance for Homeland Security." *Publius: The Journal of Federalism* 38, no. 3 (2008): 416–443.

Rucks, Andrew C., Peter M. Ginter, W. Jack Duncan, and Cindy Leisinger. "A Continuity of Operation Planning Template: Translating Public Policy into an Effective Plan." *Journal of Homeland Security and Emergency Management* 8, no. 1 (2011): 4–15.

The Serpent and the Rainbow. Directed by Wes Craven.: Universal Pictures, 1988.

Shaun of the Dead. Directed by Edgar Wright. StudioCanal, 2004

28 Days Later. Directed by Danny Boyle. Fox Searchlight Pictures, 2001.

United States. Department of Health and Human Services. "Emergency Support Functions—PHE." http://www.phe.gov/preparedness/support/esf8/Pages/default.aspx (accessed 4 March 2013).

_____. Department of Homeland Security. "Creation of the Department of Homeland Security." http://www.dhs.gov/creation-department-homeland-security (accessed 23 July 2014).

_____. "Quadrennial Homeland Security Review (QHSR)." http://www.dhs.gov/quadrennial-homeland-security-review-qhsr (accessed 23 July 2014).

_____. _____. Office of the Inspector General. *A Performance Review of FEMA's Disaster Management Activities in Response to Hurricane Katrina.* Washington, D.C.: Department of Homeland Security, 2006.

_____. Federal Emergency Management Agency. "About the Agency." Last modified 15 October 2012. http://www.fema.gov/about (accessed 25 February 2013).

_____. "Build a Kit." http://www.ready.gov/build-a-kit (accessed 28 February 2013).

_____. *Hurricane Sandy FEMA After-Action Report.* Washington, D.C.: Department of Homeland Security, 2013.

_____. *National Disaster Recovery Framework.* Washington, D.C.: FEMA, 2009.

_____. *National Recovery Framework.* Washington, D.C.: FEMA, 2008.

_____. _____. Executive Order 12127 of March 31, 1979, Federal Emergency Management Agency. http://www.archives.gov/federal-register/codification/executive-order/12127.html (accessed 25 February 2013).

The Walking Dead. Developed by Frank Darabont. AMC, 2010–present.

Webb, Jen, and Sam Byrnand. "Some Kind of Virus: The Zombie as Body and as Trope." *Body and Society* 14, no. 2 (2008): 83–98.

White House. *The Federal Response to Hurricane Katrina: Lessons Learned.* Washington, D.C.: White House, 2006.

_____. *National Strategy for Biosurveillance.* Washington, D.C.: White House, 2012.

_____. Office of the Press Secretary. *Declaration of a National Emergency with Respect to the 2009 H1N1 Influenza Pandemic.* 24 October 2009. http://www.whitehouse.gov/the-press-office/declaration-a-national-emergency-with-respect-2009-h1n1-influenza-pandemic-0 (accessed 28 February 2013).

Wilson, Lydia R., and Robert McCreight. "Public Emergency Laws and Regulations: Understanding Constraints and Opportunities." *Journal of Homeland Security and Emergency Management* 9, no. 2 (2012): 1–16.

Zombieland. Directed by Ruben Fleischer. Columbia Pictures, 2009.

A Particularly "nasty, brutish, and short" Life

Individuals, Groups and Government in the Zombie Apocalypse

ANTONIO S. THOMPSON

Life, during and after a disaster, is dangerous. In those chaotic moments when the event is unfolding survival becomes paramount. Following the disaster survivors must continue to try to live, often in desperate conditions. Vandalism, looting, violence, rape and murder may become commonplace. Often little is off-limits in the absence of law, order, and government, particularly when things seem to come down to an "us or them" scenario. Surely during a zombie apocalypse, this would be magnified. The stress of the event and of continued survival in a situation that might seem, or even be, hopeless, weighs on the survivors. The only law and order may be the notions of "might makes right" or "do unto others before they do unto me." Long-term survival depends on establishing order through the creation of a government system.

The key term following a major disaster or apocalyptic event is to establish, rather than re-establish, because post-disaster life differs from pre-disaster life. Pre-apocalyptic government is based on a social contract of written, enforced, and understood laws, both in the legal sense and regarding established norms. In the United States for instance the social contract of "life, liberty, and the pursuit of property" as espoused by John Locke, which was changed to "life, liberty, and the pursuit of happiness," guided Thomas Jefferson and other Framers the Declaration of Independence.[1]

During a major disaster, when the old order has collapsed, communication systems have failed, and, in the case of a zombie apocalypse, additional dangers lurk, the social contract must take on a new meaning. The theories of Niccolò Machiavelli and Thomas Hobbes can be applied to zombie fiction and life after disaster in the establishment of new governments.

Machiavelli was born in 1469 in Florence, Italy. He came of age during the Italian Renaissance and was a writer who worked both in history and government. His most famous work, *The Prince* (1513), described how a new ruler must be willing to resort to means both fair and foul to appease the people and secure his position. Machiavelli was writing to address the political realities and possibilities of his lifetime. Much of Italy was divided into city-states that frequently warred with each other and the use of foreign mercenaries was not uncommon. Alliances could be made and broken easily, while treachery and back-stabbing were common. Machiavelli posited that leaders may be just, unjust, or a mixture of both as long as they ruled firmly and without hesitation. His realistic but brutal theories state that the sovereign is the absolute authority and can resort to any means necessary to retain power.[2]

Hobbes was born in 1588 in Wiltshire, England. A political theorist and writer, he authored numerous works. One of his most famous is *The Leviathan* (1651), written during the English civil war (1642–1651).[3] Hobbes served for a time as the tutor to the exiled Charles, Prince of Wales, and future King Charles II. His experiences led him to further think about government foundations and rulership, a theme already explored in his book *De Cive* (1642).[4] *The Leviathan* provided a social contract theory. It described life without government as individuals who lived in a condition of constant uncertainty and insecurity as each scrambled to gain access to limited resources and to protect those from their fellow man. This condition, known as the "state of nature," is deplorable. Life in the state of nature is without "society" and therefore is marked by "continual fear, and danger of violent death; And the life of man, solitary, poor, nasty, brutish, and short."[5] The only escape from this is to form a government, but in Hobbes' estimation the new sovereign must be absolute and control all things. Those entering the contract do so willingly to escape the state of nature and therefore give power willingly to the sovereign, who may become a Leviathan. The title comes from the Biblical Leviathan, the beast from the sea mentioned in the Book of Job.

Post-apocalyptic and zombie literature provide numerous examples where the theories of Machiavelli and Hobbes can be applied. They also demonstrate how meetings between individuals and small groups can quickly become deadly.

The State of Nature in the Zombie Apocalypse

The undead in most zombie genres are presented as tireless, never needing to slow down, take a break, or sleep, and always hungry for human flesh.

Even scarier than zombies, however, is the human threat. Chuck Greene, in the *Dead Rising 2: Off the Record* video game, makes a poignant observation. He says, "These zombies ... they're annoying, sure, but they're easy to kill. The real threat is the people. Just when humanity should be coming together, they turn on each other instead."[6]

This is a theme repeated over and over in zombie lore and one that is at the core of most of George Romero's zombie movies. In *Night of the Living Dead* the survivors at the farmhouse could not get along or agree how to fight the slow-moving dead; in *Dawn of the Dead* a rogue motorcycle gang destroyed the sanctity of the mall, and in *Day of the Dead* the disagreements between the military and civilian scientists led to violence and their underground bunker being overrun.[7] In *Land of the Dead*, the corruption and factionalism led to the downfall of the post-apocalyptic society and clan feuding was a major aspect of *Survival of the Dead*.[8] Even in *Diary of the Dead* disagreements between the college student survivors, and addictions to the camcorder, disrupted the group.[9] In fact, it seems that if humankind cooperated more in these fictional situations they could learn to cope with if not overcome the zombies.

Yet before people can create groups, communities, or governments, they have to meet each other and these encounters in the state of nature can be dangerous and deadly. Since resources are scarce and, as Hobbes notes, all of mankind is in competition for them, each meeting has the potential to turn deadly. Even those individuals or groups who bear no ill will can never be sure of another group's intentions. Such is life in the state of nature; "there is always war of every one against every one."[10]

What motivates people during a disaster varies from situation to situation. Sometimes good people do bad things due to stress, fear, the desire for safety, or the "us or them argument." At other times the disaster can enable criminal or psychotic activities. In the video game *Dead Rising* and *Dead Rising 2: Off the Record*, the protagonist is Frank West.[11] Frank is best described as somewhere between a progressive muckraker, combat photographer, and the paparazzo. In the first game his goal is to find out how the zombie outbreak started; in the second game he is trying to recover some of his lost fame and gets thrown into a new zombie outbreak. In *Dead Rising 2*, motocross father Chuck Greene fights zombies in a show to support his daughter's daily need for Zombrex, a drug that prevents those bitten from turning into zombies.[12] Throughout the games Frank and Chuck encounter people that can be rescued. In a Machiavellian sense, however, the "heroes" are not doing it simply out of the goodness of their hearts as each rescued survivor forwards the story by providing clues to the zombie outbreak, supplies, money, secret passages, or experience points.

In one encounter Frank meets Cliff, a Vietnam veteran, who was shopping at the Willamette Mall when the zombies took over. The stress of the event is too much and after zombies kill his granddaughter, Cliff starts having Vietnam flashbacks and ends up killing numerous innocent survivors. In another encounter Chef Antoine is using humans as food in an attempt to make a perfect recipe to satisfy a food critic. The combined stressors of the zombie outbreak and the need for a good review lead Antoine to a psychotic break.[13]

On the other hand, in real-life situations and zombie fiction, sometimes bad people do bad things. Disasters and a lack of law and order allow people with less than noble intentions to commit crimes. In real-life situations this might include rioting, looting, stealing, and committing rape, and even murder. Add these activities to the teeming zombie hordes and life in the state of nature gets even worse. In *Dead Rising* Frank encounters three escaped convicts who are hot-rodding in an army jeep, hitting survivors with the vehicle, a baseball bat, or the mounted machine-gun. While it is unclear why they were imprisoned, their behavior cannot be completely attributed to the stress of the apocalypse. Frank also encounters the True Eye Cult, an organization busy abducting and sacrificing survivors. The concern is that an existing cult, whether fictional or real, used the disaster to move their activities from behind the scenes to front and center. After all, in the absence of law and order what is the likelihood of a crusading journalist thwarting their plans?

Yet, humans need each other. In the *Zombie Survival Guide: Complete Protection from the Living Dead*, Max Brooks argues that groups "allow mutual protection" and warns against staying alone since "security, reconnaissance, and, naturally, sleep would all be hampered."[14] Matt Mogk, author of *Everything You Ever Wanted to Know About Zombies*, argues for strength in numbers and against trying to survive alone since "you may be faster, stronger, and more confident on your own, but you may also lose your marbles."[15]

Initial meetings between individuals in zombie fiction are dangerous. It seems easiest and safest to join a group during the actual disaster or immediately after when everyone is confused, scared, looking for answers, and seeking security. Often groups are made up of seemingly random people who may not know each other or have anything in common. Enduring a trauma may bond strangers into a cohesive unit. This theme appears in the 2004 remake of *Dawn of the Dead*. Ana, a nurse, escapes her neighborhood which has been overrun by zombies. She encounters Kenneth, a police officer, and they travel together, soon encountering three other survivors, Michael, Andre, and Luda.[16] The two groups are wary of each other and guns are drawn until the situation finally de-escalates. They band together and seek refuge in the nearby mall. Part of what allows them to bond so quickly is the immediate need for safety and shelter, but also the shock and stress of the new event, that social and legal norms have not yet disappeared, and that there is still the hope of rescue and a return to normality.

As the survivors enter the mall they are greeted at gunpoint by three security guards who have thus far remained safe in the mall. While the facility has enough resources to provide for everyone, there is no trust that the "invaders" will not take it all. The security guards are not bad people, just scared. Eventually they all work together, even taking in other survivors. Over time they even create a sort of family.

Bonding is another trait of group cohesion demonstrated in zombie fiction. In the AMC show *The Walking Dead* the main group has numerous bonding times, particularly at the farm and at the prison. Community and interpersonal relationships are a clear part of the show and the quest for friendship and familial units appears even more important during the apocalypse.[17] In the *Walking Dead* novel *Fall of the Governor*, after making a dangerous supply run, Lilly comments of her group, "with enough people like these folks-people who care about each other-they just might have a shot at building a community."[18] Brooks argues that "making a group of individuals cooperate over a long period of time is the hardest task on earth. However, when successful, this group will be capable of anything."[19]

As difficult as it is to create group cohesion can fall apart for a number of reasons. The same stress that brings people together can also tear them apart, especially after the immediacy of the danger is over, when individuals lose hope of rescue, or when no greater authority oversees their actions. Interpersonal rivalries, fear, greed and betrayal can destroy groups. Brooks argues that an organized group is of the utmost importance in surviving a zombie apocalypse. He advises that "people who believe help is eventually coming are much more likely to remain loyal than those who know the future is what they make it. Discontent, mutiny, even bloodshed are always a possibility."[20]

The zombie apocalypse has its share of betrayals within groups. *In the Walking Dead: Rise of the Governor* Phillip Blake is with his daughter Penny, his brother Brian, and some of his friends when the apocalypse starts. The group is able to survive a long time thanks to Phillip's methods and the group's previous familiarity with each other. Eventually, however, the relationships within the group become somewhat dysfunctional and end with the death of everyone except Brian, including Phillip at Brian's hands.[21]

In *Dawn of the Dead*, Steve, one of the mall survivors, betrays his fellow survivors twice. He first locks them in a stairwell with a horde of zombies. He later abandons them after they wreck their escape vehicle. Steve, however, is killed by a zombie and then his reanimated corpse is killed by Ana.[22]

There are real-life examples of group cohesion and betrayal. American general Benedict Arnold, one of the heroes of the Battle of Saratoga during the American Revolution, attempted to turn West Point and its vital fortifications and communications holding the strategic Hudson River over to the

British. This is one of the most famous examples of betrayal in American history.[23] On the other hand, the shared experience and stress of combat service often tends to strengthen group cohesion. Historian Stephen Ambrose's *Band of Brothers* depicts the real-life heroes of the 101st who served in World War II.[24] Unlike the men in the movie, however, many caught in a disaster would likely not have survival or combat training, and may not have previously developed unit cohesion.

Clearly even once groups are formed there is the ever-present danger of betrayal. Both Hobbes and Machiavelli offered insight into the actions of individuals within groups. Hobbes stated that "the notions of Right and Wrong, Justice and Injustice have there no place. Where there is no common Power, there is no Law: where no Law, no Injustice. Force, and Fraud, are in war the two cardinal virtues."[25] Hobbes argued that one must put self-preservation above all else. This, however, does not necessarily indicate, that he advocated betrayal, more likely that he advocated cooperation as mutual survival is more likely to lead to individual security. Machiavelli made a clearer stance on the subject when he stated that "it cannot be called talent to slay fellow-citizens, to deceive friends, to be without faith, without mercy, without religion; such methods may gain empire, but not glory."[26]

Escaping the State of Nature and Forming Government

Hobbes' state of nature is terrible and difficult to comprehend; the experience makes one fear it and desire to escape it. Hobbes argued that fear and reason are inherent in mankind and that people can work together to escape from the state of nature. One way to do this is to form the social contract of communities and government and thus establish security and order. The formation of these governments necessarily requires a leader and, as Machiavelli argues, "there is nothing more difficult to take in hand, more perilous to conduct, or more uncertain in its success, then to take the lead in the introduction of a new order of things."[27] This is certainly true of leadership during the zombie apocalypse as survivors look to the leader to navigate the new political spectrum and provide in a time of scarcity.

The Struggles to Be Good Leaders: Shane and Rick

In both *The Walking Dead* comic and television show the main group has two leaders, first deputy Shane Walsh and second Shane's best friend and partner Rick Grimes. Both men displayed Machiavellian and Hobbesian qualities as they vacillated between doing what they thought was right and what needed to be done for the good of the group and for their own good. Shane emerged as the first leader. The group members were caught on the interstate heading for Atlanta seeking safety from the zombies when the road became clogged. Shane was a deputy sheriff, his baseball cap said police on it, and he therefore represented the legitimate authority. Shane led during a time when rescue was uncertain, goods were in short supply, and his own physical and mental stress was high. Shane was additionally having an affair with Rick's wife Lori and playing surrogate father to Carl, Rick's son.

His struggles with leadership have clear Machiavellian overtones. Shane dealt with group member Carol's troubling and abusive husband Ed by beating him unmercifully. In this sense Shane followed the Machiavellian question "is it better to be loved than feared or feared than loved?"[28] Later, Shane and Otis, a farmer's helper, went on a run to the high school for medical supplies to save Carl, who had previously been shot by Otis in a hunting accident. When they became overwhelmed by "walkers" Shane shot Otis in the leg, causing him to fall, attracting the zombies and allowing Shane to escape. Had Shane not shot Otis it was likely that both Shane and Otis would have died there, and if that happened the needed medical supplies would not have been available to save Carl. Shane sacrificed Otis to save himself and Carl and, in this Machiavellian way, "the ends justify the means."

The rivalry between Shane and Rick over leadership and Lori's love led Shane to attempt to kill Rick. As Hobbes stated when men "desire the same thing, which nevertheless they cannot both enjoy, they become enemies; and in the way to their end (which is principally their own conservation, and sometimes their delectation only) endeavor to destroy, or subdue one another."[29] Yet, it is Shane who is killed.[30]

Rick entered Atlanta and joined a small group foraging for supplies. Rick, wearing his sheriff uniform and hat, and at one time even riding a horse, clearly represented the legitimate authority and in some ways law enforcement from an earlier time in American history. Almost immediately he assumed the leadership role by locking Merle, a particularly violent and racist member of the foraging group, in handcuffs at the top of a building in Atlanta. Rick then accompanied the group back to its camp where he was reunited with his family. He continued to make decisions for the group until with the death of Shane he became the uncontested leader.

Rick's struggle with leadership varied between the comic book and the AMC series, but it is well documented in both. In the comic Rick witnessed Tyreese kill his daughter's boyfriend and even covered for him. While at the prison Dexter, the convict who was the leader of the prisoners, pulled a gun on the group, becoming a threat. Rick killed him. He also nearly beat Thomas

Richards, one of the convicts, to death when it was revealed he was actually a serial killer who had also killed members of Rick's group. He later led the group against the cannibals that had been stalking them. While on the road he and Abraham also brutally killed several bandits.

In the AMC series Rick and Glenn, another survivor, headed to the local tavern in rural Georgia to try to convince fellow group member Hershel to avoid returning to alcoholism. During this visit two men from the northeast entered the bar and a tense meeting ensued. They discussed the difficulties of survival and asked to join with Rick's group and share in the security and resources at their farm. Fearing that these tough-talking and armed northerners had bad intentions, Rick told them no. When they asked Rick what they should do Rick responded by repeating a comment from earlier in the conversation, "I hear Nebraska is nice this time of year." Rick then shot both of the men. When Rick encountered the inmates at the prison, Rick saw Tomas, the leader of the prisoners, as a threat. While clearing out walkers in the prison Tomas nearly lets Rick get killed and told Rick "shit happens." Rick later killed Tomas, repeating to him "shit happens."[31] Finally, Rick killed Shane, a decision that haunted him in both the AMC series (he saw Shane's doppelganger at Woodbury) and in the comic series (he returned to Shane's gravesite).

Some of these decisions were more difficult than others but all of them weighed on Rick. The pressure clearly got to Rick as he began speaking on a phone to his dead wife Lori. He was troubled by some of his decisions but this did not keep him from letting the group know that he was tired of catering to their whims and that "this is no longer a democracy."[32] It might appear that Rick was on his way to becoming the Leviathan that Hobbes described, but the moral dilemma and stress resulted in Rick having a psychotic break forcing him from his key role within the group. Even when Rick took a temporary break from leadership, allowing a council to run things, he still made decisions that affected everyone, like abandoning Carol during a supply run. Much of what happened to Rick can be ascribed to the stress of life in the state of nature and in the Machiavellian sense the ends justify the means. Former prisoner Axel commented, "That guy sets his mind to something—he pretty much does it no matter what, you follow me?"[33]

The Creation of a Leviathan: Brian Blake and Paul Kaufman

During the zombie apocalypse Shane and Rick struggled with leadership roles and both occasionally slipped into a dictatorial type of role. Yet, each often listened to and heeded the wishes of the group. They also recognized and wrestled with the moral ambiguity of some of their decisions. Other leaders who rose to their dictatorial Leviathan status during the zombie apocalypse seemed to never make altruistic decisions.

Groups tend to become larger and larger and newly-formed communities need to have a leadership system in place that is concerned with the common good. The social contracts formed create one way to escape the state of nature that Hobbes described as being "war of every one against every one."[34] Hobbes argued that individuals give up their rights to a leader due to the constant fear of others prevalent in the state of nature. He did not suggest that all individual rights are given to the leader, however, "as it is necessary for all men that seek peace, to lay down certain Rights of Nature; that is to say, not to have liberty to do all they list: so is it necessary for man's life, to retain some; as right to govern their own bodies; enjoy air, water, motion, ways to go from place to place; and all things else without which a man cannot live, or not live well."[35]

The leader has a duty to keep citizens safe from outside terrors and as long as this duty is kept the authority of the leader and their government should be considered legitimate. Once this authority is established, however, individual rights are difficult to regain and the power of the sovereign tends to grow and become absolute, thus a Leviathan can be formed. In the zombie apocalypse it is nearly impossible to regain rights once given and the Leviathan might have no desire to return them, since as Hobbes argues the "general inclination of all mankind, a perpetual and restless desire of power after power, that cease only in death."[36] In some ways, however, the need to leave and rejoin life in the state of nature, where even one's former community members would likely see one as competition for scarce resources and security, is something that most would prefer to avoid.

The Governor practiced a Machiavellian code of conduct. Brian transformed from the meeker Blake brother, living in the shadows of his tough-talking brother Phillip, to ultimately kill Phillip, assume his identity, and become the Governor of Woodbury, Georgia. In *The Rise of the Governor*, Philip, Brian, and their friend Nick arrived at Woodbury to find the fledgling community seeming both leaderless and rudderless. One of the survivors commented "this place is the fucking wild west ... nobody gives two pieces of a rat's ass what you do."[37] It is the escape from this lawlessness that Hobbes argued leads to the formation of the Leviathan and in this case prompted the rise of the Governor. The people struggled against both the dead and the living. The existing authority was a major and his two Marine cronies who had usurped power. Brian killed the homicidal and abusive major during a town meeting, thereby earning the respect and admiration of the folks of Woodbury. This interaction led them to willingly give up some of their individual autonomy, thus leading to the transformation from Brian into the Governor.

For Brian the killing of the Marine major was simply the ends justifying the means in the reality of the new apocalyptic world.

Woodbury, Georgia, under the Governor, on the surface resembled small-town America. The Governor seemed both a charming and disarming man smiling and waving at citizens. In this idyllic setting, the Governor had created safety and provided security; there were weapons, guards, a doctor, food, and everyone had a job to do. This was in stark contrast to life outside of Woodbury, back in the state of nature, or in Woodbury prior to the Governor's rule.

In *Land of the Dead*, Paul Kaufman, the town's leader, had also returned life relatively close to what it resembled before the apocalypse. The community was secured by a river, fences, an underground tunnel, and plenty of armed guards. The town had power, working elevators, and vehicles. The upper classes lived in the town's center, the high-rise called Fiddler's Green, while the poor lived in what seemed to be shanties and slums. Scarce supplies were available through barter thanks to those assigned to forage in the outlying regions. Since the zombie apocalypse occurred sometime before the film took place, it can be assumed that this sanctuary and Kaufman's rise to power came after numerous tribulations in the state of nature. Once again, life within the town seemed to be much preferred over life outside.

The Leviathans of Woodbury and Fiddler's Green were not as honest as their public appearances would imply. Both men were wily politicians and knew how to manipulate the citizens. In both settings they employed henchmen to do their dirty work. Gladiator-type fights provided entertainment for the community. Those few who knew the true backgrounds of the Leviathan or were dissatisfied with leadership decisions kept this to themselves as those who objected too vociferously tended to disappear. Most found it better to live within the community than outside it.

The Governor's entertainment penchant also applied to womanizing and taking what he wanted. Few knew that their leader spent hours watching the reanimated heads of his enemies bobbing in fish tanks in his living room or that he kept his zombie daughter/niece Penny chained up in his bedroom, sustaining her on the bodies of the recently deceased or killed residents. He hunted down and killed or returned those who tried to escape Woodbury as well as other groups of survivors encountered outside of Woodbury. This was likely done not only as a method of control, but of safety. If others discovered Woodbury, a stronger or more organized group could threaten the safety of the community or compete for scarce resources, both motivations discussed by Hobbes. It was his desire for safety, control, and supplies that helped lead him against Rick's group and to murder the members of the National Guard discovered after their helicopter crash. The Governor exhibited the Machiavellian traits of inducing "fear" in his subjects and inflicting "inhumane cruelty."[38] This helped deter his two remaining henchmen from turning on him after he gunned down the survivors from his own group after the first failed attempt to take the prison.[39] It seemed that life was more secure for those who lived in Woodbury than those smaller groups found by the Governor's agents outside Woodbury. When not torturing, beating, raping, plotting, killing his own people, or watching severed heads, the Governor did not seem to be that bad a guy.

In Fiddler's Green, Kaufman was revealed to be a ruthless businessman. He had a board of directors and numerous armed enforcers and security forces. He was not above playing these individuals against each other to ensure his own needs were met. He employed men like Cholo to kill troublesome individuals and to go out in the world and acquire valuable goods, such as liquor, marked just for him. He risked the safety of the groups sent to scavenge to acquire things not necessary for the survival of the community. Kaufman clearly emulated the Hobbesian thought of "profit is the measure of right."[40] He used the hope of moving into the upscale Fiddler's Green as a carrot and those that did move into the community were indebted to him. Cholo had been promised an opportunity to quit his job and move into Fiddler's Green, which Kaufman reneged on. In frustration, Cholo and a small group of mercenaries stole a tank-like machine called Dead Reckoning and threatened to use it against the town. Kaufman commissioned Riley, another of his agents, to track down Cholo. While this was happening, the zombies, another group that coveted Fiddler's Green, moved against the city. Ironically, the name Fiddler's Green has some connotations in mythology to the afterlife, which might also explain why the undead desired it. Cholo was eventually killed and turned into a zombie, but that did not keep him from hunting for Kaufman.[41] Kaufman failed to adhere to Machiavelli's advice, "men ought either to be well treated or crushed, because they can avenge themselves of lighter injuries, of more serious ones they cannot."[42]

In Hobbes' theory the Leviathan tends to be more powerful based on the number of individuals within the community being governed. The larger the size of the community the more difficult it is to reassert individual rights once freely given. While this might create the fictional dictators as seen in Kaufmann or Blake, or even have parallels to real-life leaders such as Napoleon Bonaparte or Adolf Hitler, there are alternatives to consider. For instance, in many democracies elected representatives make decisions and citizens give up some individual autonomy to the local, state, and national governments by agreeing to abide by the laws set forth. Many modern governments have strong leadership and central authority but this system makes it difficult if not impossible for individuals or smaller constituent governments to retain their autonomy. For example the Articles of Confederation, the first U.S. form of national government, favored state autonomy at the expense of central authority. The Articles were replaced with the U.S. Constitution which created a strong central government at the cost of some of the previous state autonomy. Although the Constitution has remained the law of the U.S., it has faced challenges, the most important of these instances resulted in the U.S. Civil War (1861–1865).

Real-Life Curtailment of Rights in Crisis and Disaster

While a zombie apocalypse or other disaster leading to the complete overthrow of the national or international system has not happened, the world has faced numerous epidemic diseases, large-scale disasters, and global wars. The Leviathan or Machiavellian examples of post-apocalyptic leadership in fictional Woodbury or Fiddler's Green or hostile Hobbesian communities are more realistic than one might think. Consider the genocide in Rwanda or the Balkans of the 1990s, or conflicts in parts of Africa in the 21st century. Also, few in 1920s Germany might have thought that by the 1940s they would have been involved in the dictatorship that helped usher in World War II and the Holocaust.

The curtailment of freedom and oppression of citizens often comes in times of serious crisis. The United States, championed for democratic ideals, has restricted civil liberties at different points throughout history. In the 1790s, the British and the French went to war and both nations violated what the U.S. perceived as neutral rights. Failed diplomacy led to the Quasi-War (1798–1800) between the U.S. and France. President John Adams issued the Alien and Sedition Acts of 1798 in an attempt to control French citizens in the U.S. and curtail dissent, particularly from Republicans. These acts allowed the deportation of foreigners and the arrest of anyone speaking or publishing against the government. Federalist control of the government and the crisis with France helped these laws of questionable legality get passed.[43]

During one of the greatest crises faced by the United States, the American Civil War, President Abraham Lincoln restrained rights, suspended habeas corpus, and arrested those suspected of working against the Union. This resulted in many being detained under questionable legality and sometimes for undetermined periods of time.[44]

President Woodrow Wilson passed the Espionage Act of 1917 and the Sedition Act of 1918 while the U.S. was engaged in World War I. Like the 1798 laws, these Acts were a series of laws that could be used to curtail or arrest those guilty of sedition against the American government. Sedition was broadly defined and targets of these Acts were groups like the Socialists and individuals like Eugene Debs, and immigrants, particularly Germans.[45]

During World War II, President Franklin D. Roosevelt issued Executive

Order 9066 following the attack on Pearl Harbor which led to the relocation of first and second generation Japanese-Americans; more than 100,000 were removed from their homes and relocated. Some Germans and Italians living in the U.S., particularly those who protested the war, spoke out for their home nations, or came under suspicion were also interned.[46]

While these presidents may not truly represent a Hobbesian Leviathan, their actions demonstrate that during times of crisis drastic measures can be implemented that are not fully supported by the nation's people or its laws. If the government decides that its people should give up their rights for national security, then the question that must be asked is whether the ends justify the means.

A Better Tomorrow

It is difficult to imagine how individuals or groups of people, communities, or governments might act following a disaster or crisis, particularly one that would be on the large-scale devastation such as a a potential Ebola pandemic or fictional zombie apocalypse. Some research suggests that real-life responses to disaster differ greatly from the perception of how humans are likely to respond to disaster. Disaster researcher Henry Fischer argued that after a disaster looting is typically on a much smaller scale than it was assumed it would be and that those who lived through the event are typically "found to normally behave quite rationally and are the first to respond to their needs and the needs of their neighbors."[47] A recent article supports this by suggesting that these events, which create a sense of vulnerability, led to increased cooperation as "that vulnerability is a core ingredient of social bonding."[48] Maia Szalavitz, writing for *Time*, suggested that "it's human nature to band together and be kind to one another in order to survive" as "it's when we face the toughest times that our true nature reveals itself: we're in it together."[49] Examples of cooperation during Hurricane Sandy or after the terrorist attacks of 9/11 were common. During the attack on the World Trade Center's Twin Towers in New York, police, fire fighters and other emergency personnel rushed to aid despite the potential danger and many were killed or physically and emotionally harmed. In the aftermath of the event thousands of Americans traveled to the site to help recovery efforts. Consider too how passengers aboard the 9/11 United Airlines Flight 93 worked together to try to reclaim the plane and forced the hijackers to crash before they could use the plane as a weapon.[50]

On the other hand, examples of looting and violence during disasters exist as well. Sometimes individuals put their personal needs, desires, and wants above the needs of the group. The basest of human nature can emerge when life becomes a struggle for scarce resources and immediate or long-term survival may come down to an "us or them" argument. To escape this state of nature, as Hobbes and other philosophers have suggested, government is needed. Yet sometimes liberty is sacrificed for structure in these governments, both fictional and real. During times of crisis governments often put the perceived needs of the nation above the needs of its individual constituents. The following questions can be asked. Do the ends justify the means? What is the greater good? Life is usually better under a governed society than in the state of nature. One must also consider that the measure of a "good" government is not always linear and is dictated by events and circumstances.

NOTES

1. Robert A. Goldwin, "John Locke," in *History of Political Philosophy*, 3d ed., ed. Leo Strauss and Joseph Cropsey (Chicago: University of Chicago Press, 1987); Pauline Maier, ed. *The Declaration of Independence and The Constitution of the United States* (New York: Bantam Dell, 2008).

2. Leo Strauss, "Niccolò Machiavelli," in *History of Political Philosophy*, 3d ed., ed. Leo Strauss and Joseph Cropsey (Chicago: University of Chicago Press, 1987).

3. Thomas Hobbes, *Leviathan or The Matter, Forme and Power of a Common Wealth Ecclesiastical and Civil—commonly referred to as Leviathan* (England, 1651).

4. Thomas Hobbes, *De Cive* (Paris, 1642).

5. Thomas Hobbes, *Leviathan: The Matter, Forme, & Power of a Commonwealth Ecclesiastical and Civill* (Auckland: Floating Press, 2009), 179.

6. *Dead Rising 2: Off The Record,* produced by Keiji Inafune, Capcom Vancouver, 2011, video game.

7. *Night of the Living Dead*, written by George Romero and John A. Russo, directed by George Romero, Image Ten, 1968, movie; *Dawn of the Dead*, written and directed by George Romero, United Film Distribution Company, 1978, movie; *Day of the Dead*, written and directed by George Romero, United Film Distribution Company, 1985, movie.

8. *Land of the Dead*, written and directed by George Romero, Universal Pictures, 2005, movie; *Survival of the Dead*, written and directed by George Romero, E1 Entertainment and Magnet Releasing, 2009, movie.

9. *Diary of the Dead*, written and directed by George Romero, Weinstein, 2007, movie.

10. Hobbes, *Leviathan*, 178. Translated into modern English by the author of this essay, Antonio Thompson.

11. *Dead Rising*, written by and directed by Yoshinori Kawano, produced by Keiji Inafune, Capcom Production Studio 1, 2006, video game; and *Dead Rising 2: Off The Record,* produced by Keiji Inafune, Capcom Vancouver, 2011, video game.

12. *Dead Rising 2*, produced by Keiji Inafune, Blue Castle Games/Capcom, 2010, video game.

13. The main characters are Chuck in *Dead Rising 2* and Frank in *Dead Rising 2: Off the Record* and they both encounter Antoine.

14. Max Brooks, *The Zombie Survival Guide: Complete Protection from the Living Dead* (New York: Three Rivers Press, 2003), 97.

15. Matt Mogk, *Everything You Ever Wanted to Know About Zombies* (New York: Gallery Books, 2011), 173.

16. *Dawn of the Dead*, directed by Zack Snyder, Universal Pictures, 2004, movie.

17. *The Walking Dead*, developed by Frank Darabont, based on the books by Robert Kirkman, Tony Moore, and Charlie Adlard, American Movie Classics, 2010–present, television series.

18. Robert Kirkman and Jay Bonansinga, *The Walking Dead: Fall of the Governor Part I* (New York: Thomas Dunne, 2013), 57.

19. Ibid.

20. Ibid., 159.

21. Robert Kirkman and Jay Bonansinga, *The Walking Dead: Rise of the Governor* (New York: Thomas Dunne, 2011).

22. *Dawn of the Dead*, directed by Zack Snyder, Universal Pictures, 2004, movie.

23. See James Thomas Flexner, *The Traitor and the Spy: Benedict Arnold and John Andre* (New York: Syracuse University Press, 1991); and Willard Sterne Randall, *Benedict Arnold: Patriot and Traitor* (New York: William Morrow, 1990).

24. Stephen Ambrose, *Band of Brothers: E Company, 506th Regiment, 101st Airborne from Normandy to Hitler's Eagle's Nest* (New York: Simon & Schuster, 1992); and *The Band of Brothers*, produced by Steven Spielberg, Home Box Office, 2001, miniseries.

25. Hobbes, *Leviathan*, 181.

26. Niccolò Machiavelli, *The Prince*, ed. and trans. W.K. Marriot (Auckland: Floating Press, 2008), 71.

27. Ibid., 55.

28. Ibid., 117.

29. Hobbes, *Leviathan*, 175–176.

30. In the comic book it happens rather quickly in Chapter One, "Days Gone Bye," *The Walking Dead, Compendium* 1, written by Robert Kirkman, penciler and inker Charlie Adlard, and gray tone Cliff Rathburn (Berkeley: Image Comics/Skybound Entertainment, 2009), but it does not happen in the television show until Season 2, in the episode "Better Days," *The Walking Dead*, developed by Frank Darabont, based on the books by Robert Kirkman, Tony Moore, and Charlie Adlard, American Movie Classics, 2010–present, television series.

31. "Sick," *The Walking Dead*, Season 3 developed by Frank Darabont, based on the books by Robert Kirkman, Tony Moore, and Charlie Adlard, American Movie Classics, 2010–present, television series.

32. "Beside the Dying Fire," *The Walking Dead*, Season 2 developed by Frank Darabont, based on the books by Robert Kirkman, Tony Moore, and Charlie Adlard, American Movie Classics, 2010–present, television series.

33. Chapter Six, "This Sorrowful Life," *The Walking Dead, Compendium* 1, written by Robert Kirkman, penciler and inker Charlie Adlard, and gray tones Cliff Rathburn (Berkeley: Image Comics/Skybound Entertainment, 2009).

34. Thomas Hobbes, *Leviathan: The Matter, Forme, & Power of a Commonwealth Ecclesiastical and Civill* (Auckland: Floating Press, 2009), 178.

35. Ibid., 219.

36. Ibid., 139.

37. Robert Kirkman and Jay Bonansinga, *The Walking Dead: Rise of the Governor* (New York: Thomas Dunne, 2004), 268.

38. Niccolò Machiavelli. *The Essential Writings of Machiavelli,* ed. Peter Constantine (New York: Random House, 2009), 66.

39. "Welcome to the Tombs," *The Walking Dead*, Season 3 developed by Frank Darabont, based on the books by Robert Kirkman, Tony Moore, and Charlie Adlard, American Movie Classics, 2010–present, television series.

40. Gary B. Herbert, *Thomas Hobbes: The Unity of Scientific and Moral Wisdom* (Vancouver: University of British Columbia Press, 2011), 148.

41. *Land of the Dead*, written and directed by George Romero, Universal Pictures, 2005, movie.

42. Niccolò Machiavelli, *The Prince*, ed. and trans. W.K. Marriot (Auckland: Floating Press, 2008), 35.

43. Geoffrey R. Stone. *Perilous Times: Free Speech in Wartime: From the Sedition Act of 1798 to the War on Terrorism* (New York: W.W. Norton, 2004).

44. Jonathan W. White, *Abraham Lincoln and Treason in the Civil War: The Trials of John Merryman* (Baton Rouge: Louisiana State University Press, 2011).

45. David M. Kennedy, *Over Here: The First World War and the American Society* (New York: Oxford University Press, 2004).

46. See Lawson Fusao Inada. *Only What We Could Carry: The Japanese American Interment Experience* (Berkeley: Heyday Books, 2000); and Arnold Krammer, *Undue Process: The Untold Story of America's German Alien Internees* (Lanham, MD: Rowman & Littlefield, 1997).

47. Henry W. Fischer, III, *Response to Disaster: Fact Versus Fiction and Its Perpetuation, The Sociology of Disaster* (Lanham, MD: University Press of America, 2008), 55.

48. Emma Seppala, "How the Stress of Disaster Brings People Together," *Scientific American*, November 6, 2012, http://www.scientificamerican.com/article/how-the-stress-of-disaster-brings-people-together/.

49. Maia Szalavitz, "How Disasters Bring Out Our Kindness," *Time*, October 31, 2012, http://healthland.time.com/2012/10/31/how-disasters-bring-out-our-kindness/.

50. "35 Minutes," *Flight 93 National Memorial*, http://www.honorflight93.org/remember/?fa=35-minutes, accessed August 19, 2014.

Bibliography

*In some cases I took the liberty of providing modern English equivalents to individual words in the Hobbes and Machiavelli quotes to aid the reader.

Ambrose, Stephen. *Band of Brothers: E Company, 506th Regiment, 101st Airborne from Normandy to Hitler's Eagle's Nest*. New York: Simon & Schuster, 1992.

The Band of Brothers. Produced by Steven Spielberg. Home Box Office, 2001. Miniseries.

"Beside the Dying Fire." *The Walking Dead*, Season 2. Developed by Frank Darabont, based on the books by Robert Kirkman, Tony Moore, and Charlie Adlard. American Movie Classics, 2010–present. Television series.

"Better Days." *The Walking Dead*, Season 2. Developed by Frank Darabont, based on the books by Robert Kirkman, Tony Moore, and Charlie Adlard. American Movie Classics, 2010–present. Television series.

Brooks, Max. *The Zombie Survival Guide: Complete Protection from the Living Dead*. New York: Three Rivers Press, 2003.

Diary of the Dead. Written and directed by George Romero. Weinstein, 2007. Movie.

Dawn of the Dead. Written and directed by George Romero. United Film Distribution Company, 1978. Movie.

Dawn of the Dead. Directed by Zack Snyder. Universal Pictures, 2004. Movie.

Day of the Dead. Written and directed by George Romero. United Film Distribution Company, 1985. Movie.

Dead Rising. Directed by Yoshinori Kawano and produced by Keiji Inafune. Capcom Production Studio 1, 2006. Video game.

Dead Rising 2. Produced by Keiji Inafune. Blue Castle Games/Capcom, 2010. Video game.

Dead Rising 2: Off The Record. Produced by Keiji Inafune. Capcom Vancouver, 2011. Video game.

Fischer, Henry W., III. *Response to Disaster: Fact Versus Fiction and Its Perpetuation, The Sociology of Disaster*. Lanham, MD: University Press of America, 2008.

Flexner, James Thomas. *The Traitor and the Spy: Benedict Arnold and John Andre*. Syracuse: Syracuse University Press, 1991.

Goldwin, Robert A. "John Locke." In *History of Political Philosophy*, 3d ed. Edited by Leo Strauss and Joseph Cropsey. Chicago: University of Chicago Press, 1987.

Herbert, Gary B. *Thomas Hobbes: The Unity of Scientific and Moral Wisdom*. Vancouver: University of British Columbia Press, 2011.

Hobbes, Thomas. *De Cive*. Paris, 1642.

_____. *Leviathan or The Matter, Forme and Power of a Common Wealth Ecclesiastical and Civil—commonly referred to as Leviathan*. England, 1651.

_____. *Leviathan: The Matter, Forme, & Power of a Common-Wealth Ecclesiastical and Civill* Auckland: Floating Press, 2009.

Inada, Lawson Fusao. *Only What We Could Carry: The Japanese American Interment Experience*. Berkeley: Heyday Books, 2000.

Kennedy, David M. *Over Here: The First World War and the American Society*. New York: Oxford University Press, 2004.

Kirkman, Robert, Charlie Adlard, and Cliff Rathburn. "Days Gone By." In *The Walking Dead, Compendium 1*. Berkeley: Image Comics/Skybound Entertainment, 2009.

_____, _____, and _____. "This Sorrowful Life." In *The Walking Dead, Compendium 1*. Berkeley: Image Comics/Skybound Entertainment, 2009.

Kirkman, Robert, and Jay Bonansinga. *The Walking Dead: Fall of the Governor Part I*. New York: Thomas Dunne, 2013.

_____, and _____. *The Walking Dead: Rise of the Governor*. New York: Thomas Dunne, 2011.

Krammer, Arnold. *Undue Process: The Untold Story of America's German Alien Internees*. Lanham, MD: Rowman & Littlefield, 1997.

Land of the Dead. Written and directed by George Romero. Universal Pictures, 2005. Movie.

Machiavelli, Niccolò. *The Essential Writings of Machiavelli*. Edited by Peter Constantine. New York: Random House, 2009.

_____. *The Prince*. Edited and translated by W.K. Marriot. Auckland: Floating Press, 2008.

Maier, Pauline, ed. *The Declaration of Independence and The Constitution of the United States*. New York: Bantam Dell, 2008.

Mogk, Matt. *Everything You Ever Wanted to Know About Zombies*. New York: Gallery Books, 2011.

Night of the Living Dead. Written by George Romero and John A. Russo, directed by George Romero. Image Ten, 1968. Movie.

Randall, Willard Sterne. *Benedict Arnold: Patriot and Traitor*. New York: William Morrow, 1990.

Seppala, Emma. "How the Stress of Disaster Brings People Together." *Scientific American*,

November 6, 2012. http://www.scientificamerican.com/article/how-the-stress-of-disaster-brings-people-together/.

"Sick." *The Walking Dead*, Season 3. Developed by Frank Darabont, based on the books by Robert Kirkman, Tony Moore, and Charlie Adlard. American Movie Classics, 2010–present. Television series.

Stone, Geoffrey R. *Perilous Times: Free Speech in Wartime: From the Sedition Act of 1798 to the War on Terrorism*. New York: W.W. Norton, 2004.

Strauss, Leo. "Niccolò Machiavelli." In *History of Political Philosophy*, 3d ed., edited by Leo Strauss and Joseph Cropsey. Chicago: University of Chicago Press, 1987.

Survival of the Dead. Written and directed by George Romero. E1 Entertainment and Magnet Releasing, 2009. Movie.

Szalavitz, Maia. "How Disasters Bring Out Our Kindness." *Time*, October 31, 2012. http://healthland.time.com/2012/10/31/how-disasters-bring-out-our-kindness/.

"35 Minutes." *Flight 93 National Memorial*, http://www.honorflight93.org/remember/?fa=35-minutes.

The Walking Dead. Developed by Frank Darabont, based on the books by Robert Kirkman, Tony Moore, and Charlie Adlard. American Movie Classics, 2010–present. Television series.

"Welcome to the Tombs." *The Walking Dead*, Season 3. Developed by Frank Darabont, based on the books by Robert Kirkman, Tony Moore, and Charlie Adlard. American Movie Classics, 2010–present. Television series.

White, Jonathan W. *Abraham Lincoln and Treason in the Civil War: The Trials of John Merryman*. Baton Rouge: Louisiana State University Press, 2011.

The Law and
the Living Dead

JENNIFER M. LANKFORD

"Our courts have their faults, as does any human institution, but in this country our courts are the great levelers, and in our courts all men are created equal...."[1] As he closed his defense of Tom Robinson in *To Kill a Mockingbird*, Atticus Finch spoke of an enduring truth of our legal system; it is an imperfect, but righteous attempt at justice.[2] Accurate and inspiring, Atticus's representation of American jurisprudence was also limited. He described a "human" institution where all "men" are created equal, reflecting a critical, underlying problem within our legal system. It is not yet prepared, in form or in spirit, to apply to the living dead.

The Lawless

A well-organized legal system has become as essential to American society's existence as an overpriced vanilla latte. Unfortunately, a "nation of laws is also a nation of the lawless."[3] Rules of law—whether governing the operation of automobiles or prohibiting murder—are only "rules" if there exists a means of enforcement—a Stabler and Benson or Briscoe and Logan to hold the line firm.[4] An apocalyptic disaster, such as an uprising of the living dead, will, initially, leave our nation devoid of law and order.

American jurisprudence has faced challenges before, but nothing so alarming or significant as the resurrection of the dead—instinct-driven cannibals with no concern for legal ramifications.[5] The onset of the zombie apocalypse will initially cripple our legal system. The necessity of survival will suspend judicial functioning as Americans focus not on avoiding their "just deserts," but on not becoming dessert.[6]

At present, the American legal system is based on a series of complex rules, including federal laws enacted by Congress, statutes drafted by state legislatures, and ordinances adopted by local governments, such as county commissions and city councils.[7] Officers of our legal system, including attorneys and judges, act in accordance with and apply these laws.[8] These legal officials rely on voluminous and technical rules of civil and appellate procedure, as well as evidentiary guidelines, to ensure consistent application of our laws.[9] While nuanced legal guidelines are complex enough to warrant an attorney's hourly fees, such a complicated judicial process is ill-equipped to deal with perhaps the only thing more dreaded than lawyers—flesh-eating zombies. And so, as it concerns the living dead, the law as we know it will not be practiced, but ignored.

With the dawn of the dead, and the inevitable loss of some of our best and brightest (assuming, perhaps stereotypically, that the most advanced amongst us are not necessarily the fastest runners), America will inevitably be devoid of means of mass communication. With ties to the outside world frayed, deadlines and underlying methodic formalities become impractical. As officers of the court scatter to safety, the modern-day legal system and the order it bestows will cease to exist. It will be every living man for himself. When the dust settles, however, the legal system will be one of the first forms of normality to reanimate alongside the undead, and like the undead in a cruder, more rudimentary form.[10]

Resurrecting the Law

A community justice system will arise as the first semblance of law following the zombie apocalypse. A tenuous return to normalcy, a community justice drastically differs from present day jurisprudence. In community justice forums, "the law affects everyone and everyone affects the law."[11] In other words, such "justice" systems are mercurial and depend on what the particular community needs at any given moment. In a time when there is more land than living, our legal system will be primitive, stripped down to its most basic form.[12] The living are left with what they know, need, and remember to reestablish a functioning jurisprudence, which will be a mere silhouette of past legal practice.[13]

Although formed to promote safety, reduce anxiety, and create the perception of normalcy, community justice forums will be plagued by fear and a dangerous attachment to the past.[14] The remaining population will rely on their memories of the justice system to establish rules for civilized existence.[15] Such marriage to the past typically leads to unwarranted punishment for moral, as opposed to legal, wrongs.[16] Complex rules and required procedure will be replaced by a simpler system of justice "dependent upon oaths at most stages of litigation, and permeated by both religious and superstition notions."[17] Because Americans have long believed that God passes judgment on sinful societies, superstitions will run as rampant as the living dead.[18] Just as Puritans in colonial Massachusetts created law to satisfy their godly society, Americans in the wake of a zombie apocalypse will apply the law ritualistically, with a heavy emphasis on morality.[19] A moral consciousness without a social safety net will result in a callous application of the law.[20]

The makeshift courts of an apocalyptic America are more likely to resemble the courtrooms of Salem, Massachusetts in the 1600s, than modern forums.[21] Such improvised courts are comprised of an "ugly combination of political uncertainty [and] widespread panic ... [that will lead] to a legal horror story."[22] There will be no preliminary hearings, no legal counsel, and no due process, but the ready admission of speculative evidence.[23] Influential members of the community will appoint themselves as investigators, police, judge, and jury.[24] Adjudicators will be championed not because of their experience, but because of their influence.[25] Devotion to the truth and impartiality give way to those holding the superlative "Most Likely to Keep Us from Being Eaten."

Over time, the justice system we adore will more closely resemble the Nazi regime that we abhor, in which judges and lawyers were substituted by dictators and their regimes.[26] Those caught between zombie and zealot will not be tried, but subjected to almost certain punishment.[27] Trials will not be by jury or bench, but by methods such as compurgation, during which select members of the community are called to swear to the accuracy of a person's claims.[28] These "very personal, and highly confrontative" proceedings hinge not on meritorious legal arguments, but on whether the accused can survive a popularity contest or perhaps an "ordeal," i.e., "a physical trial in which [the accused] call[s] upon God to witness his innocence by putting a miraculous sign upon his body."[29] Past American community justice systems implemented ordeals by tossing the "presumed guilty" into water to see if he or she would float. Innocence was established only if the sanctified water "accepted" the guiltless.[30]

Not only will the presentation of justice differ in the wake of the zombie apocalypse, but so will our methods of rehabilitation. Incarceration has no place in a community justice system. With less and less warm bodies, even makeshift lawmakers will know better than to bother with imprisonment; it is more important to return the law-breaking living to the community than to suffer a labor shortage or be outnumbered by the walkers.[31] Punishment will not hinge on the loss of freedom, but rather, on retribution through physical pain or shame.[32] Notwithstanding traditional tools such as a brand or whip, public degradation will serve the purpose of showing the error of one's ways, while simultaneously integrating one back into a community that cannot afford to lose him.[33] If an offender must be lost to the community, a more effective means of punishment is to make an example of the individual, e.g., by forcing the lawbreaker to confess his sins to the crowd before being turned over to the zombies beckoning outside the city gates.[34] Fear is as heavy a motivator for obedience as is imprisonment.

Eventually, the community justice system will no longer be a cure for deviance, but the cause.[35] A legal system that is the sovereignty of the community is to our disadvantage in resurrecting the justice system.[36] Yet, for all its faults, a community justice system is not without merit. As the dead arise and communications cease, chaos will ensue. Americans will suffer "not merely the breakdown of the law, but, more importantly, the breakdown of the underlying order upon which the law is based."[37] A community justice system serves to resurrect this order. By assembling even the skeleton of a legal system, our nation of laws begins to reestablish its prominence in society, showing that zombies are not the only thing that can rise from the dead.

Litigation and the Living Dead

Imagine America in the wake of a zombie apocalypse—the legal system functioning as a shadow of its former self and our government slowly rebuilding. If only Americans had the foresight of our neighbors across the pond. In 2011, when zombies were just a pop-culture phenomenon attracting millions of viewers on AMC, the British Military Defense released a plan of action: "In the event of an apocalyptic incident, e.g., zombies, any plans to rebuild and return England to its pre-attack glory would be led by the Cabinet Office."[38] While American legislatures are less progressive, at least our forefathers aided us by engaging in a little doomsday prepping of their own in the form of certain extra-constitutional measures, which are applicable to apocalyptic situations.

For instance, an issued state of emergency, the imposition and consequences of which are left to more qualified authors within this text, permits the reestablishment of law and order in chaotic situations. In an apocalyptic aftermath, our nation will require a surge of repressive laws "designed to ease our anxiety by promising greater security."[39] Recall the Patriot Act, which surpassed all *Schoolhouse Rock* promises of procedure with its 33-day journey from a bill to a law. Passed in response "to a largely undefined threat from a poorly understood source," the Patriot Act is a law of form, not substance, intended "as a symbol to reassure the country that Washington was grimly determined to step up to the fight against terrorism."[40] Americans may benefit from such a use of expediency and extrajudicial power when returning our country's legal system to normalcy following the zombie apocalypse.

To ensure community justice systems expire, thus allowing for a rehabilitation of traditional jurisprudence, the government may also enact the emergency provision within the United States Constitution permitting the suspension of *habeous corpus,* or the process of bringing a person before the court to ensure their detainment or imprisonment is valid, "when in Cases of Rebellion or Invasion the public Safety requires it."[41] Surely no definition more accurately describes the uprising of the living dead than an "invasion" threatening public safety. While this extra-constitutional measure is unorthodox, its application is not unheard of.[42] For instance, during the Civil War, President Lincoln, absent Congressional consent, issued an executive order requiring the incarceration of American citizens, specifically Confederates, and the suspension of their constitutional right to ask the courts for a writ of *habeas corpus* demanding their freedom.[43] Similarly, during the Second World War, Franklin D. Roosevelt invoked this extraordinary power and ordered more than 120,000 Japanese-American citizens into internment camps.[44]

As America cleanses itself of the primitive laws established by community justice systems, our nation must begin to establish new laws to accommodate life with the undead. Leaders in community justice systems who are reluctant to relinquish their newfound power will be a serious impediment to the rebuilding of our legal system. To deal with this resistance, our nation may look to Germany's defeat in World War II for guidance. At the end of the war, the Allied Powers assumed "supreme authority" of Germany, eradicating its new justice system and rebuilding it from the ground up.[45] Laws enacted during the Nazi regime were abolished and cleansed of ideas incompatible with German defeat. Meanwhile, former lawmakers were established to their rightful positions.[46] As time passes and community justice systems dissipate, more traditional methods of punishment to dissuade disobedience may be used. For example, those individuals continuing to act as makeshift attorneys may be prosecuted for the misdemeanor of impersonating a public servant.[47]

As Americans gain a foothold in managing a zombie crisis, the possibility of executing laws with consistency will become a possibility.[48] However, mere resurrection of our legal system is just the beginning. Our ideas of jurisprudence must adapt if there is truly to be law, and life, with the living dead.[49] Consider the term "human being" as defined by the Model Penal Code of the American Law Institute, a widespread model codification of criminal law in the United States.[50] The Model Penal Code defines a "human being" as "a person who has been born and is alive."[51] This definition highlights a crucial question facing our courts or law in the wake of the zombie apocalypse: are the undead "alive"? Certainly, zombies were born, and pre-zombiefication, operated in this world as human beings do—working, laughing, in awe of the popularity of the Kardashians.[52] But, are zombies alive or merely animated? Once our judicial and legislative branches address these basic principles of our new zombified law, i.e., what is a zombie, we can begin to incorporate our new friends into more specific provisions of the law, a few of which are addressed below.

Fight or Flight: Laws of Self-Defense

Due to the insatiable appetite of zombies, the laws of self-defense must be altered to permit self-defense in retaliation for past aggression, as well as preemptive self-defense.[53] At present, the law permits a person to "defend [themselves] against another by force [that is] intended or likely to cause death or serious bodily harm, when [that person] reasonably believes that … the other is about to inflict upon him an intentional contact or other bodily harm, and that … he is thereby put in peril of death which can safely be prevented only by the immediate use of such force."[54] Serious bodily harm includes "permanent or protracted loss of the function of any important member or organ."[55] Potential attack by a flesh-craving zombie certainly qualifies.[56]

Unfortunately, at present, the above-described privilege of self-defense does not extend to retaliatory or pre-emptive acts if the individual attacked can, in complete safety, avoid the necessity of defending himself or herself by retreating from the attacker.[57] Pre-zombie apocalypse, it was in society's best interests to retreat, i.e., avoid a bloody battle at all costs.[58] In contrast, post-zombie apocalypse, retreating serves a very limited purpose. A person may not be dinner, but there is always the possibility of becoming dessert. The law must be modified to eliminate any obligation to retreat, instead permitting pre-emptive acts of self-defense.

Similarly, the prevalence of flesh-eating zombies calls for an expansion of the "castle doctrine." Under this long-standing doctrine, a person standing in their own home may "await [an] assailant and use deadly force to repel him though he could prevent the assailant from attacking him merely by closing the door."[59] In laymen terms, the "castle doctrine" permits a person to defend his home, or "castle," with deadly force against a home invader without fear of prosecution. Persons should no longer be bound to his or her castle in order to exercise such pre-emptive self-defense. The law should be modified to make the person, not the property, the "castle."

The expansion of self-defense laws calls for a simultaneous reduction of laws pertaining to the abuse of a corpse. Certainly self-defense against zombies will, if the living are so lucky, involve some violent "end" to the undead. The law against desecration of a corpse must be modified accordingly.

Under present law, "a person who treats a corpse in a way that he knows would outrage ordinary family sensibilities commits [a] misdemeanor."[60] Desecration of a corpse, once frowned upon, should be a protected activity following the zombie apocalypse. Legislatures may even consider offering incentives for desecrating the bodies of our newly animated friends, thereby, eliminating overpopulation of the undead. Zombie hunting would perfectly fill the gap between duck and turkey season.

With the living dead walking amongst us, motivated only by the instinct to eat the living, it is essential that the law accommodate not only the basic instinct of humans to "fight," i.e., defend themselves, but also the crucial instinct of "flight." Whether a person admits it or not, when faced with a flesh-eating zombie, one is much less concerned with the old adage of "love thy neighbor." It is far more likely that said neighbor will be on his own when a small herd of zombies besieges his carefully maintained pansy garden. Pre-zombie apocalypse, a person is not required to put himself or herself in danger to save the life of another. In the spirit of adaptation, an "every man for himself" mentality may be too prevalent as people panic and use others as a means of avoiding zombification or death.

For instance, imagine a scenario in which an individual, pursued by a pack of zombies, throws his companion to the ground to gain time in his retreat. The time consumed by the zombies in devouring the fallen companion allows the individual to reach shelter. To discourage such "eat him, not me" reactions, the law must be amended to prohibit the intentional infliction upon another of substantial bodily harm in retreat from a zombie.[61] Under the amended law, the survivor would be subject to prosecution for his or her actions. In other words, no privilege exists to protect oneself from certain death by sacrificing another.[62] By the same token, the law must clarify that an individual shall not be held responsible simply for escaping a zombie attack while others perished, i.e., the law should include a provision excluding from liability a person who is merely a faster runner.[63]

Preservation of the Living

Good Samaritan laws should also be expanded following the rise of the living dead. A product of Luke 10:25–37, Good Samaritan laws protect persons who provide reasonable assistance to the injured; however, these laws have limited applications.[64] At present, absent a special relationship, an individual, viewing a stranger in dire peril, is not obligated to provide emergency assistance unless a special relationship exists between the parties.[65] For instance, the relationship between employer and employee, parent and child, or innkeeper and guest gives rise to an affirmative duty to aid and protect.[66] The tie between Good Samaritan laws and special relationships must be severed if the living are to outlast the living dead. Simply put, it is good public policy to keep as many warm bodies as possible. Our legislature should take a cue from Larry David and institute strict Good Samaritan laws for zombie attacks, even absent a special relationship.[67]

The law must also adapt to impose obligations on those brave or foolish persons who choose to become zombie caretakers, assuming a law permits such a perilous endeavor. Current law places a special duty on those individuals charged with caring for persons with "dangerous propensities." A person "who takes charge of a third person whom he knows or should know to be likely to cause bodily harm to others if not controlled is under a duty to exercise reasonable care to control the third person to prevent him from doing such harm."[68] As it pertains to zombies, the tendency to act violently is nothing if not anticipated. Legislation must be enacted that places a strict duty on persons who take charge of a zombie to exercise extraordinary care to control that zombie to prevent harm to others.

If a caretaker fails to properly restrain his or her zombie, then that individual must be subject to strict liability for engaging in an "abnormally dangerous activity." Present law defines an activity as "abnormally dangerous" if "the activity creates a foreseeable and highly significant risk of physical harm even when reasonable care is exercised by all actors; and ... the activity is not one of common usage."[69] Owning or caring for a zombie will surely be categorized as an "abnormally dangerous activity" because zombies are known to have dangerous tendencies (ahem, eating people) and, therefore, those endeavoring to care for zombies should be subject to strict liability.[70]

Strict liability principles would also allow for owners or possessors of zombies to be automatically liable for any "physical harm caused by the animal if the harm ensues from that dangerous tendency."[71] Strict liability is different from other claims of negligence because it does not require a standard of care. In other words, regardless of the care taken by the individual, he or she will be liable if their action is subject to strict liability. Such a provision to protect against a catastrophic zombie bite is distinguishable from the modern-day "first bite" rule, which holds a dog owner strictly liable for his or her pet's "first bite" only if the animal has previously bitten someone. In other words, the first bite is "free," so long as the caretaker has no knowledge of their pet's propensity to bite. No such flexibility should be given to zombie caretakers who, unlike mere dog owners, know that their "pet" or charge has the propensity to eat, let alone bite, other persons.[72]

To ensure strict liability is enforced, the law should require those choosing to take charge of a zombie to register that zombie with a national data bank. Mandatory registration encourages caretakers of the living dead to remain diligent in their charge by keeping their zombies restrained and accounted for at all times.[73] Zombie caretakers must also undertake responsibility to pay for any damage caused by their charge and the zombie's estate's wealth must be ready and willing to compensate the zombie's victims.[74] The alternative, i.e., suing a zombie for harm caused, is not only logistically difficult—imagine maintaining order in that court, but also impractical. To this end, state legislatures will need to create Zombie Reform Acts, limiting power to sue and recover from estates of a zombie's family for apocalyptic torts without certain proof.[75]

The People's Court: The Living v. the New-Age Lazarus

Just as tort law increased following the Industrial Revolution, so will it in the wake of the zombie apocalypse, where the undead, instead of machines, mangle humans at an increasingly alarming rate.[76] As noted earlier, various forms of liability exist to make a zombie's caretaker accountable; however, whether a zombie can be held responsible for its own acts of violence is a more complicated matter. To commit an intentional tort, an individual must act with intent, that this, he or she must appreciate the consequences of his or her actions. Intent exists where "the person acts with the purpose of produc[ing] the consequence; or ... acts knowing that the consequence is substantially certain to result."[77] Despite the inevitable arguments of zombie rights groups, the categorization of the undead as "persons" is highly unlikely, but, to avoid expensive litigation, the law should specifically exclude zombies from liability for intentional torts. However, as explained previously, victims of zombie violence can pursue remedies through the zombie's estate.[78]

Similar to tort law, principles of contract law will require modification post-zombie apocalypse, as contract law also rests heavily on the mental capacity of individuals. Contracts are promises for which the law gives remedy in the events of a breach, or breaking of a promise.[79] As defined prior to the zombie apocalypse, a person without legal capacity does not have the ability to enter into or be bound by a contractual agreement.[80] Legal capacity is removed where an individual has a "mental disease or defect" and is "unable to act in a reasonable manner."[81] The only way a contract may survive a party's mental defect or disease and be considered viable is if the non-affected contracting party was "without knowledge of the mental illness or defect."[82] Zombies, with their flesh-eating tendencies, certainly lack the ability to act in a reasonable manner, and thereby, the ability to contract. A person's attempt to allege an unawareness of the zombie's condition, in order to validate a contract, would inevitably fail. Even if the individual survived obtaining the zombie's signature, alleging ignorance as to zombification would be impossible, what with the rotting flesh and rabid moans.

The real litigation within contract law in a zombie-infested world will arise as persons, whether truthfully or fraudulently, claim that a person entered into the subject contract pre-zombification. In other words, the contract was agreed upon while both parties had the capacity to contract. Such claims will inevitably lead to a surge of medical expert testimony, through which physicians will opine as to when the subject party turned from human to zombie, i.e., when the person lost the requisite mental capacity to contract. So long as the contract was executed pre-zombiefication and the presence of a pulse is not necessary for the performance of the contract, then incapacity will not render the contract void.[83] The validity of the contract depends upon one of the party's having the capacity to enter into the agreement.

Finally, property and estate law will also be in flux following the zombie apocalypse. Back when the dead stayed dead, the property of a deceased family member who died intestate, that is, without a valid will, passed to the decedent's heirs. Zombies, of course, are not quite living and not quite deceased. To ensure that red tape does not prevent the passage of the legal ownership of property, the law must be amended to classify individuals as deceased upon death or zombiefication. Such an express alteration to the law would also ensure that property held by a zombie is not subject to immediate homesteading, but rather, passes to the proper parties.

Zombie Rights Movement

Although laws will certainly change to accommodate the living dead amongst us, new legislation will inevitably arise designed to protect the rights of zombies. As individuals lobby for the preservation of the undead, organizations will form to ratify zombie civil rights. Perhaps People for the Ethical Treatment of Zombies? After lobbying for zombies fundamental civil rights, PETZ must work to enforce said protections. Zombie civil rights will be limited by the abnormally dangerous tendencies of their subject cause; however, with the right lobbyist, law could be crafted to prohibit unnecessary cruelty to zombies.

Similarly, at present, cruelty to animals is illegal and punishable as a misdemeanor. Specifically, a person may be subject to legal punishment if he or she "purposely or recklessly: (1) subjects any animal to cruel mistreatment; or (2) subjects any animal in his custody to cruel neglect; or (3) kills or injures any animal belonging to another without legal privilege or consent of the owner."[84] Similar constraints might be applied to zombies. Likewise, zombie civil rights organizations may fight to prevent the illegal restraint of zombies for improper purposes. At present, one may not falsely imprison another, i.e., "knowingly restrain another unlawfully so as to interfere substantially with his liberty."[85] One might argue that zombies, as the undead, are not protected by this law. Yet, such restrictions have been held to apply to the mentally incapacitated, a category similar to that of the living dead.

Although the resurrection of our laws is essential, it is certain that new laws will breed new lawsuits.[86] The litigious nature of individuals will reanimate like the zombies preceding it, and soon court systems will function at reinvigorated rates.[87] Citizens of the new world will adapt, change, and add to the law, "in ways that suit [ ] their situation, which [is], after all, very different from the situation of the ordinary [American] man or woman."[88]

Conclusion

There is no easy transition from a legal system formed only as a "human institution" to one incorporating the living dead. Even so, as expressed by Atticus Finch, American jurisprudence need not be faultless. It need only strive to operate so as to provide equality and offer justice to those to whom it applies. An apocalyptic situation that deprives our nation of law and order will leave us as hopeless as Tom Robinson before a Georgia jury. With time, new life will be given to our legal system through community justice forums and the evolution of order therefrom. Although this resurrection will require a tumultuous transition, eventually our courts will adapt to law and the living dead.

NOTES

1. Harper Lee, *To Kill a Mockingbird* (Philadelphia: J.B. Lippincott, 1960), 218.
2. Lee, *To Kill a Mockingbird*, 51–52.
3. Charles Hoffer, *A Nation of Laws: America's Imperfect Pursuit of Justice* (Lawrence: Peter Charles Hoffer University Press of Kansas, 2010), 17.
4. Detectives Elliot Stabler and Olivia Benson headline the popular NBC crime drama *Law and Order: Special Victims Unit*, while Briscoe and Logan acted as an original partnership on the original *Law and Order*.
5. Previous disasters threatening the survival of the American way of life include the nation's conflict with the Axis Powers during World War II, which "jeopardiz[ed] our survival as an independent nation." Similarly, during the Civil War, it was "Jefferson Davis and the Confederacy [that] were aiming to destroy the Union and create a rival republic." Ronald A. Carp and Ronald Stidham, *Judicial Process in America* (Washington, D.C.: CQ Press, 1996), 5–6, 10.
6. "Just deserts" is "the punishment that a person deserves for having committed a crime." Bryan A. Garner, ed., *Black's Law Dictionary*, 9th ed. (St. Paul: West Group, 2009), accessed April 12, 2013, at https://a.next.westlaw.com/Document/I970505e3bea 311e08b05fdf15589d8e8/View/FullText.html?navigationPath=Search%2Fv3%2Fsearch

%2Fresults%2Fnavigation%2Fi0ad604060000013e0a4c9a36782cc48c%3FNav%3DBL
ACKS%26fragmentIdentifier%3DI970505e3bea311e08b05fdf15589d8e8%26startIndex
%3D1%26contextData%3D%2528sc.Search%2529%26transitionType%3DSearchItem
&listSource=Search&listPageSource=95108e061603a1631cdf3b4191b42a4a&list=BLAC
KS&rank=1&grading=na&sessionScopeId=b3cf604df3c02ddba2d366a7aa5672d7&ori
ginationContext=Search%20Result&transitionType=SearchItem&contextData=%28s
c.Search%29.

7. Carp and Stidham, *Judicial Process*, 5–6, 10. Federal laws are broad-reaching, "made up of acts of Congress, presidential orders, [and] U.S. court decisions." Carp and Stidham, *Judicial Process*, 10.

8. Carp and Stidham, *Judicial Process*, 7.

9. Lawrence M. Friedman, *Law in America: A Short History* (New York: Modern Library Chronicles, 2002), p. 97 (explaining that the American evidentiary principles are considered among the most complex in the world).

10. Hoffer, *Nation of Laws*, 20. ("Extra-legal self help may fashion its own law when written law does not reach a community or the people in a community find the strictures of the law too confining.")

11. Friedman, *Law in America*, 24.

12. Friedman, *Law in America*, 24.

13. Friedman, *Law in America*, 24.

14. Hoffer, *Nation of Laws*, 109.

15. Friedman, *Law in America*, 24.

16. Hoffer, *Nation of Laws*, 75. (Colonial courts listed blasphemy, idleness, Sabbath-breaking, and skipping church among punishable crimes.)

17. Hoffer, *Nation of Laws*, 65.

18. Friedman, *Law in America*, 76.

19. Friedman, *Law in America*, 24.

20. Friedman, *Law in America*, 46.

21. Hoffer, *Nation of Laws*, 109.

22. Hoffer, *Nation of Laws*, 109.

23. A preliminary hearing is a "criminal hearing to determine whether there is sufficient evidence to prosecute an accused person." Garner, ed., *Black's Law Dictionary*, 9th ed., accessed April 12, 2013, at https://a.next.westlaw.com/Document/Iba9e0722bea311e08b05fdf15589d8e8/View/FullText.html?navigationPath=Search%2Fv3%2Fsearch%2Fresults%2Fnavigation%2Fi0ad604060000013e0a4d445c782cc614%3FNav%3DBLACKS%26fragmentIdentifier%3DIba9e0722bea311e08b05fdf15589d8e8%26startIndex%3D1%26contextData%3D%2528sc.Search%2529%26transitionType%3DSearchItem&listSource=Search&listPageSource=d5541d61e8ca6e50fee611dd59d700e8&list=BLACKS&rank=1&grading=na&sessionScopeId=b3cf604df3c02ddba2d366a7aa5672d7&originationContext=Search%20Result&transitionType=SearchItem&contextData=%28sc.Search%29. Speculative evidence is based on matters of which there is no certain knowledge, as opposed to direct or circumstantial evidence, which tends to prove or disprove an alleged fact. Garner, ed., *Black's Law Dictionary*, 9th ed., accessed April 12, 2013, at https://a.next.westlaw.com/Document/Icb871a6ebea311e08b05fdf15589d8e8/View/FullText.html?navigationPath=Search%2Fv3%2Fsearch%2Fresults%2Fnavigation%2Fi0ad604060000013e0a4dfcfb782cc6f8%3FNav%3DBLACKS%26fragmentIdentifier%3DIcb871a6ebea311e08b05fdf15589d8e8%26startIndex%3D1%26contextData%3D%2528sc.Search%2529%26transitionType%3DSearchItem&listSource=Search&listPageSource=e1711175849426f457c1c468fd6f0943&list=BLACKS&rank=1&grading=na&sessionScopeId=b3cf604df3c02ddba2d366a7aa5672d7

&originationContext=Search%20Result&transitionType=SearchItem&contextData=%28sc.Search%29.

24. Hoffer, *Nation of Laws*, 18.

25. Hoffer, *Nation of Laws*, 84.

26. Hoffer, *Nation of Laws*, 18.

27. Hoffer, *Nation of Laws*, 18.

28. Hoffer, *Nation of Laws*, 65. An inquest involves a "verdict by a body of men from the same neighborhood who were summoned by some official, on the authority of the crown, to reply under oath to any inquiries that might be addressed to them." Leonard W. Levy, *The Palladium of Justice Origins of Trial by Jury* (Chicago: Ivan R. Dee, 1999), 4–5.

29. Levy, *Palladium of Justice*, 4–5.

30. Levy, *Palladium of Justice*, 5.

31. Friedman, *Law in America*, 77.

32. Friedman, *Law in America*, 77. Colonial criminals were whipped, fined, branded, banished, even hanged, but not imprisoned. Friedman, *Law in America*, 80.

33. Friedman, *Law in America*, 77.

34. Friedman at 77–78.

35. Hoffer, *Nation of Laws*, 18. Community justice groups have a long history in America, flourishing especially in the late 1840s in San Francisco and Oakland, California, where the groups drove lawmakers from the practice of law under promise of purging corruptness. Hoffer, *Nation of Laws*, 18.

36. Hoffer, Nation of Laws, 1 ("Americans founders rejected the monarchial law-giving of former colonial masters and substituted it for a constitution based on voter ratification").

37. Bruce Ackerman, *Before the Next Attack: Preserving Civil Liberties in an Age of Terrorism* (New Haven: Yale University Press, 2006), 170.

38. After a curious citizen submitted a Freedom of Information Act request to the British Ministry of Defense provided this tongue-in-cheek response, accessed at April 13, 2013, at http://www.telegraph.co.uk/news/newstopics/howaboutthat/9721072/Britain-is-well-prepared-to-fight-apocalyptic-zombie-invasion.html.

39. Michael Stolleis, *The Law Under the Swastika: Studies on Legal History in Nazi Germany* (Chicago: University of Chicago Press, 1998), 2.

40. Stolleis, *Swastika*, 2.

41. Ackerman, *Before the Next Attack*, 60. In Latin, habeous corpus means "You have the body." Garner, ed., *Black's Law Dictionary*, 9th ed., accessed April 12, 2013, at https://a.next.westlaw.com/Document/I896134f1bea311e08b05fdf15589d8e8/View/FullText.html?navigationPath=Search%2Fv3%2Fsearch%2Fresults%2Fnavigation%2Fi0ad6040a00000142ba138b788fe0fc84%3FNav%3DBLACKS%26fragmentIdentifier%3DI896134f1bea311e08b05fdf15589d8e8%26startIndex%3D1%26contextData%3D%2528sc.Search%2529%26transitionType%3DSearchItem&listSource=Search&listPageSource=9428441403af95aaab08ef34a9aa68cc&list=BLACKS&rank=19&grading=na&sessionScopeId=adb0a0e29b19f6b5279af2d018b45f0d&originationContext=Search%20Result&transitionType=SearchItem&contextData=%28sc.Search%29.

42. Friedman, *Law in America*, 171–172.

43. Stolleis, *Swastika*, 20.

44. Stolleis, *Swastika*, 20.

45. Stolleis, *Swastika*, 3.

46. Stolleis, *Swastika*, 3.

47. Stolleis, *Swastika*, 3.

48. Hoffer, *Nation of Laws*, 80–81.

49. Hoffer, *Nation of Laws*, 17.

50. Herbert Wechsler, Chief Reporter, et al., MODEL PENAL CODE (American Law Institute: 1985), accessed April 12, 2013, at https://a.next.westlaw.com/Document/ N25A3CA9004C211DC9A5BEF4472C1061E/View/FullText.html?originationContext =documenttoc&transitionType=CategoryPageItem&contextData=(sc.Default).

51. Wechsler, MODEL PENAL CODE, § 210, accessed April 12, 2013, at https://a. next.westlaw.com/Document/N25A3CA9004C211DC9A5BEF4472C1061E/View/ FullText.html?originationContext=documenttoc&transitionType=CategoryPageItem &contextData=(sc.Default).

52. The Kardashian family, comprised of a klan of "Ks," including Kris, Kim, Kourtney, Khloe, Kendall, and Kylie, have appeared on their own E! network reality show, *Keeping Up with the Kardashians*, and are unofficial pop culture royalty. Wikipedia, The Free Encyclopedia, accessed November 3, 2013, at http://en.wikipedia.org/ wiki/Keeping_Up_with_the_Kardashians.

53. Herbert Wechsler, Chief Reporter, et al., RESTATEMENT (SECOND) OF TORTS (American Law Institute: 1965), § 65(a)–(b), accessed at https://a.next.westlaw.com/ Document/I4722d6bb662911dca5lecfdfa1ed2cd3/View/FullText.html?origination Context=documenttoc&transitionType=CategoryPageItem&contextData=(sc. Default).

54. Herbert Wechsler, Chief Reporter, et al., RESTATEMENT (SECOND) OF TORTS (American Law Institute: 1965), § 65(a)–(b), accessed at https://a.next.westlaw.com/ Document/I4722d6bb662911dca5lecfdfa1ed2cd.3/View/FullText.html?origination Context=documenttoc&transitionType=CategoryPageItem&contextData=(sc.Default).

55. Wechsler, RESTATEMENT (SECOND) OF TORTS, § 65(a)–(b) cmt. (b), accessed at https://a.next.westlaw.com/Document/I4722d6bb662911dca5lecfdfa1ed2cd3/View/ FullText.html?originationContext=documenttoc&transitionType=CategoryPageItem &contextData=(sc.Default).

56. Wechsler, RESTATEMENT (SECOND) OF TORTS, § 65(a)–(b) cmt. (b), accessed at https://a.next.westlaw.com/Document/I4722d6bb662911dca5lecfdfa1ed2cd3/View/ FullText.html?originationContext=documenttoc&transitionType=CategoryPageItem &contextData=(sc.Default).

57. Wechsler, RESTATEMENT (SECOND) OF TORTS, § 65(a)–(b) cmt. (j), accessed at https://a.next.westlaw.com/Document/I4722d6bb662911dca5lecfdfa1ed2cd3/View/ FullText.html?originationContext=documenttoc&transitionType=CategoryPageItem &contextData=(sc.Default).

58. Wechsler, RESTATEMENT (SECOND) OF TORTS, § 65(a)–(b) cmt. (j), accessed at https://a.next.westlaw.com/Document/I4722d6bb662911dca5lecfdfa1ed2cd3/View/ FullText.html?originationContext=documenttoc&transitionType=CategoryPageItem &contextData=(sc.Default).

59. Wechsler, RESTATEMENT (SECOND) OF TORTS, § 65(a)–(b) cmt. (i), accessed at https://a.next.westlaw.com/Document/I4722d6bb662911dca5lecfdfa1ed2cd3/View/ FullText.html?originationContext=documenttoc&transitionType=CategoryPageItem &contextData=(sc.Default).

60. Wechsler, MODEL PENAL CODE, § 250.10, accessed April 12, 2013, at https:// a.next.westlaw.com/Document/N70971822025C11DD8320AE42787FBF1D/View/ FullText.html?navigationPath=Search%2Fv3%2Fsearch%2Fresults%2Fnavigation%2 Fi0ad604060000013e0a54ff0c782ccf16%3FNav%3DANALYTICAL%26fragmentId entifier%3DN70971822025C11DD8320AE42787FBF1D%26startIndex%3D1%26con

textData%3D%2528sc.Search%2529%26transitionType%3DSearchItem&listSource=S earch&listPageSource=8c6c724c605a5ef6ee8141f11d80fc43&list=ANALYTICAL&rank =1&grading=na&sessionScopeId=b3cf604df3c02ddba2d366a7aa5672d7&origination Context=Search%20Result&transitionType=SearchItem&contextData=%28sc.Search%29.

61. Wechsler, RESTATEMENT (SECOND) OF TORTS, § 73, accessed at https://a. next.westlaw.com/Document/Ia62d216c662911dca51ecfdfa1ed2cd3/View/FullText. html?originationContext=documenttoc&transitionType=CategoryPageItem&context Data=(sc.Default).

62. Wechsler, RESTATEMENT (SECOND) OF TORTS, § 73 (Illustrations 1–4), accessed at https://a.next.westlaw.com/Document/Ia62d216c662911dca51ecfdfa1ed2 cd3/View/FullText.html?originationContext=documenttoc&transitionType=Cate goryPageItem&contextData=(sc.Default).

63. Wechsler, RESTATEMENT (SECOND) OF TORTS § 83, accessed at https://a.next. westlaw.com/Document/I47223a7c662911dca51ecfdfa1ed2cd3/View/FullText.html? originationContext=documenttoc&transitionType=CategoryPageItem&contextDa ta=(sc.Default).

64. The scripture states, in pertinent part: "'You shall love the Lord your God with all your heart, with all your soul, with all your strength, and with all your mind,' and 'your neighbor as yourself.'"

65. Wechsler, RESTATEMENT (SECOND) OF TORTS, § 314, accessed April 12, 2013, at https://a.next.westlaw.com/Document/I7937bb01662911dca51ecfdfa1ed2cd3/View/ FullText.html?originationContext=documenttoc&transitionType=CategoryPageItem &contextData=(sc.Default).

66. Wechsler, RESTATEMENT (SECOND) OF TORTS, § 314, accessed April 12, 2013, at https://a.next.westlaw.com/Document/I7937bb01662911dca51ecfdfa1ed2cd3/View/ FullText.html?originationContext=documenttoc&transitionType=CategoryPageItem &contextData=(sc.Default).

67. In the May 1998 series finale of *Seinfeld*, Jerry and friends, while visiting a small Massachusetts town, were all prosecuted and sentenced to one year in jail for making fun of, as opposed to helping, a man who was being robbed at gunpoint. Wechsler, RESTATEMENT (SECOND) OF TORTS, § 314, accessed April 12, 2013, at https:// a.next.westlaw.com/Document/I7937bb01662911dca51ecfdfa1ed2cd3/View/FullText. html?originationContext=documenttoc&transitionType=CategoryPageItem&context Data=(sc.Default).

68. Wechsler, RESTATEMENT (SECOND) OF TORTS, § 319, accessed April 12, 2013, at accessed April 12, 2013 at https://a.next.westlaw.com/Document/I7937bb01662911 dca51ecfdfa1ed2cd3/View/FullText.html?originationContext=documenttoc&transit ionType=CategoryPageItem&contextData=(sc.Default).

69. Wechsler, RESTATEMENT (SECOND) OF TORTS, § 20, accessed April 12, 2013, at https://a.next.westlaw.com/Document/I793793f5662911dca51ecfdfa1ed2cd3/View/ FullText.html?originationContext=documenttoc&transitionType=CategoryPageItem &contextData=(sc.Default).

70. Wechsler, RESTATEMENT (SECOND) OF TORTS, § 23, accessed April 12, 2013, at https://a.next.westlaw.com/Document/I15e147d3662911dca51ecfdfa1ed2cd3/View/ FullText.html?originationContext=documenttoc&transitionType=CategoryPageItem &contextData=(sc.Default).

71. Wechsler, RESTATEMENT (SECOND) OF TORTS, § 23, accessed April 12, 2013, at https://a.next.westlaw.com/Document/I15e147d3662911dca51ecfdfa1ed2cd3/View/ FullText.html?originationContext=documenttoc&transitionType=CategoryPageItem &contextData=(sc.Default).

72. Wechsler, Restatement (Second) of Torts, § 14 cmt. (c), accessed April 12, 2013, at https://a.next.westlaw.com/Document/Ia62d6f63662911dca51ecfdfa1ed2 cd3/View/FullText.html?originationContext=documenttoc&transitionType=Cate goryPageItem&contextData=(sc.Default).

73. Wechsler, Restatement (Second) of Torts, § 319 cmt. 4, accessed April 12, 2013, at accessed April 12, 2013 at https://a.next.westlaw.com/Document/I7937 bb01662911dca51ecfdfa1ed2cd3/View/FullTexthtml?originationContext=documenttoc &transitionType=CategoryPageItem&contextData=(sc.Default).

74. Wechsler, Restatement (Second) of Torts, § 319 cmt. 3, accessed April 12, 2013, at accessed April 12, 2013 at https://a.next.westlaw.com/Document/I7937bb 01662911dca51ecfdfa1ed2cd3/View/FullText.html?originationContext=documenttoc &transitionType=CategoryPageItem&contextData=(sc.Default).

75. Hoffer, Nation of Laws, 17 (discussing reform following mass litigation).

76. Friedman, Law in America, 43.

77. Herbert Wechsler, Chief Reporter, et al., Restatement (Third) of Torts: Phys. and Emotional Harm (American Law Institute: 2010), § 1, accessed April 12, 2013, at https://a.next.westlaw.com/Document/I3f41a4e4662211dca51ecfdfa1ed2cd3/ View/FullText.html?originationContext=documenttoc&transitionType=Category PageItem&contextData=(sc.Default).

78. Herbert Wechsler, Chief Reporter, et al., Restatement (Third) of Torts: Phys. and Emotional Harm (American Law Institute: 2010), § 1, accessed April 12, 2013, at https://a.next.westlaw.com/Document/I3f41a4e4662211dca51ecfdfa1ed2cd3/ View/FullText.html?originationContext=documenttoc&transitionType= CategoryPageItem&contextData=(sc.Default).

79. Herbert Wechsler, Chief Reporter, et al. Restatement (Second) of Con-tracts (American Law Institute: 1981), § 1, accessed at April 12, 2013, https://a.next. westlaw.com/Document/I4d6ce0714a5511de9b8c850332338889/View/FullText. html?originationContext=documenttoc&transitionType=CategoryPageItem&conte-xtData=(sc.Default).

80. Wechsler, Restatement (Second) of Contracts, § 12, accessed at April 12, 2013, https://a.next.westlaw.com/Document/I27a7d94e4a5511de9b8c8503323388 89/View/FullText.html?originationContext=documenttoc&transitionType=Category PageItem&contextData=(sc.Default).

81. Wechsler, Restatement (Second) of Contracts § 15, accessed at April 12, 2013, https://a.next.westlaw.com/Document/I27a7d8fc4a5511de9b8c850332338889/ View/FullText.html?originationContext=documenttoc&transitionType=Category PageItem&contextData=(sc.Default).

82. Wechsler, Restatement (Second) of Contracts § 15, accessed at April 12, 2013, https://a.next.westlaw.com/Document/I27a7d8fc4a5511de9b8c850332338889/ View/FullText.html?originationContext=documenttoc&transitionType=Category PageItem&contextData=(sc.Default).

83. Present law holds a valid contract voidable upon a contracting party's illness only "[i]f the existence of a particular person is necessary for the performance of a duty [and] his death or such incapacity [ ] makes performance impracticable...." Wechsler, Restatement (Second) of Contracts § 262, accessed at April 12, 2013, https://a.next.westlaw.com/Document/I4d6ce00b4a5511de9b8c850332338889/View/ FullText.html?originationContext=documenttoc&transitionType=CategoryPageItem &contextData=(sc.Default.

84. Wechsler, Model Penal Code, § 250.11, accessed at April 12, 2013, https:// a.next.westlaw.com/Document/I11a6d7fd4a4c11de9b8c850332338889/View/FullText.

html?originationContext=documenttoc&transitionType=CategoryPageItem&context Data=(sc.Default).

85. Wechsler, Model Penal Code, § 212.3, accessed at April 12, 2013, https:// a.next.westlaw.com/Document/N334EC530025C11DD8320AE42787FBF1D/View/ FullText.html?originationContext=documenttoc&transitionType=CategoryPageItem &contextData=(sc.Default).

86. Hoffer, Nation of Laws, 17. Similar to the War on Drugs in the 1980s and Prohibition in the 1920s, life with the living dead will result in an increased desire for litigation.

87. Hoffer, Nation of Laws, 15 (explaining "Every period of economic downturn [brings] an upturn in litigation, driven by creditors trying to recover from debtors").

88. Hoffer, Nation of Laws, 15 (explaining "Every period of economic downturn [brings] an upturn in litigation, driven by creditors trying to recover from debtors").

Bibliography

Ackerman, Bruce. *Before the Next Attack: Preserving Civil Liberties in an Age of Terrorism*. New Haven: Yale University Press, 2006.

Carp, Ronald A., and Ronald Tidham. *Judicial Process in America*. Washington, D.C.: CQ Press, 1996.

Friedman, Lawrence M. *Law in America: A Short History*. Modern Library Chronicles, 2002.

Garner, Bryan A., ed., *Black's Law Dictionary*, 9th ed. St. Paul: West, 2009. Accessed April 12, 2013. https://a.next.westlaw.com/Document/I970505e3bea311e08b05fdf1 5589d8e8/View/FullText.html?navigationPath=Search%2Fv3%2Fsearch%2Fre sults%2Fnavigation%2Fi0ad604060000013e0a4c9a36782cc48c%3FNav%3DBLA CKS%26fragmentIdentifier%3DI970505e3bea311e08b05fdf15589d8e8%26startIn dex%3D1%26contextData%3D%2528sc.Search%2529%26transitionType%3DSea rchItem&listSource=Search&listPageSource=95108e061603a1631cdf3b4191b42a4a &list=BLACKS&rank=1&grading=na&sessionScopeId=b3cf604df3c02ddba2d36 6a7aa5672d7&originationContext=Search%20Result&transitionType=SearchIte m&contextData=%28sc.Search%29.

Hoffer, Charles. *A Nation of Laws: America's Imperfect Pursuit of Justice*. Lawrence: University Press of Kansas, 2010.

Lee, Harper. *To Kill a Mockingbird*. Philadelphia: J.B. Lippincott, 1960.

Stolleis, Michael. *The Law Under the Swastika: Studies on Legal History in Nazi Germany*. Chicago: University of Chicago Press, 1998.

The Telegraph, Accessed April 13, 2013. http://www.telegraph.co.uk/news/newstopics/ howaboutthat/9721072/Britain-is-well-prepared-to-fight apocalyptic-zombie-invasion.html.

Wechsler, Herbert, Chief Reporter, et al., Model Penal Code. American Law Institute (1985). Accessed April 12, 2013. https://a.next.westlaw.com/Document/ N25A3CA9004C211DC9A5BEF4472C1061E/View/FullText.html?origination Context=documenttoc&transitionType=CategoryPageItem&contextData=(sc. Default).

_____. Restatement (Second) of Contracts. American Law Institute, 1981. Accessed April 12, 2013. https://a.next.westlaw.com/Document/I4d6ce0714a5511de9b8c85 0332338889/View/FullText.html?originationContext=documenttoc&transition Type=CategoryPageItem&contextData=(sc.Default).

_____. Restatement (Second) of Torts. American Law Institute, 1965. Accessed April 12, 2013. https://a.next.westlaw.comDocument/I4722d6bb662911dca51ecfd

fa1ed2cd3/View/FullText.html?originationContext=documenttoc&transition
Type=CategoryPageItem&contextData=(sc.Default).

_____. RESTATEMENT (THIRD) OF TORTS: PHYS. AND EMOTIONAL HARM. American
Law Institute, 2010. Accessed April 12, 2013. https://a.next.westlaw.com/Docu
ment/I3f41a4e466221ldca51ecfdfa1ed2cd3/View/FullText.html?originationCont
ext=documenttoc&transitionType=CategoryPageItem&contextData=(sc.De
fault).

Wikipedia, The Free Encyclopedia. Accessed November 3, 2013. http://en.wikipedia.
org/wiki/Keeping_Up_with_the_Kardashians.

A Sociologist Responds
to Zombies

David F. Steele

"Zombies scare me." This is what my wife said when I told her I was writing the sociology essay of this book. Given the popularity of zombies, I do not think she is the only one to fear them. You may be unfamiliar with the discipline of sociology and wondering what it contributes to a discussion about zombies. Being a sociologist, it should not surprise you that my response is "quite a lot."

A comprehensive definition of sociology by Jon Witt, a contemporary sociologist, states that sociology is "the systematic study of the relationship between the individual and society and of the consequence of difference."[1] One aspect of sociology examines how humans, past and present, have collectively created current social institutions (examples include the economy, religion, and the media), organizations, groups, and rules for interacting with each other. A corresponding aspect of sociology examines how these same social institutions, organizations, groups, and accepted social rules for interacting with each other influence the social actions of people. Sociologists are also concerned with social inequality which is identified in Witt's definition as the "consequence of difference."[2]

One can learn a great deal about society from zombies. Sociologically, the idea of zombies can be viewed as what Emile Durkheim, a classical sociologist, identified as a "collective representation." In *The Rules of Sociological Method and Selected Texts on Sociology and Its Method*, Durkheim explains that collective representations like "myths, popular legends, religious conceptions of every kind, moral beliefs, etc., express a different reality from individual reality."[3] Durkheim claims this different reality requires that "several individuals at the very least must have interacted together and the resulting combination must have given rise to some new production" which sociologists call a social fact or collective representation.[4] Collective representations are not reducible to any one individual and are an example of the existence of a collective, social reality.

The wedding ring provides an example of a collective representation. Many spouses expect their significant other to wear a wedding ring. They do not expect their significant other to wear it simply because it is a ring. They expect their significant other to wear it because of the cultural rules and values the "wedding" ring symbolizes. The cultural rules and values symbolized by the wedding ring exist beyond just one couple and have been socially agreed upon by numerous people to constitute a social reality.[5]

This leads to the question: Are zombies a collective representation? There is evidence to suggest that the idea of zombies is a collective representation as attributed to the work of Durkheim. Does this collective representation have a large permeation throughout society? Evidence suggests that it does. Currently, a popular example of zombies is the AMC television series *The Walking Dead* which originated from the Image Comics comic book of the same title.[6] According to Rob Moynihan in the February 11–24, 2013, issue of *TV Guide*, *The Walking Dead*'s "Season 3 premiered last October [2012] to a series high 10.9 million viewers and continued to break records for AMC during its eight-week run, until the show was eventually crowned the top scripted series on all of TV in the coveted 18–49 demographic."[7] According to the March 1, 2013, issue of *Entertainment Weekly*, episodes of *The Walking Dead* television series held nine slots on the iTunes Top 25 list.[8] In the March 2013 issue of *Previews* magazine *The Walking Dead #106* comic book was ranked number 19 of the "Best Top 100 Comics: Sellers" for January 2013 and is the first non–DC or non–Marvel comic book listed in the "Top 100."[9] *The Walking Dead* comic books in "trade paperback" form (collections of issues) held four slots on the "Top 10 Graphic Novels & Trade Paperbacks" list and *The Walking Dead Season 3 Zombie Poster* was number five of the "Top 5 Prints & Posters" list.[10]

Other examples of the current idea and meaning of zombies as a collective representation exist as well. Marvel Comics have produced the *Marvel Zombies* comic books.[11] The *Resident Evil* movie series, based on a video game, consists of five movies to date.[12] Max Brooks' books *The Zombie Survival Guide* and *World War Z: An Oral History of the Zombie War* (both a book and a movie) are also examples of the pervasiveness of the idea of zombies.[13]

The evidence presented suggests that zombies are indeed a collective representation for society as the idea of zombies and the cultural rules and values the idea of zombies embodies exist beyond any one individual and belong to society, making it a collective representation. If you were to randomly stop someone on the street in the United States and ask them about the examples presented, most people could provide information about at least one of them or general information about zombies. This chapter examines three questions associated with the cultural rules and values presented by the collective representation of zombies: (1) Do zombies represent a threat to individuality? (2) Could the zombie threat become real? and (3) How would people respond to a zombie threat?

Do Zombies Represent a Threat to Individuality?

Many sociologists and social scientists examine the relationship between the individual and society. One key question in sociology is if society is "making" people into whom they are and controlling their actions, do people really possess individuality? In other words, could the collective representation of zombies be a response to the fear of people becoming zombies? Or are people zombies already? The works of George Ritzer, Emile Durkheim, Erving Goffman, Robin Leidner, and Benjamin R. Barber are presented to examine these issues.

The discipline of sociology emerged during a period of great social upheaval in the late 1700s and early 1800s. Ritzer, a contemporary sociologist, identifies the social factors of "political revolutions," "the industrial revolution," "the rise of capitalism," "the rise of socialism," "feminism," "urbanization," "religious change," "the growth of science," "the enlightenment," and "the conservative reaction to the enlightenment" as elements leading to the emergence of sociology.[14] Zombies reflect a similar concern or fear of the possibility of social upheaval in the modern world as it would again bring so many of these cultural elements into question. The fear of social upheaval can be found in Durkheim's work related to mechanical and organic solidarity.[15] In a period of mechanical solidarity, people are basically all doing the same thing. This reduces one's dependency on others but requires everyone to think and act the same to avoid conflict. In a period of organic solidarity, people are not held together as much by thinking and acting the same way as they are by an interdependency created by the modern day division of labor. With organic solidarity people depend on others for survival. People expect food to be available for purchase in the grocery store and clothes to be available at the mall. The modern day division of labor affords people more individuality. Since everyone performs different types of specialized work people do not have to think the same way everyone else does, at least not to the degree one would if living during a period of mechanical solidarity. People working in different fields have different needs and wants than others. This allows people to formulate different ideas about the world without disrupting society. But what happens if the division of labor that people depend on in the modern world fails? This is a key concern raised by the idea of zombies. If it did fail the survivors would be forced to revert to a period of mechanical solidarity and would most likely find it difficult to give up the individuality they have attained during this period of organic solidarity. From movies such as George Romero's 1968 classic *Night of the Living Dead* to currently AMC's *The Walking Dead*, a common theme is the presentation of small groups of people, often thrown together by happenstance, who are left to their own means for survival.[16] With the breakdown of services from social institutions, the groups face difficulty surviving due not only to being pursued by zombies but because they lack the knowledge necessary for survival since modern culture is not held by any one individual but by society collectively. The connections people have to other groups are cut off and therefore the group to which one belongs is the only connection to others and the only means of survival. This is important as Georg Simmel, a classical sociologist, suggests individuality is increased in the modern world because people belong to many groups which weakens the power any one group has over a person.[17] Being limited to only one group provides the group greater influence over a person's actions and reduces individuality. In zombie stories expulsion from the group potentially equates to death as no other groups may be readily available to join.

Are people already zombies? Do people have individuality now? Goffman, a contemporary sociologist, in *The Presentation of Self in Everyday Life*, likened the action of individuals to a theatrical performance.[18] Even what Goffman terms a "cynical" performance, which is a performance conducted by a person who does not believe in their role, may become real. Goffman states an "illustration may be found in the raw recruit who initially follows army etiquette in order to avoid physical punishment and eventually comes to follow the rules so that his organization will not be shamed and his officers and fellow soldiers will respect him."[19] From Goffman's perspective most people spend a significant amount of time doing what has come to be expected of them and working to avoid embarrassment from a spoiled performance by conducting the roles they occupy as best as possible. Leidner, a contemporary sociologist, in *Fast Food, Fast Talk: Service Work and the Routinization of Everyday Life*, examines how the growth of service oriented jobs affect people.[20] There is a difference among the growing service sector jobs compared to others. Employers of "interactive service workers" do not want to just control the work process but "claim authority over many more aspects of workers' lives ... seeking to regulate workers' appearance, moods, demeanors, and attitudes."[21] This is done by imposing routines that workers must follow, particularly scripts on what to say and how to act toward customers. This situation also pressures customers to adopt scripts and act in a routine way.[22] Employers attempt to control not only how employees and customers act but also what they think and feel. For Leidner, this issue "presents dilemmas of identity."[23] There is concern that social forces, particularly the rules by which people live, turn people into zombies thus reducing individuality. Is work, which individuals spend a significant time doing, and other roles people perform day-to-day leading them to become mindless zombies?

Individuality appears to be threatened on both sides. If there is a reversion from organic solidarity to mechanical solidarity people's individuality will be diminished, however, if the routines and rules used to keep modern society functioning continue to expand into more and more aspects of people's lives individuality will be diminished as well. These two diametrical threats to individuality are presented by Barber, a contemporary political scientist, as they relate to democracy in his 1992 *Atlantic Monthly* article, "Jihad vs. McWorld" and correlates to the fear of individuality diminishing as portrayed by the idea of zombies.[24] According to Barber:

> Just beyond the horizon of current events lie two possible political futures—both bleak, neither democratic. The first is a retribalization of large swaths of humankind by war and bloodshed: a threatened Lebanonization of national states in which culture is pitted against culture, people against people, tribe against tribe—a Jihad in the name of a hundred narrowly conceived faiths against every kind of interdependence, every kind of artificial social cooperation and civic mutuality. The second is being borne in on us by the onrush of economic and ecological forces that demand integration and uniformity and that mesmerize the world with fast music, fast computers, and fast food—with MTV, Macintosh, and McDonald's, pressing nations into one commercially homogenous global network: one McWorld tied together by technology, ecology, communications, and commerce. The planet is falling precipitately apart *AND* coming reluctantly together at the very same moment.[25]

The world of zombies provides an eerily strong reflection to the tension between the forces identified by Barber. Zombies can be viewed as the representation of ultimate uniformity, literally consuming individuality, while the remaining groups of humans struggle against the zombie hoard and possibly other groups for survival.

Could the Zombie Threat Become Real?

Could the zombie threat become real? It is improbable, but in the modern world it cannot be ruled out. To address this question the work of Anthony Giddens, a contemporary sociologist, is examined. The possibility of the social creation of a catastrophic problem, as exemplified by the idea of a zombie threat, illustrates how risk is an important and difficult part of modern society.[26] Giddens suggests modern society should be viewed as a juggernaut-a runaway engine of enormous power which, collectively as human beings, we can drive to some extent but which also threatens to rush out of our control and which could rend itself asunder. The juggernaut crushes those who resist it, and while it sometimes seems to have a steady path, there are times when it veers away erratically in directions we cannot foresee.... But, so long as the institutions of modernity endure, we shall never be able to control completely either the path or the pace of the journey. In turn, we shall never be able to feel entirely secure, because the terrain across which it runs is fraught with risks of high consequence.[27]

During this period of modernity people live in a constant state between "trust and risk, opportunity and danger—these polar, paradoxical features of modernity permeate all aspects of day-to-day life."[28] Risk can never be fully resolved because modern society operates with a focus on science, not tradition. This leads to a state of reflexivity where "social practices are constantly examined and reformed in the light of incoming [new] information" and "we can never be sure that any given element of knowledge will not be revised."[29] In the modern world, there is the possibility of something getting out of control that is "life-threatening for millions of human beings and potentially for the whole of humanity."[30]

The world has experienced pandemics or global epidemics that have killed millions. The 1918 Influenza Pandemic is one of the most deadly of the 20th century as it is estimated to have killed "between 30 and 50 million" people around the world.[31] The documentary *The Truth Behind Zombies* identified other recent pandemics such as HIV, avian flu, and SARS.[32] The documentary also examined the possibility of a zombie virus. While the documentary finds that a zombie virus is unlikely, it does indicate that it is not impossible. Samita Andreansky, a virologist at the University of Miami's Miller School of Medicine, discusses the possibility of the rabies virus mutating into a deadly airborne virus or the rabies virus being genetically modified with the influenza virus. Andreansky states that "hypothetically, one day we can create a zombie virus on a dish."[33]

What problems with modern society lead to the possibility of catastrophe? Giddens identified four general factors that could lead to modern society veering out of control: (1) "design faults"—the possibility that there is a flaw in the system, (2) "operator failure"—the possibility that the people who operate the systems may make a mistake, (3) "unintended consequences"—modern systems are so complex no government, organization, group, or person can foresee all of the consequences of initiating an action or how other parts of the system will respond, and (4) "reflexivity or circularity of social knowledge"—as researchers study the social world new knowledge continues to change the world in new ways.[34]

The four factors identified by Giddens are presented in zombie stories. The graphic novel *28 Days Later: The Aftermath* explains that the creation of the zombie creating rage virus was due to the efforts of two scientists to control anger and aggression in humans.[35] A design fault occurred when they combined the treatment with Ebola genomes to act as a delivery agent which created the opposite effect of what they were trying to produce.[36] *The Return of the Living Dead* and *Return of the Living Dead Part II* movies are examples of operator failure as it occurs when someone is infected due to an accidental release of a zombie creating chemical.[37] In *The Return of the Living Dead* canisters of the zombie creating chemical are incorrectly sent to a medical supply company due to an Army administrative error and then accidentally released years later by warehouse employees.[38] In *Return of the Living Dead Part II* the Army is moving the zombie creating chemical through a town and the drivers of a truck containing the zombie creating chemical fail to notice when three barrels fall out of the truck, which eventually results in a zombie outbreak.[39] Another example of operator failure is how the rage virus is released by protesters who are able to overcome security measures in a lab to free infected animals which bite people leading to the spread of the rage virus in the movie *28 Days Later* and the graphic novel *28 Days Later: The Aftermath*.[40] There is often the unintended consequence of a large number of people becoming infected with a zombie creating agent, turning them into zombies who attack people which overwhelms all social systems. An example of this unintended consequence is the release of the zombie creating t-virus in Raccoon City in the first *Resident Evil* movie in which the social systems cannot stop the t-virus from eventually spreading across the world.[41] The circularity of social knowledge occurs as the reality of zombies changes how the social world is defined. For example, in the movie *Resident Evil: Retribution* it is explained that the Umbrella Corporation "derived its primary income from the sale of viral weaponry" and created a new arms race with countries wanting to purchase the biological weapons created.[42] Once world leaders gained knowledge of the existence of the t-virus and its ability to create biological weapons, it changed their perception of the world and they wanted access to the virus.

When disasters occur it is often the result of one or a combination of the factors identified by Giddens. The documentary *The Truth Behind Zombies* states the Zombie Research Society "use[s] the zombie apocalypse idea as a metaphor for any ... disaster."[43] This leads to the question: how would people respond to a zombie threat?

How Would People Respond to a Zombie Threat?

While the zombie threat would affect people on an individual level (individual trauma), it would also affect people on a social level (cultural trauma/

collective trauma).[44] Cultural trauma is different than individual trauma as it affects the majority of people in a community and disrupts the system of social support. Contemporary sociologists Johanna Nurmi, Pekka Räsänen, and Atte Oksanen examined the issue of cultural trauma related to a high school shooting in Jokela, Finland.[45] They identified four factors associated with cultural trauma: (1) the incident is sudden, (2) the event is perceived by residents as "unexpected, shocking and repulsive," (3) "everyday social interactions, especially people's sense of security," are disrupted, and (4) the event usually is associated with a social origin instead of an accident or natural disaster.[46]

These factors are common in zombie stories. For example, in the movie *World War Z* the outbreak of a zombie attack suddenly occurs in Philadelphia, Pennsylvania.[47] It is perceived as shocking and repulsive as zombies run through the streets randomly attacking people. The outbreak disrupts social interaction and people's sense of security as they do not understand why it is occurring or what to do. It is also somewhat associated with a social origin as people are often attacked by family members who have turned into zombies, the very people they depend on most for security.[48]

Social scientists have found evidence that social solidarity or strong community focus is common during times of emergency, catastrophe, or disaster.[49] Examining natural disasters contemporary sociologists E.L. Quarantelli and Russell R. Dynes identified "seven factors that are associated with the absence of community conflict in a natural disaster situation."[50] The seven factors are (1) "natural disasters involve an external threat," (2) "in almost all natural disaster situations, the disaster agent can generally be perceived and specified," (3) "there is high consensus on priorities in natural disasters ... the saving of lives takes precedence over anything else," (4) "natural disasters almost by definition create community-wide problems that need to be quickly solved," (5) "disasters lead to a focusing of attention on the present, at least in the emergency period, as the past and the future are temporarily laid aside," (6) "there is a leveling of social distinctions in disaster situations," and (7) "disasters strengthen community identification" by "(a) creating a dramatic event in the life history of the community; and (b) allowing wide opportunities for participation in community-relevant activities."[51]

These seven factors that support an absence of community conflict in natural disasters would be expected in a zombie outbreak. The one factor that might not function is the first factor, viewing the zombie outbreak as an "external threat" since people in the community would be turned into zombies, however, a case can be made that zombies would quickly be defined as an external threat since they are dead and attacking the living. The other factors fit readily with establishing social solidarity in a zombie outbreak. The zombies would be easily identified as the disaster agent. Keeping people from being killed or bitten would take priority, and the community would directly recognize stopping the zombie threat as the problem that needs to be solved as quickly as possible. The focus would be on working together to survive to the next moment, requiring social distinctions and old rivalries to be put aside, at least for the short term, leading to a strengthening of community identification.

Zombie stories often include examples of conflicts before the zombie threat and how they are set aside during the zombie threat. For example, Magneto, a Marvel Comics mutant character often presented as viewing humans as inferior to mutants, forms a group with humans and protects them from the superheroes who have become zombies.[52] If someone does not set aside a conflict they are usually punished for it in a zombie story to serve as a cultural cautionary tale. In *The Walking Dead* comic book, Shane puts his position in the group and his attraction to Rick's wife Lori ahead of the interests of the group, which leads to Shane's death.[53] Zombie stories do not often continue with the efforts to rehabilitate the community and the conflicts that emerge in the restoration period after the crisis is over. Quarantelli and Dynes indicate that a common pattern is "the absence of conflict in the emergency period and its presence in the post-emergency period" as at some point a "new normal" is established which "provides the setting for the emergence of conflict."[54] Social conflict can occur as a result of an emergency period, and organizations, particularly nonlocal organizations, and the people involved may be labeled with the social identity of "outsiders" with the potential for conflict.[55] An emergency period may also lead to the development of new social divisions.[56] *In the Flesh* is a television series on BBC America that examines the social conflicts created after a zombie apocalypse is quelled.[57] The show examines such issues as zombies now being medically treated and labeled as suffering from "partially deceased syndrome," the prejudice and discrimination encountered due to the stigma of this condition, the efforts of the government to regain social control, the reluctance of the "human volunteer force" to give up its power, and some of the undead viewing themselves as a liberation force.[58]

What would make it possible for groups to resolve conflicts during and after a zombie threat? Contemporary sociologists John D. DeLamater and Daniel J. Myers acknowledge that the "insider versus outsider" identification is difficult to overcome but recognize four techniques that may lead to resolving conflicts: (1) "superordinate goals," (2) "intergroup contact," (3) "mediation/third-party intervention," and (4) "unilateral conciliatory initiatives."[59] Superordinate goals are needs that both groups value but cannot achieve alone so they must work together, which restructures relationships, reduces stereotypes, and may lead to the emergence of a common superordinate identity.[60] Intergroup contact in itself may reduce conflict as people begin to communicate with each other and see one another as a person instead of as a stereotype.[61] Mediation may be possible if a mediator or arbitrator that each group trusts could be found.[62] Unilateral conciliatory initiatives is the idea that if one group concedes to the other group on an issue the other group will come to see the first group with less aggression and be willing to reciprocate, which could lead to other concessions by each group and the reduction of hostilities.[63] The two types of conflict resolution most common in zombie movies are (1) the superordinate goal of surviving the zombies and (2) intergroup contact as people are "thrown together during the crisis" and come to know each other beyond a stereotype. In *The Walking Dead* comic book, Hershel Green and survivors at his farm joined Rick and his companions living within a prison and helped with its maintenance as living and working together provided greater protection against the zombies indicating a superordinate goal.[64] The movie *Detention of the Dead* illustrates intergroup contact as the smart kid, cheerleader, bully, athlete, stoner, and goth must work together against a zombie threat and come to see each other as unique individuals.[65]

Conclusion

This chapter examined the idea of zombies as a collective representation in social reality and how it invokes questions of individuality, the possibility of catastrophe, and how people might respond to a catastrophe. I started this chapter with my wife's comment that zombies scare her. The idea of zombies scares me too. The reason why is that disasters, especially the type of disaster represented by zombies, would have the potential to destroy communality or the close network of relationships people share as Kai T. Erikson, a contemporary sociologist, discovered with the 1972 Buffalo Creek Flood disaster.[66] Erikson finds communality important as he explains:

> The difficulty is that when you invest so much of yourself in that kind of social arrangement you become absorbed by it, almost captive to it, and the larger collectivity around you becomes an extension of your own personality, an extension of your own flesh. This means that not only are you diminished as a person when that surrounding tissue is stripped away, but that you are no longer able to reclaim as your own the emotional resources you invested in it. To "be neighborly" is not a quality you can carry with you into a new situation like negotiable emotional currency; the old community was your niche in the classic ecological sense, and your ability to relate to that niche is not a skill easily transferred to another setting.[67]

Joseph A. Tainter, a contemporary anthropologist, explained that societies solve problems by becoming more complex.[68] As a society solves previous problems it confronts new ones that are more difficult and costly to overcome. According to Tainter, a society may eventually be confronted with diminishing returns in regards to problem solving and without adequate energy and resources, collapse.[69] Previous historical collapses have resulted in "a great decline in population (up to 90 percent)."[70] If a problem such as a zombie apocalypse could not be solved many survivors may experience a chronic condition of disaster where they are constantly in a state of emergency and are not able to regain a sense of communality.[71] While older adults might never recover a sense of communality given their cultural awareness of a time before zombies, children would recognize this new "cultural order as a natural order" since it is the only world they have ever known.[72] From my cultural background and social experiences, I would consider this a difficult and frightening world, but the children of this new world would make it their own.

NOTES

1. Jon Witt, *SOC*, 2d ed. (New York: McGraw-Hill, 2011), 4.

2. Ibid., 8.

3. Emile Durkheim, Preface to the second edition to *The Rules of Sociological Method and Selected Texts on Sociology and Its Method* by Emile Durkheim, trans. W.D. Halls (1895; rpt., New York: Free Press, 1982), 41.

4. Ibid., 45.

5. Michael Mayerfeld Bell uses a similar example of the wedding ring in explaining the Hau in his book, *An Invitation to Environmental Sociology*, 4th ed. (Thousand Oaks: Pine Forge Press, 2012), 53.

6. *The Walking Dead*, 2010– , New York: AMC Studios, television series. *The Walking Dead*, 2003– , Berkeley: Image Comics, comic book series.

7. Rob Moynihan, "To Live and Die in GA.," *TV Guide*, February 11–24, 2013, 18.

8. "The Chart," *Entertainment Weekly*, March 1, 2013, 57.

9. "Best Top 100 Comics: Sellers," *Previews*, March 2013, FS-8.

10. "Top 10 Graphic Novels & Trade Paperbacks," *Previews*, March 2013, FS-9; "Top 5 Prints & Posters," *Previews*, March 2013, FS-9.

11. For brevity, Marvel zombie stories are cited using the *hardcover* editions by first year published when available: Robert Kirkman and Sean Phillips, *Marvel Zombies*, Collecting *Marvel Zombies* #1–5 (New York: Marvel, 2006); Reginald Hudlin and Francis Portella, *Black Panther: Four the Hard Way* (trade paperback—not hardcover), Collecting *Black Panther* #26–30 (#27–30 include zombies), (New York: Marvel, 2007); John Layman and Fabiano Neves, *Marvel Zombies Vs. Army of Darkness*, Collecting *Marvel Zombies Vs. Army of Darkness* #1–5 (New York: Marvel, 2007); Mark Millar, Greg Land and Mitch Breitweiser, *Ultimate Fantastic Four Vol. 3*, Collecting *Ultimate Fantastic Four* #21–32 (#21–23 & #30–32 include zombies), (New York: Marvel, 2007); Robert Kirkman and Sean Phillips, *Marvel Zombies 2*, Collecting *Marvel Zombies 2* #1–5 (New York: Marvel, 2008); Fred Van Lente and Kev Walker, *Marvel Zombies 3*, Collecting *Marvel Zombies 3* #1–4 (New York: Marvel, 2009); Fred Van Lente and Kev Walker, *Marvel Zombies 4*, Collecting *Marvel Zombies 4* #1–4 (New York: Marvel, 2009); Fred Van Lente and Kano, *Marvel Zombies 5*, Collecting

Marvel Zombies 5 #1–5 (New York: Marvel, 2010); Fred Van Lente, David Wellington, Jonathan Maberry, Seth Grahame-Smith, Nick Dragotta, Andrea Mutti, Jason Shawn Alexander, Richard Elson, and Wellinton Alves, *Marvel Zombies Return*, Collecting *Marvel Zombies Return* #1–5 (New York: Marvel, 2010); Jim McCann, David Baldeon, and Jeremy Treece, *Marvel Zombies Christmas Carol*, Collecting *Marvel Zombies Christmas Carol* #1–5 (New York: Marvel, 2011); Frank Marraffino and Fernando Blanco, *Marvel Zombies Supreme*, Collecting *Marvel Zombies Supreme* #1–5 (New York: Marvel, 2011); Frank Marraffino, Peter David, Mirco Pierfederici and Al Barrionuevo, *Marvel Zombies Destroy!*, Collecting *Marvel Zombies Destroy!* #1–5 (New York: Marvel, 2012); Single issue Marvel zombie comic book stories by first year published include Robert Kirkman and Sean Phillips, *Marvel Zombies: Dead Days* (New York: Marvel, 2007); Karl Kesel and Roberto Di Salvo, *Marvel Zombies: Evil Evolution* (New York: Marvel, 2009); Fred Van Lente and Alessandro Vitti, *Marvel Zombies: Halloween* (New York: Marvel, 2012).

12. *Resident Evil* movies presented by first year released: *Resident Evil*, directed by Paul W.S. Anderson, 2002, Screen Gems, motion picture; *Resident Evil: Apocalypse*, directed by Alexander Witt, 2004, Screen Gems, motion picture; *Resident Evil: Extinction*, directed by Russell Mulcahy, 2007, Screen Gems, motion picture; *Resident Evil: Afterlife*, directed by Paul W.S. Anderson, 2010, Screen Gems, motion picture; *Resident Evil: Retribution*, directed by Paul W.S. Anderson, 2012, Screen Gems, motion picture. The *Resident Evil* movies are based on the *Resident Evil* video games by Capcom.

13. Max Brooks, *The Zombie Survival Guide: Complete Protection from the Living Dead* (New York: Broadway Paperbacks, 2003); Max Brooks, *World War Z: An Oral History of the Zombie War* (New York: Broadway Paperbacks, 2006); *World War Z*, directed by Marc Foster, 2013, Paramount Pictures, 2013, DVD.

14. George Ritzer, *Classical Sociological Theory*, 5th ed. (New York: McGraw Hill, 2008), 6 (quotation); ibid., 7 (quotation); ibid., 7 (quotation); ibid., 7 (quotation); ibid., 8 (quotation); ibid., 8 (quotation); ibid., 9 (quotation); ibid., 9 (quotation); ibid., 10 (quotation); ibid., 11 (quotation).

15. Emile Durkheim, *The Division of Labor in Society*, trans. W.D. Halls (1893; rpt., New York: Free Press, 1997), 85.

16. *Night of the Living Dead*, directed by George A. Romero, 1968, Image Ten, motion picture; *The Walking Dead*, AMC Studios, television series.

17. Georg Simmel, "Group Expansion and the Development of Individuality," in *Georg Simmel on Individuality and Social Forms*, ed. Donald N. Levine (1908; rpt., Chicago: University of Chicago Press, 1971), 252.

18. Erving Goffman, *The Presentation of Self in Everyday Life* (Garden City, NY: Doubleday Anchor Books, 1959).

19. Ibid., 20.

20. Robin Leidner, *Fast Food, Fast Talk: Service Work and the Routinization of Everyday Life* (Berkeley: University of California Press, 1993).

21. Ibid., 2.

22. Ibid., 6.

23. Ibid., 12.

24. Benjamin R. Barber, "Jihad vs. McWorld," *The Atlantic*, March 1, 1992, accessed April 14, 2013, http://www.theatlantic.com/magazine/archive/1992/03/jihad-vs-mcworld/303882/; Barber also published the book *Jihad vs. McWorld: How Globalism and Tribalism Are Shaping the World* (New York: Ballantine, 1996).

25. Barber, "Jihad vs. McWorld," para. 1.

204 ...But If a Zombie Apocalypse *Did* Occur

26. Anthony Giddens, *The Consequences of Modernity* (Stanford: Stanford University Press, 1990).

27. Ibid., 139.

28. Ibid., 148.

29. Ibid., 38 (quotation); ibid., 39 (quotation).

30. Ibid., 131.

31. "The Great Pandemic: The United States—The Pandemic," United States Department of Health and Human Services, accessed November 11, 2013, http://www.flu.gov/pandemic/history/1918/the_pandemic/index.html, para. 6.

32. *The Truth Behind Zombies*, directed by Michael Wafer, 2010, National Geographic Channels (Produced by Zig Zag Productions for National Geographic Channels), documentary.

33. Ibid.

34. Giddens, *The Consequences of Modernity*, 151 (quotation); ibid., 152 (quotation); ibid., 152–153 (quotation); ibid., 153 (quotation).

35. Steve Niles, Dennis Calero, Diego Olmos, and Nat Jones, *28 Days Later: The Aftermath* (New York: Fox Atomic Comics released through HarperCollins, 2007).

36. Ibid.

37. *The Return of the Living Dead*, directed by Dan O'Bannon, 1985, Metro-Goldwyn-Mayer, motion picture; *Return of the Living Dead Part II*, directed by Ken Wiederhorn, 1988, Warner Bros., motion picture.

38. *The Return of the Living Dead*, directed by Dan O'Bannon.

39. *Return of the Living Dead Part II*, directed by Ken Wiederhorn.

40. *28 Days Later*, directed by Danny Boyle, 2002, Fox Searchlight Pictures, motion picture; Niles et al., *28 Days Later: The Aftermath*.

41. *Resident Evil*, directed by Paul W.S. Anderson; *Resident Evil: Apocalypse*, directed by Alexander Witt; *Resident Evil: Extinction*, directed by Russell Mulcahy.

42. *Resident Evil: Retribution*, directed by Paul W.S. Anderson.

43. *The Truth Behind Zombies*, directed by Michael Wafer.

44. Jeffery C. Alexander, "Toward a Theory of Cultural Trauma," in *Cultural Trauma and Collective Identity*, eds. Jeffery C. Alexander, Ron Eyerman, Bernard Giesen, Neil J. Smelser, and Piotr Sztompka (Berkeley: University of California Press, 2004); Kai T. Erikson, *Everything in Its Path: Destruction of Community in the Buffalo Creek Flood* (New York: Simon & Schuster, 1976).

45. Johanna Nurmi, Pekka Räsänen, and Atte Oksanen, "The Norm of Solidarity: Experiencing Negative Aspects of Community Life After a School Shooting Tragedy," *Journal of Social Work* 12, no. 3 (2012), doi:10.1177/1468017310386426.

46. Ibid., 304; ibid., 304 (quotation); ibid., 304 (quotation); ibid., 304.

47. *World War Z*, directed by Marc Foster.

48. Ibid.

49. James Hawdon, John Ryan, and Laura Agnich, "Crime as a Source of Solidarity: A Research Note Testing Durkheim's Assertion," *Deviant Behavior* 31, no. 8 (2010), doi:10.1080/01639620903415901; E.L. Quarantelli and Russell R. Dynes, "Community Conflict: Its Absence and Its Presence in Natural Disasters," *Mass Emergencies* 1, no. 2 (1976), http://massemergencies.org/v1n2/Quarantelli_v1n2.pdf; Nurmi, Räsänen, and Oksanen, "The Norm of Solidarity: Experiencing Negative Aspects of Community Life After a School Shooting Tragedy."

50. Quarantelli and Dynes, "Community Conflict: Its Absence and Its Presence in Natural Disasters," 141.

51. Ibid., 141 (quotation); ibid., 142 (quotation); ibid., 142 (quotation); ibid., 142

(quotation); ibid., 142 (quotation); ibid., 143 (quotation); ibid., 143 (quotation); ibid., 143 (quotation).

52. Mark Millar and Greg Land, "Crossover: Part 2," *Ultimate Fantastic Four* no. 22 (2005).

53. Robert Kirkman and Tony Moore, *The Walking Dead, Volume 1: Days Gone Bye*, Collecting *The Walking Dead* #1–6 (Berkeley: Image Comics, 2012).

54. Quarantelli and Dynes, "Community Conflict: Its Absence and its Presence in Natural Disasters," 140–141 (quotation); ibid., 145 (quotation).

55. Matthew Carroll, Lorie L. Higgins, Patricia J. Cohn, and James Burchfield, "Community Wildfire Events as a Source of Social Conflict," *Rural Sociology* 71, no. 2 (2006), doi:10.1526/003601106777789701; Quarantelli and Dynes, "Community Conflict: Its Absence and Its Presence in Natural Disasters"; Nurmi, Räsänen, and Oksanen, "The Norm of Solidarity: Experiencing Negative Aspects of Community Life After a School Shooting Tragedy"; Carroll et al., "Community Wildfire Events as a Source of Social Conflict," 276 (quotation); Quarantelli and Dynes, "Community Conflict: Its Absence and Its Presence in Natural Disasters," 144 (quotation).

56. Quarantelli and Dynes, "Community Conflict: Its Absence and Its Presence in Natural Disasters," 145; 147; Nurmi, Räsänen, and Oksanen, "The Norm of Solidarity: Experiencing Negative Aspects of Community Life After a School Shooting Tragedy," 311–313.

57. *In the Flesh*, 2013– , BBC America, television series.

58. Ibid.

59. John D. DeLamater and Daniel J. Myers, *Social Psychology*, 6th ed. (Belmont, CA: Thomson Higher Education, 2007), 424 (quotation); ibid., 425 (quotation); ibid., 426 (quotation); ibid., 429 (quotation).

60. Ibid., 424–425.

61. Ibid., 425–426.

62. Ibid., 426–429.

63. Ibid., 429–430.

64. Robert Kirkman, Tim Daniel, Charlie Adlard, and Cliff Rathburn, "Hershel," *The Walking Dead Survivor's Guide* (Berkeley: Image Comics, 2012).

65. *Detention of the Dead*, directed by Alex Craig Mann, 2012, Gala Films, motion picture.

66. Erikson, *Everything in Its Path: Destruction of Community in the Buffalo Creek Flood*, 187.

67. Ibid., 191.

68. Joseph A. Tainter, "Sustainability of Complex Societies," *Futures* 27, no. 4 (1995), doi:10.1016/0016-3287(95)00016-P.

69. Ibid., 399.

70. Ibid.

71. Erikson, *Everything in Its Path: Destruction of Community in the Buffalo Creek Flood*, 255.

72. Tainter, "Sustainability of Complex Societies," 397.

Bibliography

Alexander, Jeffery C. "Toward a Theory of Cultural Trauma." In *Cultural Trauma and Collective Identity*, edited by Jeffery C. Alexander, Ron Eyerman, Bernard Giesen, Neil J. Smelser, and Piotr Sztompka, 1–30. Berkeley: University of California Press, 2004.

Barber, Benjamin R. "Jihad vs. McWorld." *The Atlantic*, March 1, 1992. http://www.theatlantic.com/magazine/archive/1992/03/jihad-vs-mcworld/303882/.

_____. *Jihad vs. McWorld: How Globalism and Tribalism Are Shaping the World*. New York: Ballantine, 1996.

Bell, Michael Mayerfeld. *An Invitation to Environmental Sociology*, 4th ed. Thousand Oaks: Pine Forge Press, 2012.

"Best Top 100 Comics: Sellers." *Previews*, March 2013.

Brooks, Max. *World War Z: An Oral History of the Zombie War*. New York: Broadway Paperbacks, 2006.

_____. *The Zombie Survival Guide: Complete Protection from the Living Dead*. New York: Broadway Paperbacks, 2003.

Carroll, Matthew, Lorie L. Higgins, Patricia J. Cohn, and James Burchfield. "Community Wildfire Events as a Source of Social Conflict." *Rural Sociology* 71, no. 2 (2006): 261–280. doi:10.1526/003601106777789701.

"The Chart," *Entertainment Weekly*, March 1, 2013.

DeLamater, John D., and Daniel J. Myers. *Social Psychology*, 6th ed. Belmont, CA: Thomson Higher Education, 2007.

Detention of the Dead. Directed by Alex Craig Mann. 2012, Gala Films. Motion picture.

Durkheim, Emile. *The Division of Labor in Society*. 1893. Translated by W.D. Halls. Reprint, New York: Free Press, 1997.

_____. Preface to the second edition to *The Rules of Sociological Method and Selected Texts on Sociology and Its Method*, by Emile Durkheim, 34–47. 1895. Translated by W.D. Halls. Reprint, New York: Free Press, 1982.

Erikson, Kai T. *Everything in Its Path: Destruction of Community in the Buffalo Creek Flood*. New York: Simon & Schuster, 1976.

Giddens, Anthony. *The Consequences of Modernity*. Stanford: Stanford University Press, 1990.

Goffman, Erving. *The Presentation of Self in Everyday Life*. Garden City, NY: Doubleday Anchor Books, 1959.

"The Great Pandemic: The United States in 1918–1919: The Pandemic." United States Department of Health and Human Services. Accessed November 11, 2013. http://www.flu.gov/pandemic/history/1918/the_pandemic/index.html.

Hawdon, James, John Ryan, and Laura Agnich. "Crime as a Source of Solidarity: A Research Note Testing Durkheim's Assertion." *Deviant Behavior* 31, no. 8 (2010): 679–703. doi:10.1080/01639620903415901.

Hudlin, Reginald, and Francis Portella. *Black Panther: Four the Hard Way*. Paperback. Collecting *Black Panther* #26–30. New York: Marvel, 2007.

In the Flesh. 2013– . BBC America. Television series.

Kesel, Karl, and Roberto Di Salvo. *Marvel Zombies: Evil Evolution*. New York: Marvel, 2009.

Kirkman, Robert, Tim Daniel, Charlie Adlard, and Cliff Rathburn. "Hershel." *The Walking Dead Survivor's Guide*. Berkeley: Image Comics, 2012.

_____, and Tony Moore. *The Walking Dead, Volume 1: Days Gone Bye*. Collecting *The Walking Dead* #1–6. Berkeley, CA: Image Comics, 2012.

_____, and Sean Phillips. *Marvel Zombies*. Hardcover. Collecting *Marvel Zombies* #1–5. New York: Marvel Publishing, Inc., 2006.

_____. *Marvel Zombies: Dead Days*. New York: Marvel, 2007.

_____. *Marvel Zombies 2*. Hardcover. Collecting *Marvel Zombies 2* #1–5. New York: Marvel, 2008.

Layman, John, and Fabiano Neves. *Marvel Zombies Vs. Army of Darkness.* Hardcover. Collecting *Marvel Zombies Vs. Army of Darkness* #1–5. New York: Marvel, 2007.

Leidner, Robin. *Fast Food, Fast Talk: Service Work and the Routinization of Everyday Life.* Berkeley: University of California Press, 1993.

Marraffino, Frank, and Fernando Blanco. *Marvel Zombies Supreme.* Hardcover. Collecting *Marvel Zombies Supreme* #1–5. New York: Marvel, 2011.

_____, Peter David, Mirco Pierfederici and Al Barrionuevo. *Marvel Zombies Destroy!* Hardcover. Collecting *Marvel Zombies Destroy!* #1–5. New York: Marvel, 2012.

McCann, Jim, David Baldeon, and Jeremy Treece. *Marvel Zombies Christmas Carol.* Hardcover. Collecting *Marvel Zombies Christmas Carol* #1–5. New York: Marvel, 2011.

Millar, Mark, and Greg Land. "Crossover: Part 2." *Ultimate Fantastic Four* no. 22. New York: Marvel, 2005.

_____, _____, and Mitch Breitweiser, *Ultimate Fantastic Four Vol. 3.* Hardcover. Collecting *Ultimate Fantastic Four* #21–32. New York: Marvel, 2007.

Moynihan, Rob. "To Live and Die in GA." *TV Guide*, February 11–24, 2013.

Night of the Living Dead. Directed by George A. Romero. 1968. Image Ten. Motion picture.

Niles, Steve, Dennis Calero, Diego Olmos, and Nat Jones. *28 Days Later: The Aftermath.* New York: Fox Atomic Comics released through HarperCollins, 2007.

Nurmi, Johanna, Pekka Räsänen, and Atte Oksanen. "The Norm of Solidarity: Experiencing Negative Aspects of Community Life After a School Shooting Tragedy." *Journal of Social Work* 12, no. 3 (2012): 300–319. doi:10.1177/1468017310386426.

Quarantelli, E. L., and Russell R. Dynes. "Community Conflict: Its Absence and Its Presence in Natural Disasters." *Mass Emergencies* 1, no. 2 (1976): 139–152. http://massemergencies.org/v1n2/Quarantelli_v1n2.pdf.

Resident Evil. 1996. Capcom. Video game.

Resident Evil. Directed by Paul W.S. Anderson. 2002. Screen Gems. Motion picture.

Resident Evil: Afterlife. Directed by Paul W.S. Anderson. 2010. Screen Gems. Motion picture.

Resident Evil: Apocalypse. Directed by Alexander Witt. 2004. Screen Gems. Motion picture.

Resident Evil: Extinction. Directed by Russell Mulcahy. 2007. Screen Gems. Motion picture.

Resident Evil: Retribution. Directed by Paul W.S. Anderson. 2012. Screen Gems. Motion picture.

The Return of the Living Dead. Directed by Dan O'Bannon. 1985. Metro-Goldwyn-Mayer. Motion picture.

Return of the Living Dead Part II. Directed by Ken Wiederhorn. 1988. Warner Bros. Motion picture.

Ritzer, George. *Classical Sociological Theory*, 5th ed. New York: McGraw Hill, 2008.

Simmel, Georg. "Group Expansion and the Development of Individuality." 1908. In *Georg Simmel on Individuality and Social Forms*, edited by Donald N. Levine, 251–293. Reprint, Chicago: University of Chicago Press, 1971.

Tainter, Joseph A. "Sustainability of Complex Societies." *Futures* 27, no. 4 (1995): 397–407. doi:10.1016/0016-3287(95)00016-P.

The Truth Behind Zombies. Directed by Michael Wafer. 2010. National Geographic Channels (Produced by Zig Zag Productions for National Geographic Channels). Documentary.

"Top 5 Prints & Posters." *Previews*, March 2013.

"Top 10 Graphic Novels & Trade Paperbacks." *Previews*, March 2013.

28 Days Later. Directed by Danny Boyle. 2002. Los Angeles: Fox Searchlight Pictures. Motion Picture.

Van Lente, Fred. *Marvel Zombies 4*. Hardcover. Collecting *Marvel Zombies 4 #1–4*. New York: Marvel, 2009.

_____, and Kano. *Marvel Zombies 5*. Hardcover. Collecting *Marvel Zombies 5 #1–5*. New York: Marvel, 2010.

_____, and Alessandro Vitti. *Marvel Zombies: Halloween*. New York: Marvel, 2012.

_____, and Kev Walker. *Marvel Zombies 3*. Hardcover. Collecting *Marvel Zombies 3 #1–4*. New York: Marvel, 2009.

_____, David Wellington, Jonathan Maberry, Seth Grahame-Smith, Nick Dragotta, Andrea Mutti, Jason Shawn Alexander, Richard Elson, and Wellinton Alves, *Marvel Zombies Return*. Hardcover. Collecting *Marvel Zombies Return #1–5*. New York: Marvel, 2010.

The Walking Dead. 2003– . Berkeley, CA: Image Comics. Comic book series.

The Walking Dead. 2010– . AMC Studios. Television series.

Witt, Jon. *SOC*, 2d ed., 2011 ed. New York: McGraw Hill, 2011.

World War Z. Directed by Marc Foster. 2013. Paramount Pictures, 2013. DVD.

The Psychology of Surviving the Zombie Apocalypse

Scott Mirabile

Zombies and the zombie apocalypse have been portrayed in almost as many ways as there are books, comics, and movies on the topic. Zombie outbreaks have been depicted as geographically limited, such as Romero's *Night of the Living Dead*, and as full-blown world-wide pandemics, as in *World War Z* and *The Walking Dead*.[1] Likewise, zombies themselves have ranged from slow, unintelligent, and shuffling in *Night of the Living Dead* to highly mobile and energetic in *28 Days Later* and *Dawn of the Dead*.[2] Across most depictions of a zombie apocalypse, there are four common themes: the survivors are isolated from contact or assistance from the outside world, if one even still exists; the survivors are confined to an area of uncertain safety; the survivors face lethal threats to their well-being, sometimes from one another in addition to the zombies; and finally the survivors must attempt to cope with these stressors and adapt their behavior to the demands of their new reality.

The skills and preparation likely needed to survive the initial outbreak and find shelter have already been speculated on by others, notably by Max Brooks in *The Zombie Survival Guide*.[3] In a sense, this essay picks up where other authors leave off by assuming that humanity is reduced to isolated groups of humans living on the road or in questionably-secure compounds with limited contact between such groups. This chapter describes the probable psychological and interpersonal needs, stressors, and consequences of the zombie apocalypse for individuals fortunate (or unfortunate) enough to find themselves as survivors, grouped in tentative safety.

Given that humanity has never faced a zombie apocalypse, it may seem presumptuous to speculate on psychological and interpersonal challenges faced by survivors. Surely the zombie apocalypse is not the only setting in which humans have faced deadly, prolonged threats in relative or absolute isolation from one another. If one is to speculate about how humans will adapt to the zombie apocalypse threat, it would be wise to consider how humans have adapted to similar threats in the past. Indeed, human history is replete with examples of individuals or groups struggling to survive in hostile environments—notably soldiers at war, victims of natural disasters, and explorers of extreme environments. Each of these groups would seem to be reasonable analogs to zombie apocalypse survivors.

On Choosing a Reasonable Analog

Soldiers

Are soldiers reasonable analogs for zombie apocalypse survivors? Soldiers seem a natural choice, as they are likely to face deadly adversaries on a regular basis while living in confined, semi-secure conditions. Further, warfare requires that soldiers constantly adapt to changing ground conditions and enemy tactics. Similar to popular depictions of survivors of a zombie apocalypse, soldiers often experience many negative psychological consequences. Rates of depression, suicide, and post-traumatic stress disorder among active and retired military personnel are strikingly high.[4]

Soldiers, however, may not be ideal analogues for human adaptation in a zombie apocalypse. Unlike zombie apocalypse survivors, soldiers often have the support of numerous individuals and organizations wielding truly awesome technology such as long-range artillery, air support, surveillance drones, and the like. They also know that should they be injured, every comrade is trained in first aid and life-saving medical evacuation may be available. Additionally, whereas a hypothetical rag-tag band of survivors may have only limited training in first aid, firearms, and military tactics, soldiers are extensively conditioned and trained to combat the enemies they face on the battlefield. Finally, although soldiers often experience the same forms of mental health problems commonly depicted in zombie apocalypse survivors, soldiers have access to various forms of social and institutional support that are unlikely to exist in a zombie apocalypse.

Soldiers are not ideal analogues for zombie apocalypse survivors, but there are still lessons to be learned from this population. Many active and non-active duty soldiers struggle with post-traumatic stress disorder (PTSD) and depression, suggesting that these mental health issues are likely to be prevalent among less prepared and less well supported individuals facing daily threats to their safety. Indeed, popular depictions of zombie apocalypse survivors often include highly traumatized individuals experiencing multiple PTSD and depressive symptoms. In Romero's classic *Night of the Living Dead*, a female survivor of a zombie attack experiences intense horror and help-lessness when facing the undead and experiences numbness, shock, and hys-teria in response to losing her brother and seeing another survivor maul the undead attackers.[5] Unlike many popular depictions of a zombie apocalypse, AMC's *The Walking Dead* follows survivors long enough for the acute stress of the emerging zombie apocalypse to develop into *post*-traumatic stress dis-order and depression, as illustrated by Beth Greene's numbness, hopelessness, and suicidal behavior.[6]

Natural Disaster Survivors

Might natural disaster survivors be reasonable analogs for zombie apoc-alypse survivors? It would seem that survivors of extreme weather and geo-logical events would be obvious analogs for survivors of a zombie apocalypse. Indeed, natural disaster victims often experience the themes common in zombie lore: they are initially separated from outside sources of assistance; they may be confined to a small area, restrained by natural barriers or debris; they are faced with potentially life-threatening conditions such as limited food, water, and shelter; and they must struggle to adapt to such conditions until help arrives. Considering their struggle to adapt, survivors of large-scale natural disasters may experience various psychological problems, with PTSD being the most likely and most commonly studied outcome.[7] Unfortu-nately, relatively little is known about the long-term course of PTSD after nat-ural disasters.[8]

Unlike a zombie apocalypse, natural disasters are typically single-time-point events. Hurricanes do not linger indefinitely; earthquakes typically sub-side within minutes or hours; and floodwaters often recede within days or weeks. Although the economic, environmental, and sociopolitical impact of natural disasters may last for years, the period of time in which such disasters pose deadly threats to survivors has greatly diminished in developed coun-tries with large, sophisticated emergency response agencies. While the scale of the devastation of a large natural disaster such as Hurricane Katrina most closely approximates the systemic failures that would likely accompany a zombie apocalypse, the time-limited nature of such events and the fact that neighboring communities are available to assist such survivors rule them out as analogs for survivors of a zombie apocalypse. Though it is likely that the long term threats to health and safety posed by natural disasters in less devel-oped countries may closely resemble the stressors of a zombie apocalypse, too little research on the psychological consequences has been conducted in such populations.

It is, however, possible to draw lessons on the zombie apocalypse from survivors of natural disasters. Natural disaster survivors may closely approximate the average zombie apocalypse survivor in terms of their pre-disaster mental health and coping ability. This similarity suggests that rates of PTSD and other mental health problems among zombie apocalypse survivors may be at least as high as those experienced by natural disaster survivors.

Explorers

Can one learn about surviving a zombie apocalypse from pioneering explorers and scientists who willingly expose themselves to separation and isolation in deadly environments? The exploration of novel, often hostile environments may ultimately be the greatest ambition of humanity; and the study of such explorers may prove the most fruitful window into how humans might cope with a zombie apocalypse. Modern-day explorers such as astronauts and researchers living near the South Pole are the best analogs for zombie apocalypse survivors, as they share all four themes common in zombie lore.

First, modern explorers are separated from outside contact and material support. While explorers have the comfort of knowing that civilized society still exists and may even have the opportunity to regularly contact outside individuals, they are largely unable to gain physical or material help such as food and supplies from outside sources. Zombie apocalypse survivors may not know whether civilized society still exists and, in any event, are unlikely to receive aid from outside sources.

Second, many modern explorers live in confined settings. Accommodations in polar research stations vary from tents and huts to fairly modern quarters with bedrooms and kitchens; and astronauts in the International Space Station experience few familiar comforts while enduring relatively small living and working quarters. Though such settings are smaller than the settings depicted in zombie lore such as farms, malls, walled estates, and prisons, all such settings share the common feature of confinement—the relatively safe zone is restricted to a known area; and egress from that area is either impossible or highly hazardous.

Third, the modern explorer lives in an extreme and dangerous environment, one not typically experienced by humans or particularly suited to our survival. Astronauts and polar scientists understand that they are completely surrounded by a foreign, deadly environment and that their safety is not guaranteed.[9] Likewise the zombie apocalypse survivor understands that even in areas of relative safety, they are still surrounded by hordes of undead.

Finally, concerning the need for adaptation, modern explorers and scientists are not a perfect analog. Modern explorers are typically selected for their intelligence and adaptability, with the understanding that working in extreme environments demands flexibility and creative thinking.[10] Because such individuals are a highly selected group, it is likely that inferences drawn from research on such individuals may overestimate the adaptability and survivability of the average zombie apocalypse survivor. Despite this shortcoming, it is worth considering how humans function in isolated, confined, and extreme environments.

Human Performance in Isolated, Confined and Extreme Environments

Explorers have long endured and even sought out isolated, confined, and extreme (ICE) environments in the pursuit of prestige and knowledge; for example, Sir Ernest Henry Shackleton's failed voyage to the South Pole. Unfortunately, many such journeys provide mere anecdotal accounts of human adaptation to the stressors of ICE environments. Historically, researchers have neglected to study the psychological side of such enterprises as space flight and exploration for a variety of reasons ranging from the perceptions of engineers that psychology and psychiatry are "soft sciences" to our trusting in "the right stuff" of airmen and astronauts.[11] Until relatively recently most space flights have been of fairly short duration, which likely minimizes the frequency and severity of psychological stressors, thus limiting the need for research on human adaptation to space exploration.[12] More recently both anecdotal and empirical evidence from studies of the Shuttle-Mir Space Program, long-term Russian missions, and other ICE environments like research stations in Antarctica and space simulators have prompted an upsurge of interest in how humans adapt and interact in such conditions.[13] Indeed, the consistent consequences of depression, hostility, poor leadership, failures of communication, and simple human error have forced scientists to probe our ability to live and work in groups in ICE environments. Further, psychological aspects of human performance and coping can promote the health and well-being of the explorers or contribute to the utter failure of the mission. Given that in a zombie apocalypse, the "mission" may be the survival of the human race, studying how humans cope in ICE environments may ultimately prove to be incredibly important.

Researchers have identified three different domains of behavior that are affected by ICE environments: the intrapersonal domain, the interpersonal domain, and the organizational domain. Research in the intrapersonal domain has focused largely on stressors faced by individuals, individual differences in how persons cope with such stressors, and various mental health outcomes relevant to those stressors. Research in the interpersonal domain has considered social problems such as isolation and conflict, changing social dynamics, leadership, and communication. The organizational domain will not be considered here as it is primarily concerned with communication between the individuals in the ICE environment and their control base, such as NASA's Mission Control in Houston; and it seems unlikely that such communications would either be possible or of a similar nature in a zombie apocalypse.

Intrapersonal Domain

Basic Needs

Basic human needs such as food and shelter and higher-level needs such as affiliation and self-determination are clearly powerful motivators behind many common behaviors. It is safe to assume that humans in a zombie apocalypse will still strive to meet many of the same needs. While the drive to secure shelter and food is necessary to keep survivors alive, other higher-level motivations may actually impair positive adaptation. Indeed, the desire for affection from others, which is natural in every-day settings, is linked to lower task ability, poorer emotional stability, worse social compatibility, and worse overall performance in ICE environments. Many individuals in ICE environments find it difficult to balance their own need for affection and affirmation with others' need for personal space.[14] Further, the need to create order by establishing routines and rules is also inversely related to individuals' emotional stability and leadership quality. Likewise, individuals' need for achievement may interfere with their social compatibility. In ICE scenarios, environmental conditions likely determine the degree to which individuals' needs for order and achievement can be met; thus, people who adjust their expectations and behavior based on current conditions are most likely to be well adapted.

Unfortunately for zombie apocalypse survivors in an ICE environment, many of their higher-order human needs are unlikely to be satisfied. Survivors in ICE environments will likely experience separation from or loss of loved ones and limited or no communication with other survivors. Further, individuals in prolonged ICE environments such as space missions are likely to experience restricted privacy and personal space, territoriality of others, and boredom and social monotony arising from interactions with the same individuals every day.[15] Such findings based on the experiences of astronauts suggest that zombie apocalypse survivors also will experience similar issues. These stressors are commonly included in multiple literary and film depictions of the zombie apocalypse. In the popular television drama *The Walking Dead*, many of the survivors lost loved ones in the initial zombie outbreak, and the fear of losing additional loved ones is a major factor driving the decisions of group leaders. Additionally, the inability of Rick to contact other survivors using a walkie-talkie was highlighted regularly throughout season one. Clearly, the need to contact others is psychologically powerful, and significant stress is likely when that and other needs cannot be met.

Psychological Functioning and Outcomes

In response to such stress, zombie apocalypse survivors, much like explorers and scientists in ICE environments, are likely to suffer anxiety and depression, insomnia, irritability and anger, and reduced cognitive performance.[16] Whereas such outcomes are obviously problematic for individuals in non–ICE environments, they may pose critical threats to the safety, survival, and mission success of individuals in ICE environments. Indeed, zombie lore is full of examples of characters who "break" under the stress, often committing acts that endanger themselves and other survivors. For example, in the final episode of season one of *The Walking Dead*, Dr. Edwin Jenner and Jacqui commit suicide rather than risk facing a difficult life and a gruesome death. Though such self-destructive actions may be exaggerated for dramatic effect in zombie apocalypse lore, approximately half of individuals in ICE environments report serious, impairing (though technically "sub-clinical") problems with sleep, depression, irritability, memory, and concentration. Participants in the studies finding such high rates of problems at least have the comfort of knowing that their time in an ICE environment is limited; yet survivors of a zombie apocalypse are left with no such comfort. It is likely safe to assume that indefinite exposure to such stressors will have even stronger negative effects.

Some researchers have failed to find these effects or suggest that the impairments are minimally important to the health and safety of crews or success of missions in ICE environments.[17] There are three possible ways to reconcile the conflicting research findings. First, ICE environments are not uniquely stressful.[18] Second, the all-volunteer nature of scientific or exploratory ICE environment crews means that the individuals are uniquely capable of maintaining optimal performance.[19] And third, individual differences in personality, motivation, and other factors account for the discrepancy in individuals' outcomes.[20] While the research literature on ICE environment scientists and astronauts allows for these multiple explanations, some of the explanations simply do not apply to a zombie apocalypse ICE scenario. One can exclude the argument that a zombie apocalypse ICE environment is not "uniquely stressful," and one can argue that findings from all volunteer teams are likely to produce results substantially better than would be found from an average group of survivors. Thus one must consider what kinds of individual differences may account for success or failure in zombie apocalypse ICE environments.

Individual Differences

Although researchers had historically ignored the role of individual differences in how humans adapt to ICE environments, often trusting the "right stuff" of airmen and astronauts, the past few decades have seen increased interest in determining just what the "right stuff" is that enables some individuals to thrive and succeed under the pressures of an ICE environment. Researchers have grouped characteristics that predict astronaut effectiveness into three clusters: the "right stuff," the "wrong stuff," and "no stuff."[21] Individuals with the "right stuff" are highly goal-oriented and independent, high on interpersonal warmth and sensitivity, and have a strong desire to work hard and excel at their job. Individuals with the "right stuff" also are unlikely to be arrogant, hostile, fussy, or exhibit social vulnerability. It is no coincidence that the hero characters across many different examples of zombie lore also exhibit many of these strengths and lack most of these weaknesses. For example, Glenn in AMC's *The Walking Dead* is highly motivated to perform well as a group member and is still able to exhibit warmth and sensitivity in his romantic relationship with Maggie. In sharp contrast, individuals with the "wrong stuff" are highly competitive, arrogant, hostile, impatient and irritable, and are unlikely to demonstrate warmth and sensitivity. As a case-study in the "wrong stuff," consider Shane, best friend and former police partner to Rick, in *The Walking Dead*. Although Shane attempts to lead, his competitive relationship with Rick and his arrogance, hostility, and lack of sensitivity to the needs of the larger group ultimately undermine his bid for power. Finally, some individuals have "no stuff"; they exhibit low levels of warmth, sensitivity, goal-orientation and independence and high levels of self-subordination, subservience, and verbal aggressiveness. The *Walking Dead* character Carol, wife of abusive husband Ed, illustrates many of these traits in season one.

It would seem from this analysis that adapting post-zombie apocalypse is simply a matter of one's personality characteristics; but as with virtually every psychological outcome, both the individual and the context in which that individual is situated must be considered. Unfortunately, researchers have found that relatively stable traits of individuals such as their personality are much poorer predictors of performance than are the ICE conditions themselves.[22] That is, although researchers can identify what kind of person has the best chance of adapting well in an ICE environment, the specific challenges and conditions of that ICE environment may ultimately have a larger impact on individuals' adaptation, mission success, and even survival. Furthermore, the coping skills and resources that one uses outside of ICE environments are typically not transferrable to ICE environments.

Essentially, no matter how well adapted or successful an individual is in her daily life, her personality is a worse predictor of her performance than are the stressors she would face in a zombie apocalypse ICE environment; and the skills she possesses to cope with her every-day stressors may do little for her in such a scenario. A possible extension of this finding is that some zombie apocalypse survivors who thrive and adapt well to their new ICE conditions may have been fairly poorly adjusted or unsuccessful pre-apocalypse. As a humorous example of this counterintuitive possibility, consider Shaun and Ed, the two main characters from *Shaun of the Dead*, both of whom are not particularly successful or well-adjusted in their pre-apocalypse lives but who manage to adapt and (mostly) survive until their eventual rescue.[23] Consider also the success of viewer-favorite Daryl in AMC's *The Walking Dead*; his abrasive, loner personality and survivalist skill set are not particularly well suited to modern society, yet they are among his greatest assets in his new zombie apocalypse ICE setting.

Interpersonal/Social Domain Issues

Although individual-level variables may predict how humans would cope with a zombie apocalypse ICE scenario, it is imperative to look beyond the individual to the social context of survival. Indeed, the social context of survival may be even more important than individual-level variables.

Stages of Social Interaction in ICE Environments

Researchers studying the social dynamics of ICE environments have identified three common stages of social interaction which are similar to communication stages observed in non–ICE environments. Initially, group members interact openly; this stage is characterized by social comparison and the identification of individual differences and preferences. This stage is well illustrated in multiple examples of zombie lore in which characters disclose interests, share stories, and build relationships, as occurred frequently in the multiple seasons of *The Walking Dead* and as survivors interact during the dinner scene in Romero's *Dawn of the Dead*.[24]

In stage two, individuals begin to form subgroups, typically based on common interests, political orientation, and job demands. These groups may even become exclusive and develop into cliques. Stage two is well illustrated in *The Walking Dead* when Rick's group finds Hershel's farm; characters with shared responsibilities such as household chores or interests such as perimeter security tend to spend more time with one another, at times even exhibiting hostility toward members of other sub-groups.

In stage three, most individuals begin to unite around a shared social identity. While interactions among members may still vary according to friendships and groups already formed, the social core and attendant norms are widely supported. Individuals who fail to follow group norms may be ostracized or exhibit self-imposed isolation. Zombie lore contains many such examples of ostracized individuals. Daryl from AMC's *The Walking Dead* frequently imposes such isolation upon himself. Likewise, Shane and Andrea feel an increasing sense of "otherness" or an inability to fit into their pre-scribed roles and act in line with group expectations. Finally, Carol's covert teaching of children how to use weapons and her euthanizing two sick individuals places her at odds with the group's standards, ultimately leading to her exile.

Patterns of Group Cohesion in ICE Environments

Importantly, at each stage of group development, the group may be characterized as more cohesive or more conflicted, likely due to differences in individual personalities and the specifics of the ICE environment. By analyzing multiple years of data on the crew structure of South Pole research stations, researchers have identified three distinct patterns of social dynamics in ICE environments.[25] In one pattern, cliques are the dominant group formats, often formed around members' leisure activities such as playing cards or watching movies. This pattern also includes a few "isolates" who are not part of any clique, for example, Daryl in season one of *The Walking Dead* and Michonne for much of season three. The second pattern is a more complex "core-periphery" structure. Most individuals identify as members of the core group; some individuals remain somewhat more independent of the core; and some "periphery" individuals remain completely independent or unaffiliated in their social interactions. Finally, the third configuration is a hybrid of the first two in which the crew forms a relatively cohesive core group with clear subgroups as well. The "core-periphery" structure is clear in Woodbury under the leadership of the Governor in season three of *The Walking Dead*. Woodbury is fairly cohesive, with most residents seemingly strongly identifying as members of the town. At the periphery there are semi-independent members such as Andrea and Michonne and militant sub-groups which protect the town and venture out to secure additional resources. How individuals in ICE environments form groups and the structure of those groups matters, as crews that form cliques demonstrate significantly more anger, tension, anxiety, and depression than groups which develop a core-periphery structure.

Social Causes of Conflict in ICE Environments

A Russian astronaut aboard the space station Mir for 211 days once estimated that 30 percent of his time was spent in crew conflict. Indeed, the prolonged isolation and confinement characteristic of most ICE environments often leads to a variety of interpersonal problems ranging from decreased cohesiveness to increased isolation, withdrawal, antagonism, and tension. These findings extend not just to lengthy space station missions but also to shuttle missions, space simulators, submarines, and polar research stations. No matter the ICE environment, conflict is likely to occur for a variety of reasons including personality, gender, and cultural differences.

In an ICE environment, individuals who are domineering do not work well with one another.[26] An obvious example of this kind of personality conflict is illustrated by the repeated, violent confrontations between Rick and Shane in season two of *The Walking Dead* and between Rick and the Governor in seasons three and four. Interestingly, introverts who have less need for social interaction and affection are more likely to adapt successfully to ICE environments than are extroverts with greater needs for social interactions. Daryl in AMC's *The Walking Dead* well illustrates this principle, as his loner proclivities frequently insulate him from the interpersonal conflict and dysfunction of the rest of the group.

Gender is another source of conflict in ICE environments, with researchers finding that both outright and subtle sexual stereotyping leads to misunderstandings and strained social relationships in ICE environments.[27] While popular zombie lore depictions of male survivors include their desire for sexual encounters with other survivors, as in *28 Days Later*, research suggests that these relationships may be a source of additional conflict.[28] In AMC's *The Walking Dead*, Andrea pushes against the gender norms of women serving only support roles, often actively seeking participation in more male-dominated activities like search and rescue and perimeter defense; but her efforts to resist sex stereotyping cause considerable conflict between her and other, mostly female, survivors. For example, Andrea's avoidance of domestic chores in favor of securing the camp earns her harsh words and angry glances from Lori.

Finally, cultural differences among individuals in an ICE environment also contribute to group conflict and poor responses to dangerous situations. Such differences are often mitigated by the "microculture" of shared training and common expectations of professionals living in ICE environments, but it is unlikely that a random group of individuals facing a zombie apocalypse ICE situation would have the advantage of such shared skills and expectations.[29] Further, as group size increases to include individuals with different abilities, expectations, and backgrounds, the development of a unifying microculture is likely to be impaired.[30] The implications for zombie apocalypse ICE contexts is clear: survivors of diverse backgrounds will likely suffer miscommunications and conflict and may have difficulty establishing shared expectations and group cohesion. These problems will likely only intensify as additional survivors join the group. These issues are well illustrated in multiple seasons of *The Walking Dead* in which characters struggle to integrate with and trust other bands of survivors.

Leadership in ICE Environments

Researchers have extensively studied the leadership style of individuals in a variety of ICE environments such as winter-over crews in polar research stations, space flights and long-duration simulators, and submarines, largely concluding that leaders must, above all, be flexible to suit the demands of the crew and the mission. Beyond flexibility, ideal leadership involves four domains: personal traits, task management style, interpersonal style, and group maintenance style.[31] Personally, the ideal leader must be positive, optimistic, confident, experienced, and focused on the achievement of mission goals. As a task-manager, the leader should delegate to and seek the advice of others when possible and appropriate, should actually engage in routine work with others, and should take command in crises. In group maintenance, the leader should attempt to reduce conflicts among cliques and sub-groups while appearing impartial in all decision making. Interpersonally, the leader should adopt a largely democratic approach to leadership, be sensitive to the needs and well-being of others, and recognize them for accomplishments. Finally, the leader must emphasize discipline and be a clear communicator.

Additionally, ideal leadership varies according to the length and focus of the mission. Task-based leadership is especially important during the early stages of a mission, while interpersonal supportiveness is more important later.[32] In a zombie apocalypse ICE setting, these findings suggest the necessity of a strong leader focused initially on the task of group security but who is also capable of providing social and emotional support to others as basic needs are met and more complicated social problems take precedent. Popular depictions of a zombie apocalypse sometimes allow for enough time to pass that immediate survival concerns are met, allowing for other kinds of problems to dominate the leader's focus. This shift in focus is clearly depicted in Romero's *Dawn of the Dead* after the survivors clear and secure the mall and in multiple seasons of *The Walking Dead* once the group reaches the relatively safe zones of the Centers for Disease Control, Hershel's farm, and the prison. At these points, the group leaders shift from primarily security concerns such as shelter seeking to largely social issues like new relationships, leadership issues, and sexual conflicts.

Although researchers suggest that adopting a more democratic approach to certain aspects of leadership is adaptive, such a transition also may be problematic. The process by which the formal or informal chain of command is dropped for a more egalitarian approach to group governance is known as "status leveling."[33] Status leveling can also occur when leadership roles are informally adopted by additional individuals. While such a shift may seem preferable to those being lead, status leveling poses a serious danger to mission success and group survival during crisis scenarios when clear command is needed. Zombie lore is replete with examples of crises in which individual and group safety is threatened. Death and disaster often follow when such crises are not met with an organized, well-lead response, for example the zombies entering the mall in *Dawn of the Dead* and the attack on Hershel's farm at the end of season two of *The Walking Dead*.

Communication in ICE Environments

Communication in ICE environments is often stressed as crucial to interpersonal relations and even mission success.[34] Despite its importance, even trained professionals separated from safety and loved ones for limited amounts of time are likely to communicate less intensely and less comprehensively than in normal conditions, a process sometimes referred to as "psychological closing."[35] Much research on psychological closing was conducted using professionals selected and trained for their ICE missions, suggesting that in civilian populations not prepared for living in ICE conditions, the decrease in communication quality may be even more severe. The development of this more constrained communication style has clear implications for the mental health of zombie apocalypse survivors. This particular facet of ICE-context communication is poorly represented in popular zombie lore, perhaps because psychological closing and constricted communication among characters does not make for entertaining dialogue or plot development.

Interestingly, research conducted using both trained individuals and untrained civilians communicating with strangers demonstrated that the decrements in communication quality experienced by trained pairs may not be present in untrained pairs of strangers facing ICE-like conditions. Specifically, communication between strangers actually deepened when they were placed in ICE-like conditions. While their frequency of disclosure did not reach that typically experienced between good friends, the depth of their disclosure did reach such levels. These findings suggest that strangers facing an ICE scenario may actually communicate and connect in ways that promote strong social bonds. Indeed, the formation and strengthening of friendships among strangers is a common theme in much zombie lore.

Implications for Human Survival in a Zombie Apocalypse

The demands of ICE contexts are extreme, and the consequences of individual or group failure are dire. For these reasons only the best adapted and best trained civilians and military scientists are selected for missions in such challenging settings. Research conducted on these highly selected and highly trained individuals largely supports the hypothesis that prolonged exposure to ICE environments is deleterious to individual mental health, interpersonal relationships, and group cohesion and function. It is also clear that individuals living and working in ICE environments require highly skilled leaders possessing a host of personality characteristics and interpersonal skills.

Whereas most professionals in ICE environments have some expectation that their lives will return to normal when their mission ends, zombie apocalypse survivors have no such comfort. Further, many professionals working in ICE environments such as space missions have the comfort of regular communication via email and video-chat with support staff and loved ones. Survivors of a zombie apocalypse not only will not have such options, they are likely to be burdened by the additional knowledge that their loved ones are casualties of the zombie apocalypse. Because there are no actual studies of human performance in a zombie apocalypse ICE setting, it is impossible to know for sure how the average survivor would function and what the limits of human endurance and adaptability truly are. Unfortunately, it seems that the level of adaptive function and group cohesion that well supported and well trained professionals are able to achieve in ICE environments is unlikely to be matched by a group of traumatized survivors who find themselves together in one of the most isolated, confined, and extreme environments ever conceived of: the zombie apocalypse.

Notes

1. George A. Romero, *Night of the Living Dead*, Image Ten, 1968. Max Brooks, *World War Z* (New York: Crown, 2007). Frank Darabont, *The Walking Dead*, AMC, 2010. Robert Kirkman and Tony Moore, *The Walking Dead* (Berkeley: Image Comics, 2003).

2. Romero, *Night of the Living Dead*. Danny Boyle, *28 Days Later*, Berkeley: DNA Films, 2002. Zack Snyder, *Dawn of the Dead*, Strike Entertainment, 2004.

3. Max Brooks, *The Zombie Survival Guide* (New York: Three Rivers Press, 2003).

4. Coady Lapierre, Andria Schwegler, and Bill LaBauve, "Posttraumatic stress and depression symptoms in soldiers returning from combat operations in Iraq and Afghanistan," *Journal of Traumatic Stress* 20 (2007): 933–43. Matthew K. Nock, Charlene A. Deming, Carol S. Fullerton, Stephen E. Gilman, Matthew Goldenberg, Ronald C. Kessler, James E. McCarroll, et al., "Suicide among soldiers: A review of psychosocial

risk and protective factors," *Psychiatry: Interpersonal & Biological Processes* 76 (2013): 97–125. Timothy S. Wells, Shannon C. Miller, Amy B. Adler, Charles C. Engel, Tyler C. Smith, and John A. Fairbank, "Mental health impact of the Iraq and Afghanistan conflicts: a review of U.S. research, service provision, and programmatic responses," *International Review of Psychiatry* 23 (2011): 144–152.

     5. Romero, *Night of the Living Dead*.

     6. Darabont, *The Walking Dead*.

     7. Naomi Breslau, G.A. Chase, and J.C. Anthony, "The uniqueness of the DSM definition of post-traumatic stress disorder: implications for research," *Psychological Medicine* 32 (2002): 573–576. Sandro Galea, Arijit Nandi, and David Vlahov, "The epidemiology of post-traumatic stress disorder after disasters," *Epidemiologic Reviews* 27 (2005): 78–91. Fran H. Norris, Matthew J. Friedman, Patricia J. Watson, Christopher M. Byrne, Eolia Diaz, and Krzysztof Kaniasty, "60,000 disaster victims speak: Part I. An empirical review of the empirical literature, 1981–2001," *Psychiatry: Interpersonal and Biological Processes* 65 (2002): 207–239.

     8. Yuval Neria, Arijit Nandi, and Sandro Galea, "Post-traumatic stress disorder following disasters: a systematic review," *Psychological Medicine* 38 (2008): 467–480.

     9. Lawrence A. Palinkas, "Psychosocial issues in long-term space flight: Overview," *Gravitational and Space Biology* 14 (2001): 25–33.

     10. Patricia A. Santy, *Choosing the Right Stuff: The Psychological Selection of Astronauts and Cosmonauts* (Westport, CT: Praeger, 1994).

     11. Santy, *Choosing the Right Stuff*. Albert Harrison, "On resistance to the involvement of personality, social and organizational psychologists in the U.S. Space Program," *Journal of Social Behavior and Personality* 1 (1986): 315–324.

     12. Robert Helmreich, "Applying psychology to outer space: unfulfilled promises revisited," *American Psychologist* 38 (1983): 445–450.

     13. Bryan Burrough, *Dragonfly: NASA and the Crisis Aboard Mir* (New York: HarperCollins, 1998). Jerry M. Linenger, *Off the Planet* (New York: McGraw-Hill, 2000). Valentin Lebedev, *Diary of a Cosmonaut: 211 Days in Space* (College Station: Phytoresource Research, 1988). James E. Oberg, *Red Star in Orbit* (New York: Random House, 1981). E.K. Eric Gunderson, "Psychological studies in Antarctica," in E.K. Eric Gunderson, ed., *Human Adaptability to Antarctic Conditions* (Washington, D.C.: American Geophysical Union, 1974): 115–131. Gro M. Sandal, R. Vaernes, and H. Ursin, "Interpersonal relations during simulated space Missions," *Aviation, Space, and Environmental Medicine* 66 (1995): 617–624.

     14. Lawrence A. Palinkas, "Going to extremes: the cultural context of stress, illness and coping in Antarctica," *Social Science & Medicine* 35 (1992): 651–664.

     15. H.S. Cooper, Jr., "The loneliness of the long-duration astronaut," *Air Space* 11 (1996): 37–45.

     16. Julien Christensen and John Talbot, "A review of the psychosocial aspects of space flight," *Aviation, Space, and Environmental Medicine* 57 (1986): 203–212. Gunderson, "Psychological studies in Antarctica," 115–131. Nick Kanas, "Psychological factors affecting simulated and actual space missions," *Aviation, Space, and Environmental Medicine* 56 (1985): 806–811.

     17. Gloria R. Leon, Carl McNally, and Yossef S. Ben-Porath, "Personality characteristics, mood, and coping patterns in a successful North Pole expedition team," *Journal of Research in Personality* 23 (1989): 162–179.

     18. Peter Suedfeld and G. Daniel Steel, "The environmental psychology of capsule habitats," *Annual Review of Psychology* 51 (2000): 227–253.

     19. Lawrence A. Palinkas, Peter Suedfeld, and G. Daniel Steel, "Psychological

functioning among members of a small polar expedition," *Aviation, Space, and Environmental Medicine* 66 (1995): 1591–1596.

20. Lawrence A. Palinkas, E.K. Eric Gunderson, Albert W. Holland, Christopher Miller, and Jeffrey C. Johnson, "Predictors of behavior and performance in extreme environments: the Antarctic space analogue program," *Aviation, Space, and Environmental Medicine* 71 (2000): 619–625.

21. Thomas Chidester, Robert Helmreich, Steven Gregorich, and Craig Geis, "Pilot personality and crew coordination: Implications for training and selection," *International Journal of Aviation Psychology* 1 (1991): 25–44. Janet T. Spence, and Robert L. Helmreich, *Masculinity & Femininity: Their Psychological Dimensions, Correlates, and Antecedents* (Austin: University of Texas Press, 1978).

22. Charles Carver and Michael Scheier, "Situational coping and coping dispositions in a stressful Transaction," *Journal of Personality and Social Psychology* 66 (1994): 184–195.

23. Edgar Wright, *Shaun of the Dead*, StudioCanal, 2004.

24. Darabont, *The Walking Dead*. George Romero, *Dawn of the Dead*, Laurel Group, 1978.

25. Lawrence A. Palinkas, E.K. Eric Gunderson, Jeffrey C. Johnson, and Albert W. Holland, "Behavior and performance on long-duration spaceflights: evidence from analogue environments," *Aviation, Space, and Environmental Medicine* 71 (2000): A29–36.

26. Kanas, "Psychological factors," 806–811.

27. Lebedev, *Diary of a Cosmonaut*. Oberg, *Red Star in Orbit*.

28. Boyle, *28 Days Later*.

29. Mary Connors, Albert Harrison, and Faren Akins, *Living Aloft: Human Requirements for Extended Spaceflight* (Washington, D.C.: NASA, 1985).

30. Committee on Space Biology and Medicine, *A Strategy for Research in Space Biology and Medicine in the Next Century* (Washington, D.C.: National Academy Press, 1998).

31. John M. Nicholas and Larry W. Penwell, "A proposed profile of the effective leader in human spaceflight based on findings from analog environments," *Aviation, Space, and Environmental Medicine* 66 (1995): 63–72.

32. E.K. Eric Gunderson and Paul Nelson, "Adaptation of small groups to extreme environments," *Aerospace Medicine* 34 (1963): 1111–1115. Paul D. Nelson, "Similarities and differences among leaders and followers," *The Journal of Social Psychology* 63 (1964): 161–167.

33. Gro M. Sandal, G.R. Leon, and Lawrence A. Palinkas, "Human challenges in polar and space Environments," in *Life in Extreme Environments* (The Netherlands: Springer, 2007): 281–296.

34. Nicholas and Penwell, "Profile of the effective leader." Palinkas, "Psychosocial issues."

35. Vadim Gushin, Nina Zaprisa, Tatiana Kolinitchenko, Vladimir Efimov, Tatiana Smirnova, Alla Vinokhodova, and Nick Kanas, "Content analysis of the crew communication with external communicants under prolonged isolation," *Aviation, Space, and Environmental Medicine* 68 (1997): 1093–1098.

Bibliography

Boyle, Danny. *28 Days Later*. DNA Films, 2002.

Breslau, Naomi, G.A. Chase, and J.C. Anthony. "The uniqueness of the DSM definition of post-traumatic stress disorder: implications for research." *Psychological Medicine* 32 (2002): 573–576.

Brooks, Max. *The Zombie Survival Guide*. New York: Three Rivers Press, 2003.
_____. *World War Z*. New York: Crown, 2007.
Burrough, Bryan. *Dragonfly: NASA and the Crisis Aboard Mir*. New York: Harper-Collins, 1998.
Carver, Charles, and Michael Scheier. "Situational coping and coping dispositions in a stressful Transaction." *Journal of Personality and Social Psychology* 66 (1994): 184–195.
Chidester, Thomas, Robert Helmreich, Steven Gregorich, and Craig Geis. "Pilot personality and crew coordination: Implications for training and selection." *International Journal of Aviation Psychology* 1 (1991): 25–44.
Christensen, Julien, and John Talbot. "A review of the psychosocial aspects of space flight." *Aviation, Space, and Environmental Medicine* 57 (1986): 203–212.
Committee on Space Biology and Medicine. *A Strategy for Research in Space Biology and Medicine in the Next Century*. Washington, D.C.: National Academy Press, 1998.
Connors, Mary Albert Harrison, and Faren Akins. *Living Aloft: Human Requirements for Extended Spaceflight*. Washington, D.C.: NASA, 1985.
Cooper, H.S., Jr. "The loneliness of the long-duration astronaut." *Air Space* 11 (1996): 37–45.
Darabont, Frank. *The Walking Dead*. AMC, 2010.
Galea, Sandro, Arijit Nandi, and David Vlahov. "The epidemiology of post-traumatic stress disorder after disasters." *Epidemiologic Reviews* 27 (2005): 78–91.
Gunderson, E.K. Eric. "Psychological studies in Antarctica," in E.K. Eric Gunderson, ed., *Human Adaptability to Antarctic Conditions*. Washington, D.C.: American Geophysical Union, 1974: 115–131.
_____, and Paul Nelson. "Adaptation of small groups to extreme environments." *Aerospace Medicine* 34 (1963): 1111–1115.
Gushin, Vadim, Nina Zaprisa, Tatiana Kolinitchenko, Vladimir Efimov, Tatiana Smirnova, Alla Vinokhodova, and Nick Kanas. "Content analysis of the crew communication with external communicants under prolonged isolation." *Aviation, Space, and Environmental Medicine* 68 (1997): 1093–1098.
Harrison, Albert. "On resistance to the involvement of personality, social and organizational psychologists in the U.S. Space Program." *Journal of Social Behavior and Personality* 1 (1986): 315–324.
Helmreich, Robert. "Applying psychology to outer space: unfulfilled promises revisited." *American Psychologist* 38 (1983): 445–450.
Kanas, Nick. "Psychological factors affecting simulated and actual space missions." *Aviation, Space, and Environmental Medicine* 56 (1985): 806–811.
Kirkman, Robert, and Tony Moore. *The Walking Dead*. Berkeley: Image Comics, 2003.
Lapierre, Coady, Andria Schwegler, and Bill LaBauve. "Posttraumatic stress and depression symptoms in soldiers returning from combat operations in Iraq and Afghanistan." *Journal of Traumatic Stress* 20 (2007): 933–43.
Lebedev, Valentin. *Diary of a Cosmonaut: 211 Days in Space*. College Station: Phytoresource Research, 1988.
Leon, Gloria R., Carl McNally, and Yossef S. Ben-Porath. "Personality characteristics, mood, and coping patterns in a successful North Pole expedition team." *Journal of Research in Personality* 23 (1989): 162–179.
Linenger, Jerry M. *Off the Planet*. New York: McGraw-Hill, 2000.
Nelson, Paul D. "Similarities and differences among leaders and followers." *The Journal of Social Psychology* 63 (1964): 161–167.

Neria, Yuval, Arijit Nandi, and Sandro Galea. "Post-traumatic stress disorder following disasters: a systematic review." *Psychological Medicine* 38 (2008): 467–480.

Nicholas, John M. and Larry W. Penwell. "A proposed profile of the effective leader in human spaceflight based on findings from analog environments." *Aviation, Space, and Environmental Medicine* 66 (1995): 63–72.

Nock, Matthew K., Charlene A. Deming, Carol S. Fullerton, Stephen E. Gilman, Matthew Goldenberg, Ronald C. Kessler, James E. McCarroll, et al. "Suicide among soldiers: A review of psychosocial risk and protective factors." *Psychiatry: Interpersonal & Biological Processes* 76 (2013): 97–125.

Norris, Fran H., Matthew J. Friedman, Patricia J. Watson, Christopher M. Byrne, Eolia Diaz, and Krzysztof Kaniasty. "60,000 disaster victims speak: Part I. An empirical review of the empirical literature, 1981–2001." *Psychiatry: Interpersonal and Biological Processes* 65 (2002): 207–239.

Oberg, James E. *Red Star in Orbit.* New York, NY: Random House, 1981.

Palinkas, Lawrence A. "Going to extremes: the cultural context of stress, illness and coping in Antarctica." *Social Science & Medicine* 35 (1992): 651–664.

_____. "Psychosocial issues in long-term space flight: Overview." *Gravitational and Space Biology* 14 (2001): 25–33.

_____, E.K. Eric Gunderson, Albert W. Holland, Christopher Miller, and Jeffrey C. Johnson. "Predictors of behavior and performance in extreme environments: the Antarctic space analogue program." *Aviation, Space, and Environmental Medicine* 71 (2000): 619–625.

_____, _____, Jeffrey C. Johnson, and Albert W. Holland. "Behavior and performance on long-duration spaceflights: evidence from analogue environments." *Aviation, Space, and Environmental Medicine* 71 (2000): A29–36.

_____, Peter Suedfeld, and G. Daniel Steel. "Psychological functioning among members of a small polar expedition." *Aviation, Space, and Environmental Medicine* 66 (1995): 1591–1596.

Romero, George A. *Dawn of the Dead.* Laurel Group, 1978.

_____. *Night of the Living Dead.* Image Ten, 1968.

Sandal, Gro M., G.R. Leon, and Lawrence A. Palinkas. "Human challenges in polar and space Environments." In *Life in Extreme Environments.* The Netherlands: Springer, 2007: 281–296.

_____, R. Vaernes, and H. Ursin. "Interpersonal relations during simulated space Missions." *Aviation, Space, and Environmental Medicine* 66 (1995): 617–624.

Santy, Patricia A. *Choosing the Right Stuff: The Psychological Selection of Astronauts and Cosmonauts.* Westport, CT: Praeger, 1994.

Snyder, Zack. *Dawn of the Dead.* Strike Entertainment, 2004.

Spence, Janet T., and Robert L. Helmreich. *Masculinity & Femininity: Their Psychological Dimensions, Correlates, and Antecedents.* Austin: University of Texas Press, 1978.

Suedfeld, Peter, and G. Daniel Steel. "The environmental psychology of capsule habitats." *Annual Review of Psychology* 51 (2000): 227–253.

Wells, Timothy S., Shannon C. Miller, Amy B. Adler, Charles C. Engel, Tyler C. Smith, and John A. Fairbank. "Mental health impact of the Iraq and Afghanistan conflicts: a review of U.S. research, service provision, and programmatic responses." *International Review of Psychiatry* 23 (2011): 144–152.

Wright, Edgar. *Shaun of the Dead.* StudioCanal, 2004.

"Three men, and the place is surrounded"

Reel Women in the Zombie Apocalypse

LuAnne Roth *and* Kate Shoults

"My call to action has nothing to do with stocking up on food and guns; I prefer instead to consider how we might shake up patriarchy, how we might address power imbalances that will carry over into the post-apocalyptic world if they are not dealt with now."— Jen Rinaldi[1]

In the original *Night of the Living Dead,* a group of survivors devise a plan to escape from the farmhouse besieged by zombies.[2] In ticking off their assets and liabilities, Harry Cooper, the cranky husband hiding in the basement, loudly proclaims: "How could we possibly get away from here? We've got a sick child, two women, one woman out of her head, three men, and the place is surrounded." Although there are six able-bodied adults present, the men consider only themselves relevant in assuring the group's survival. Even worse, the women allow the men to consider them incapable, a problem that, according to our survey, befalls far too many zombie narratives.

This essay begins with the understanding that zombie films provide a useful point from which to gauge the borders of "appropriate" gender relations in society—that is, they are useful cultural texts with which to debate the politics of gender representation.[3] Often judged as misogynistic, critical film theory exonerates parts of the horror genre from this demonization, revealing it as one of the few genres that actually affords its female characters real power.[4] In fact, women comprise over 40 percent of the zombie genre's fan base, according to some observers.[5] Despite the large number of women in its fan base, the sub-genre has remained notoriously male dominated since its inception.[6]

Perhaps the reason for this fan base lies in Carol J. Clover's seminal "Final Girl" theory. Having no one else to rely upon, the Final Girl has to save herself. For Clover, "the character of stature who does live to tell the tale is in fact the Final Girl…. She is intelligent, watchful, levelheaded; the first character to sense something amiss and the only one to deduce from the accumulating evidence, the pattern and extent of the threat; the only one, in other words, whose perspective approaches our own privileged understanding of the situation."[7] The Final Girl emerges from degradation, bodily harm, and emotional trauma to wreak havoc upon the typically male coded villain. This courageous figure epitomizes female empowerment. She is the person that viewers want to be victorious.

This article argues that, while female characters in zombie films have gradually become more capable of defending themselves, unlike the savvy female celebrated in the horror genre proper, the women of zombie films, with a few notable exceptions addressed below, are either required to transform into masculine roles or devolve into weak females reliant on the patriarchy for survival. Close textual examination supports this thesis—from the apocalyptic films of George A. Romero's *Dead* series, the *Resident Evil* franchise, *28 Days Later*, *World War Z*, and television's *The Walking Dead* series to the more positive outlook taken by such recent "rom-com-zoms" as *Shaun of the Dead*, *Zombieland*, and *Warm Bodies*.

Romero's Zombie Apocalypse: Stereotypes and Binaries

The first real cinematic instance of the zombie apocalypse setting appears in Romero's seminal *Dead* series, which reveals a range of female character types.[8] The 1968 cult classic *Night of the Living Dead* introduces the infamous character of Barbra. Whereas the character's actions and inactions have enraged feminist critics for decades, some have attempted to reconcile embedded meanings. "It is ultimately the symbol of patriarchy that destroys Barbara [sic] One," says gender studies theorist Natasha Patterson, "as she is unable to distinguish between living Johnny and zombified Johnny, and her lack of participation and awareness in the face of danger (i.e., behaving in ways appropriate to her gender) ultimately causes her death."[9] Representing the stereotypically submissive woman, Barbra's character functions as a critique of patriarchal systems: Barbara's survival in the 1990 remake—and her enactment of real agency—might demonstrate Romero's growing empathy toward women.[10] Film scholar Barry Keith Grant concludes that the remake "attempts to reclaim the horror genre for feminism, and for all those female victims in such movies who attempt to resist patriarchal containment."[11] Beyond presenting women in progressive ways, Romero's zombies may enable them to be "demonstrably aggressive and unapologetic."[12] For example, in *Dawn of the Dead* (1978), the audience meets Fran, a female character light years ahead of Barbra. Fran is pregnant at the start of the narrative. When the men discover this their behavior towards her changes to "a chauvinistic attitude that seems oddly out of place and somewhat ridiculous in light of the circumstances surrounding them."[13] Fran confronts the men, and their patriarchal assumptions, saying: "I don't want to be treated any differently than you treat each other … and I'm not going to be den mother for you guys and I want to know what's going on, and I want to have something to say about the plans. There's four of us, okay?" By insisting that women be counted, Fran seems to respond to Cooper's aforementioned line from *Night of the Living Dead*. Despite her insistence, Fran does eventually become "den-mother" to the surviving men, cooking and serving meals while also nursing a wounded male protagonist. Yet throughout the film Fran's gender coding becomes more complex as she navigates between masculine and feminine gender divides. Fran seems to understand the reality of the situation better than others and, to become self-sufficient, she demands to be taught how to pilot a helicopter and shoot a gun. The fact remains, however, that Fran must perform something other than female roles to survive, while the men are not required to assume traditionally feminine traits.

The remakes of *Night of the Living Dead* (1990) and *Dawn of the Dead* (2004) greatly improve upon the initial films' gender politics. Both Barbra and Fran's counterparts are rendered even more androgynous in oscillation between female and male coding. In the former, Barbara ends up becoming a fearless fighter; in the latter, Ana employs quick thinking gleaned from her career as an emergency room nurse. This change results from Romero's attempt to demonstrate the futility of gender binaries in the zombie apocalypse. Patterson suggests that "[t]he zombie film restores pleasure to the female viewer through the zombie's ambivalence toward gender (and genre), while the male viewer has a space opened up for him that allows for a relationship to the body that may not otherwise be realized with other kinds of horror films," largely because the eating of flesh ignores binaries of male and female.[14] At first this interpretation seems tantalizing, but stripping women of their gender may be more troubling.[15] Romero's films require females to assume typically masculine traits, such as firing weapons or becoming ruthless, but the supposed shucking of binary norms rarely works the other way (forcing men to cook, clean, or assume caretaking roles). Performing masculinity is assumed to be enough to ensure survival.

Sexualization, Domestication and Marginalization

One solution to Romero's tendency of destroying femininity is to celebrate the female in all her glory, much like the protagonist in Paul W.S. Anderson's film adaptations of the survival horror video game series *Resident Evil* and *Resident Evil 2*. The film introduces Alice and a band of commandos as they attempt to contain the outbreak of the T-Virus at a top-secret research facility owned by the Umbrella Corporation. In this franchise, which currently spans five films—*Resident Evil* (2002), *Resident Evil: Apocalypse* (2004), *Resident Evil: Extinction* (2007), *Resident Evil: Afterlife* (2010), *Resident Evil: Retribution* (2012)—Alice joins other survivors, as well as her own clones, to fight against zombies and the evil Umbrella Corporation. At first Alice appears to be a strong character capable of reversing the male gaze, a dynamic reinforcing gender imbalance in which filmmakers use the camera to put the audience into a heterosexual male point of view.[16] Naturally, Alice would seem to be the go-to zombie killing action heroine, especially when outfitted with her trusty and numerous female sidekicks, but the unwarranted achievement of near super-human status and the constant cinematic exploitation of her body render Alice problematic as a feminist role model.

For example, consider her costuming. During the first film she combats zombies in a flimsy red cocktail dress and biker boots, but what this film tries to present as empowerment ends up being demeaning. Actress Milla Jovovich reports that she designed the costume, fancying herself a young Ripley, in reference to the sci-fi icon's underwear attire during the last scene in *Alien*.[17] With this connection, Jovovich forgets that Final Girl Ripley sports the costume only during the last few minutes of the film's final showdown, reminding the audience that the ass-kicker is also a vulnerable human being. Given no narrative reasons for her attire, the glamorization of Alice's body over practicality and survival undercuts the impact of the character's representation as a powerful force. Despite the fetishization of women in the *Resident Evil* franchise, there is something oddly fascinating about watching Alice in combat. She is lithe, athletic, and graceful, more like a dancer than a fighter. The fact remains, however, that Alice's advantageous body is genetically engineered by Umbrella Corporation. As such, notes Jen Rinaldi, a scholar of disability studies, "she does not contradict patriarchy since medical control and manipulation of female bodies are common patriarchal practices."[18] Because Alice was engineered to be perfect—strong, smart, and never missing her mark—identifying with such a flawless character can be difficult for female audience members.

More current zombie films seem to negate the issue of female sexuality, but by the time the credits role, women typically end up being the requisite damsels in distress. Danny Boyle's *28 Days Later* (2002) typifies this dynamic. Following a group of survivors searching for safe haven in zombie-ravaged Britain, the film depicts Selena from the start as a pragmatic character, exemplifying the Final Girl in almost every way. Selena makes cutthroat decisions with her own survival in mind: "If someone gets infected," she says to the protagonist Jim, "you have between 10 and 20 seconds to kill them. It might be your brother or your sister or your oldest friend, makes no difference. And just so you know … if it happens to you, I'll do it in a heartbeat." The statement recalls the unfortunate demise of Barbra, in *Night of the Living Dead*, whose fear renders her incapable of making such decisions and leads to death at the hands of her zombified brother. Despite her deductive reasoning and comments to the contrary, Selena's stony exterior starts to slip and she begins to care for her companions, finding kinship with Frank and Hannah, the father/daughter survivors they encounter, and even developing a romantic interest in Jim. When Selena's emotional side is revealed, she is not painted as hysterical or irrational. After her strong, independent introduction, however, both Selena and Hannah are preyed upon with the threat of sexual violence.

Once the gang has found and settled in with the male soldiers at the fortress, dialogue reveals that the military group has been promised women with whom to procreate. Just because the end of the world is nigh, sexuality will probably not cease to exist. In fact, with the possible necessity of repopulation, it may become more important than before. Here, though, the goal of reproduction is a thin veil used to mask the soldiers' violent sexual desires: "Rape becomes a means of domestication because—through reproduction— it establishes the mother the soldiers have lacked in this 'home.'"[19] Beyond the threat of sexual violence, Selena and Hannah are forced by their attackers to wear red ballroom gowns. The clothing not only sexualizes the women, but it also restricts their survival abilities. In a sense, the filmmakers seem to punish Selena for performing masculinity by exploiting her henceforth forgotten sexuality. The dress and sexual violence are harsh reminders of her gender role, so that despite the independence and strength she earlier displays, Selena seems unable to save either herself or Hannah and must rely on Jim to end the patriarchal violence.[20] Also, despite the critical eye the attempted rape places on ideas of reproduction and motherhood, the film still shoves Selena into a familial role through what purports to be a happy ending. Selena becomes Hannah's de facto mother figure when the girl's father becomes infected. Upon rescuing Selena and Hannah, Jim assumes the role of protector.[21] This leaves Jim as the patriarch, Hannah as the child, and Selena as the matriarch. To further this point, in the theatrical ending, Jim awakens in a cottage to find Selena sewing her red dress into a large cloth sign with the letters "HELL" (the "O" not yet finished) for airplanes passing overhead to see. Since sewing is considered "women's work," this cinematic ending marks Selena's return to stereotypical womanhood.[22] Altogether, it also suggests that Selena can only survive if she assumes her gender-dictated place in a traditional nuclear family. Christopher Williams puts it bluntly: "The film's goal appears to be the re-domestication of Selena," reinforcing Selena's proper place in society before the final credits roll.[23]

Based loosely on Max Brooks' similarly titled novel, the film *World War Z* attempts to provide a realistic rundown of the zombie apocalypse on a global scale, mostly through the perspective of male protagonist Gerry Lane, a United Nations employee.[24] While the film shows the pandemic toppling governments and armies, it suggests that the patriarchal family is key to humanity's survival and offers a fairly passive picture of women's roles in the face of such a disaster. Although Gerry's family functions as his anchor throughout the film, they do little beyond acting as a liability. For example, as soon as Karin is out of her husband's sight, she is attacked and in need of his assistance. Both of the pre-teen daughters, Constance and Rachel, are similarly passive and incapable of taking care of themselves. Compare this to Gerry asking a young boy the family saves to "take care of the ladies" until he returns. In contrast to the novel, there is only one named female character of importance in the film. Segen, an Israeli soldier who is masculine in appearance, is tasked with bringing Gerry to his getaway plane. With very few lines of dialogue, Segen proves to be level headed under pressure, even as her unit is attacked. When Segen is suddenly bitten on the hand by a zombie, Gerry acts quickly to save her from being infected by severing the hand with his machete. Despite this traumatic setback, Segen remains active throughout the rest of the film. While refreshing to find such a capable female, Segen's character rarely offers her own input and seems to exist in the narrative as Gerry's sidekick and to remind us of the protagonist's humanity. In fact, the film's ratio of male-to-female characters in power is staggeringly uneven. The only other female of note is a scientist introduced toward the end and, although she works in a male-dominated profession, she has no name and little screen time or dialogue.[25]

The popular AMC television show *The Walking Dead*, praised as the inevitable outlook of daily life under the constant threat of zombies, is criticized for the way in which it rearticulates patriarchy, defends hypermasculinity, and reinforces sexist gender roles. As with *World War Z*, there is a notable absence of female leaders in both the comic book series and television show. In the first few seasons, "the roles of men and women are clearly demarcated by a feminist's nightmare: men 'hunt' while women 'gather'—traditional gender roles are almost organically claimed. Men assert roles of leadership, while the women remain behind fulfilling domestic chores, namely laundry and childcare."[26] Women hold no leadership positions until the start of Season 4. A small-town sheriff's deputy, Rick, quickly emerges as the leader of one group, while psychopath, Phillip (aka the Governor), becomes leader of another. When Rick relinquishes his leadership role, the "Prison Council" purportedly shifts political power toward something resembling democracy. While the five-member team is comprised of two females—Carol and Sasha—the latter is absent due to illness.[27] While this composition (two females / three males) strikes more balance than the previous male dictator model, the television audience sees the Council function only once by the end of Season 4 (to respond to the deadly flu outbreak) and, despite the Council's existence, Rick still acts as the group's leader.[28] Theologian Ashley Barkman claims that the women in *The Walking Dead* universe concur with this leadership arrangement, citing this as support for her argument that women are biologically predestined to take subservient positions in the zombie apocalypse. That the *fictional* characters seem to support this system, however, does not render it unproblematic.[29] Since the story starts within a patriarchal society before the outbreak, note Danee Pye and Peter O'Sullivan, "it's no surprise that they begin to recreate patriarchal norms as they rebuild new societies."[30] Yet, female characters do express a desire for something different. For instance, Jacqui complains about this presumed "natural" division of labor, "Could someone explain to me how all the women ended up doing all the Hattie McDaniel work?" to which Amy responds, "The world ended. Didn't you get the memo?" and Carol closes with "It's just the way it is."[31] In the comic series, another character, Donna, even adds, "I just don't understand why we're the ones doing the laundry while they go off and hunt. When things get back to normal, I wonder if we'll still be allowed to vote." Judging Donna's complaint to be unwarranted, Lori rebuts with "This isn't about women's rights. It's being about realistic and doing what needs to be done."[32] In Season 2, Lori even chastises Andrea, the group's best sharpshooter, for taking guard-duty shifts rather than engaging in traditional domestic activities.[33]

Despite its overall conservative gender politics, *The Walking Dead* does offer some tentatively redeeming roles for women—Carol, Andrea, Sasha, Maggie, and Michonne—however, each female character possesses major flaws that qualify this assertion: Carol transitions from cowering victim of domestic abuse to decisive woman capable of killing both zombies and people, while being a surrogate mother to Judith (Rick's infant daughter) as well as two orphaned girls (Lizzie and Mika). By Season 4, however, Carol's rising strength corresponds to her loss of humanity, a development made clear when, trying to protect others from a deadly flu epidemic, she callously burns alive two innocent people. In response, Rick banishes Carol from the group without even consulting the Council formed to handle such matters.[34] Andrea, a former civil rights attorney, eventually learns to operate a gun and becomes an excellent shot. Although smart enough to survive the winter with Michonne (and no help from men), Andrea ends up making extremely poor decisions. She seems to search for power by sleeping with powerful men who are revealed to be psychopaths (first Shane and then the Governor). Dismissing Michonne's accurate intuition about the Governor's evil ways, Andrea chooses to stay with her new lover at Woodbury rather than leave with Michonne, whose friendship, trustworthiness, and wisdom was established. Sasha and Maggie demonstrate excellent fighting skills and sound judgment. Introduced in Season 3, Sasha's character shows promise thus far. Maggie begins Season 2 as a confident and outspoken woman yet, existing as a love interest, she gradually devolves to become deferential to Glenn. After being terrorized and sexually assaulted by the Governor, the show focuses more on how the threatened rape impacts Glenn's insecure masculinity than on Maggie's trauma.[35] Many commenters identify Michonne as the show's promise of redemption. Indeed, the extraordinarily fearless character, introduced at the end of Season 2 as Andrea's savior, is smart, resilient, and skilled with a sword, having survived for at least eight months before joining the others.[36] Unlike a number of other survivors, who are quickly taken into the fold, Michonne is strangely regarded as an outsider viewed with suspicion, despite her obvious strengths. Even the child Carl is consulted as to whether or not Michonne should be accepted as "one of them."[37] Then, after finally proving herself to the group, Rick and the Governor—two southern white men fighting over territory—attempt to negotiate a truce by using the African-American woman's body as an object for barter.[38] To help make this decision, Rick consults with three other white men—Hershel, Daryl, and Meryl (the explicitly racist and sexist southerner)—about whether he should give Michonne to the Governor. In the comics, Michonne is brutally raped by the Governor and, while gratefully this does not occur on the television show, the fact that her fate is determined by men demonstrates that, in *The Walking Dead* universe, women are denied real power.[39]

As this brief discussion illustrates, there is much feminist critique to be made of *The Walking Dead*. Some might question whether feminism is relevant in a zombie apocalypse. In such a scenario, Barkman argues, leadership should remain in the hands of men.[40] A series of assumptions are cited in support of this position, including the ideas that men and women are biologically and spiritually different, that men tend to be stronger than women, and that masculinity "naturally" lends itself to leading while femininity lends to being led. If two potential leaders—one male and one female—are equally competent, according to this trajectory, the man should be the leader.[41] Underlying this flawed argument is the notion that men and women are different in terms of how they assert authority or respond to problems. On a biological level, Barkman claims that women are disadvantaged by their inferior strength and size, "their lack of testosterone-driven aggression," and their vulnerability as females: "A woman using her strength to brutally rape a man is unheard of, if it is even possible. But the temptation of a man to rape a woman is a popular motif in any zombie tale."[42] Indeed, several female characters are threatened with rape (e.g., Michonne and Maggie), but both emerge as fighters and survivors. This assumption that men are naturally better leaders is the sort of problematic and dangerous reasoning that reinforces learned passivity, and the fact that men try to rape women is no valid reason to exempt women from leadership roles. If anything, this flawed sense of humanity provides one reason why men might *not* be good leaders.

The Rom-Com-Zom Development

In searching for capable female characters in the zombie apocalypse, additional glimmers of hope can be found as well in the "rom-com-zom" sub-genre. These films marry elements from horror, zombie films, and the romantic comedy. Although they borrow from the female focused romance genre, the following films—Edgar Wright's *Shaun of the Dead* (2004), Richard Fleischer's *Zombieland* (2009), and Jonathan Levine's *Warm Bodies* (2013)—feature male leads alongside fairly positive representations of female characters.

Shaun of the Dead tells the story of a slacker in his late-twenties, who is forced to learn adult responsibilities during a zombie apocalypse taking place in the UK. Although the film is a coming-of-age story that propels Shaun into becoming an unlikely hero, the cast brims with female characters, including Shaun's girlfriend Liz, her roommate Dianne, Shaun's friend Yvonne, and Shaun's mother Barbara, a direct homage to Barbra from the original *Night of the Living Dead*. Aside from Barbara, these women are capable and savvy. In fact, the women are often depicted as being better off without their men. The start of the zombie outbreak coincides with Liz breaking up with the immature Shaun. Liz and Dianne are coaxed out of their secure third floor flat by Shaun, who brings them to a more dangerous situation at the Winchester Pub. Barbara and Dianne's downfall results from the trust they place in the patriarchy, while Liz survives as girlfriend/surrogate mother for Shaun.[43] Yvonne, a minor character who appears only three times as foil to Shaun, is far more capable than Shaun. During her limited screen time, Yvonne expresses polite and subtle disapproval at Shaun's choice in safe houses ("Oh, good luck") and later shows that her wise choices saved her group. Yvonne is clearly the most effective player in the film, begging the question of why she is not the film's protagonist or, as one astute female audience member remarked, why the film was not instead refocused to be "Yvonne of the Dead." The film slyly implies that women are as strong as they allow themselves to be and that waiting on patriarchy for permission and guidance is inadequate—a lesson appropriate to both reel life and real life.

On the other side of the Atlantic, *Zombieland* is structured around a list of survival rules developed by the timid, obsessive-compulsive shut-in named Columbus. The film follows four strangers who meet and then head west together across the zombie-ravaged America. Sister con artists, twenty-something Wichita and her twelve-year-old sister Little Rock, are formidable forces wary of trusting the male protagonists. From their first introduction, the sisters are depicted as smart, brave, perfectly capable of surviving on their own and, in fact, able to gain the upper hand in every encounter.[44] Despite their well-established smarts and survival skills, near the end Wichita and Little Rock inexplicably decide to abandon the group, and the safety and security of Bill Murray's mansion, and continue alone to Pacific Playland. Upon their arrival, they power up the amusement park—rides, lights, music, and all—an act that attracts hordes of zombies. To escape, the sisters run to the towering "Blast Off" and ride it to the top, however, a horde of zombies await them below. The climax of the film, therefore, becomes the classic showdown between hero (Columbus) and monster over the distressed damsels in a tower.[45] While the females begin as strong, stereotype-repelling, and empowered, in the end they prove unable to survive without men.[46]

In *Warm Bodies*, the women demand to be heard but, like Wichita and Little Rock, the female protagonist relies on men to save her in the end. Based on the book *Warm Bodies: A Novel* and set in the post-apocalyptic world, the film follows a zombie known as R, who falls in love at first sight with Julie, a living woman.[47] Through loving Julie, R becomes capable of rudimentary speech, but despite his favorite phrase, "Keep you safe," Julie needs no one to fret over her safety. She remains calm under pressure and yet can be vulnerable without signaling weakness. During the first attack, Julie and her friend Nora are the first to sense that something is amiss. As the zombies crash through the doors, the action is portrayed through R's eyes as he watches Julie wield a shotgun against zombie aggressors, her long blonde hair flowing in slow-motion glory. It is her strength and vibrancy, in addition to her beauty, that draws R to Julie, but all of these initial positive steps take a giant leap backwards with Julie's lack of judgment regarding her new zombified love interest.[48] Unbelievably, a girl who came of age during the apocalypse believes her best chance at survival is to follow the rotting corpse who just ate her boyfriend, simply because he manages to growl her name. Julie did not previously need a zombie bodyguard. Once Julie is in R's domain, however, she diverts to his lead. At the climax of the film, in fact, Julie becomes another damsel in distress. She trips while running away from the fast and ferocious "Boneys," which have become the true villains, and R must rescue her. This inability to effectively run does not affect Nora, Julie's friend, the only other named female character in the film. Throughout *Warm Bodies*, Nora shows pragmatism and capability. She manages to survive without being saved by either zombie or men. She also has no qualms about pointing a gun at Julie's father Grigio, leader of the human settlement, while informing him that she will "totally" pull the trigger. Yet none of this detracts from Nora's femininity. She dreams about becoming a nurse, a profession typically dominated by females, and reveals a stockpile of makeup she was "saving for a special occasion." Through these "feminine" behaviors, Julie, R, and Nora collaborate to ensure the survival of the entire human race. Using Nora's makeup, the women help disguise R as a living human while the gang tries to convince Grigio that the zombies are evolving for the better. In the end, Julie and a now-human R manage to persuade Grigio that the zombies are indeed being cured and that the real enemies are the Boneys. Grigio must not only listen to his daughter to survive, but he must also adapt to this new worldview. In other words, the militaristic male must adopt a stereotypically feminine trait (empathy) to save humanity.

So What Now?

While this article focuses on the role of women in reel life, it is worth drawing comparison with studies of women and children in real-life disasters. Unfortunately, a dearth of data exists on the subject.[49] From what can be discerned, women and children suffer more during disasters than their adult male counterparts, despite the Victorian adage of "women and children first." The sinking of the *Titanic* in April 1912 provides one shocking case study. According to popular legend, "the men stood in dinner jackets watching the women and children get into the lifeboats," yet, according to J.P.W. Rivers, one-third of the women and children did not gain access to the partly empty lifeboats. As a result, deaths of women and children were particularly high, especially of those from the steerage class.[50] Countries where women have fewer rights suffer increased female causalities, such as the restrictive dress women are required to wear in Bangladesh, which inhibited their ability to swim during the 2004 tsunami, and the tradition that considers it improper for women to leave the house unescorted by a male relative.[51] During the evacuation of the World Trade Center on September 11, 2001, women were twice as likely to be injured due to their high-heeled shoes and then, upon being discarded, the shoes causing others to trip over them in the staircases.[52]

Because women comprise a large portion of paid caregivers, they are also more likely to be at home with children, placing them at higher risk during plagues and earthquakes.[53] The fact that gender norms differently impact the experience of women and men in disaster situations does not, however, detract from the capacity of women to make positive contributions. "Disaster researchers are accumulating clear evidence," write B.H. Morrow and B.D. Phillips, "that, as a group, women are likely to respond, experience, and be affected by disasters in ways that are qualitatively different."[54] In Haiti, Marlene Richard, a 23-year-old nursery assistant, made sure to evacuate her building before retreating to safety during the 2010 earthquakes that devastated the country.[55] After studying tsunamis in class, 10-year-old Abby Wutlzer ran down the beaches of Lalomanu, Samoa, to warn the village of the impending wave.[56] When the tsunami of 2011 hit the shores of Japan, public worker Miki Endo saved thousands of lives by broadcasting the message for residents to "run away, run away fast," electing to stay behind and continue urging others to evacuate until the studio was destroyed and she perished.[57] During the aftermath of Hurricane Katrina in 2005, reports of female volunteers rescuing victims proliferate, including women helping to relocate the injured and sick to different hospitals, Coast Guard swimmer Sara Faulkner doing excellent rescue work, and female employees of Louisiana's Department of Wildlife and Fisheries saving lives.[58] The European Commission's Humanitarian Aid and Civil Protection Department has already taken notice and specifically targets women to teach life-saving techniques and strategies in Vietnam.[59] In a time of increasing diplomatic and economic disaster, Serbia lauds its female politicians for "steering their countries through the storms, even going so far as to require that every third candidate in elections must be female.[60] Yet, even as these contemporary women change the world, they continue to be judged according to gender, as Vlora Citaku, Minister for European Integration of Kosovo reflects: "[It] is almost impossible to forget even for a moment that I am a woman—I've been reminded of it since I became minister.... First of all they ask you, 'are you married?' 'What does your dad think of you traveling alone surrounded by all men?'"[61] Or worse, women are simply erased and forgotten. When the British invaded Connecticut in 1777, 16-year-old Sybil Ludington rode approximately 40 miles on horseback to rally the militia, nearly 10 miles more than the much-celebrated Paul Revere.[62]

A study that asked men and women to complete various tasks found that, while both genders had the same success, women consistently underestimated their performance, suggesting that women can be a factor in their own subordination by assuming inability.[63] Making capable women more visible in fiction can alleviate the confusion met by women in power and society's reticence to acknowledge and remember their contributions. Simply watching women react effectively will bolster courage and help to demolish the so-called confidence gap that prevents women's full participation in leadership roles.[64] In its present incarnation, the on-screen zombie apocalypse discourages women from taking leadership roles or trying to save themselves. Any apocalyptic society that presumes females to be second-class citizens would likely result in a higher rate of death. Women need art to start mimicking real life by proving that reliance on patriarchy will only decrease the chance for survival.

Zombie apocalypse narratives possess an opportunity to highlight the social construction of gender. For starters, there is an absence of gender amongst the zombies, who have no need to distinguish sex roles, while the living continue to find themselves clinging blindly to the patriarchal systems of their pre-apocalyptic world. Many zombie scholars interpret post-apocalyptic worlds as "fantasies of liberation."[65] "It is possible that patriarchy would cease to exist if we had such a drastic paradigm shift as a zombie outbreak, author Jen Rinaldi reflects. "Indeed, as long as there are no human beings to reinforce patriarchy, since patriarchy is social, it would cease to exist.... However, I have a feeling that both patriarchy and cockroaches will survive the apocalypse, at least as long as human beings live."[66] If the fictional worlds of film and television are any indication, Rinaldi's suspicion is probably correct—even post-apocalyptic zombie worlds might manage to retain the patriarchal status quo. While some say zombie films open up narrative spaces "that create genderless identificatory viewing positions," because zombies do not recognize gender, female fans of zombie films are still left with insufficient representations.[67] To levy this, perhaps women have created something similar to feminist theorist bell hooks' "oppositional gaze."[68] Growing up as a black female lacking strong, realistic on-screen role models, hooks developed a sort of oppositional gaze in which she reminds herself that the onscreen character did not represent her. A similar scenario inheres here. Because women cannot trust female representations in extant zombie films, they will need to maintain an oppositional gaze to remedy the shortcomings of female characters in the zombie apocalypse. If film genres exist as a sort of social contract between the filmmakers and the audience, audiences need to pressure filmmakers to give them strong female representations. With every new era of the zombie genre, as the "rom/com/zom" movement reveals, the female characters become more multifaceted, but they still fall short as role models. The representation of women in zombie apocalypse films may be lacking, but this is something that audiences can confront and push to change. Audiences deserve to see the strength and tenacity of real life women reflected to foster this empowerment. Perhaps watching *reel* women ensure their own survival in the zombie apocalypse will empower *real* women to follow suit in real-life disaster scenarios.

NOTES

1. Jen Rinaldi, "What Feminism Has to Say About World War Z," in *Braaaii-innnsss! From Academics to Zombies,* ed. Robert Smith (Ottawa: University of Ottawa Press, 2011), 15.

2. *Night of the Living Dead,* DVD, directed by George A. Romero, 1968, GT Media, 2007.

3. Barry Keith Grant, "Taking Back the *Night of the Living Dead*: George Romero, Feminism, and the Horror Film," in *The Dread of Difference: Gender and the Horror Film,* ed. Barry Keith Grant (Austin: University of Texas Press, 1996), 200–212. Natasha Patterson, "Becoming Zombie Grrrls On and Off Screen," in *Braaaii-innnsss! From Academics to Zombies,* ed. Robert Smith (Ottawa: University of Ottawa Press, 2011), 227.

4. For reflections on the problematic handling of gender in horror films, see Cynthia A. Freeland, "Feminist Frameworks for Horror Films," in *Post-Theory: Reconstructing Film Studies* (Madison: University of Wisconsin Press, 1996): 195–218; Linda Williams, "When the Woman Looks," and Barbara Creed, "Horror and the Monstrous-Feminine: An Imaginary Abjection," in Barry Keith Grant, ed., *The Dread of Difference: Gender and the Horror Film* (Austin: University of Texas Press, 1996): 15–34 and 35–65.

5. Brett Shenker, "Facebook Fandom Spotlight: Zombies," *Graphic Policy,* July 17 (2013), accessed 1/4/2014. Jennifer Rutherford, *Zombies* (New York: Routledge, 2013).

6. "The majority of zombie fare is produced by relatively privileged White men and what limits this entails on zombies' itinerary of terror," writes sociologist Todd Platts. Todd K. Platts, "Locating Zombies in the Sociology of Popular Culture," *Sociology Compass* 7 (2013): 553.

7. Clover draws from such films as *Psycho* (1960), *The Texas Chainsaw Massacre* (1974), and *Halloween* (1978) to support this theory. Carol J. Clover, "Her Body, Himself," in *Critical Visions in Film Theory,* ed. Timothy Corrigan, Patricia Wright, and Meta Mazaj (Boston: Bedford, 2011), 518.

8. Based on Richard Matheson's novel *I Am Legend* (1954), the apocalyptic world depicted in the films *Last Man on Earth* (1964) and *The Omega Man* (1971) is populated with vampires, not zombies, although the lumbering creatures that feed on the living served as inspiration for the flesh-eating zombies of Romero's films. *Last Man on Earth*, DVD, directed by Ubaldo Ragona and Sidney Salkow, 1964, 20th Century–Fox Home Entertainment, 2007. *The Omega Man*, DVD, directed by Boris Sagal, 1971, Warner Bros., 2000. Richard Matheson, *I Am Legend* (New York: Gold Medal, 1954).

9. Natasha Patterson, "Cannibalizing Gender and Genre: A Feminist Re-Vision of George Romero's Zombie Films," in *Zombie Culture: Autopsies of the Living Dead,* ed. Shawn McIntosh and Marc Leverette (Lanham, MD: Scarecrow Press, 2008), 110.

10. Patterson, "Cannibalizing Gender and Genre," 110. The character's name is spelled "Barbra" in the original and "Barbara" in the remake.

11. Grant, "Taking Back the *Night of the Living Dead*: George Romero, Feminism, and the Horror Film," 206–7.

12. Patterson, "Cannibalizing Gender and Genre," 111.

13. Patterson, "Cannibalizing Gender and Genre," 111.

14. Patterson, "Cannibalizing Gender and Genre," 114.

15. For discussion about the flat, genderless representation of women in *Land*

of the Dead, see Kim Paffenroth, *"Land of the Dead*: The Deepest Abyss of Hell and the Final Hope," in *Gospel of the Dead: George Romero's Visions of Hell on Earth* by Kim Paffenroth (Waco: Baylor University Press, 2006), 125.

16. Laura Mulvey, "Visual Pleasure and Narrative Cinema." *Screen* 16.3 (1975): 6–18. Stephen Harper, "'I could kiss you, you bitch': Race, Gender, and Sexuality in *Resident Evil* and *Resident Evil 2: Apocalypse*," *Jump Cut* 49 (2007): 1.

17. *Alien*, DVD, directed by Ridley Scott, 1979, 20th Century–Fox Home Entertainment, 1999.

18. Rinaldi, "What Feminism Has to Say About World War Z," 15.

19. Christopher G. Williams, "Birthing an Undead Family: Reification of the Mother's Role in the Gothic Landscape of *28 Days Later*," *Gothic Studies* 9.2 (2007): 41.

20. Williams draws attention to the film's title. Literally, Jim wakes up from a coma, naked in a hospital bed, 28 days after the outbreak began, but the film's preoccupation with abject blood and bodily fluids points to a deeper meaning, that is, 28 days is the length of the average women's menses. Williams, "Birthing an Undead Family," 42.

21. Williams, "Birthing an Undead Family," 43.

22. The original ending of the film, on the DVD special features, has gun-toting Selena and Hannah leaving Jim's corpse behind as they set off into the world. Williams suggests that the "contradiction could have been rectified with Boyle's 'proper ending' and the ultimate 'rescue' of the women through resistance to all collectives." Aside from the problematic use of rape as a reminder of their female vulnerability, this ending may have given a more positive outlook on the fluidity of gender roles, but a film should not be judged according to what it "could" have been. Williams, "Birthing an Undead Family," 42.

23. Williams, "Birthing an Undead Family," 38.

24. *World War Z*, DVD, directed by Marc Forster, 2013, Paramount Pictures, 2013.

25. A similar situation ensues in the 2007 cinematic adaptation of *I Am Legend*, where female character Anna shows her meddle by saving the male protagonist while retaining aspects of her feminine side (nursing the protagonist's injuries, caring for a young boy, and maintaining her spirituality). Despite these positive attributes, the minor character appears late in a narrative otherwise centered on the male protagonist.

26. Barkman, "Women in a Zombie Apocalypse," in *The Walking Dead and Philosophy: Zombie Apocalypse Now*, ed. Wayne Yuen (Chicago: Open Court, 2012), 99.

27. "Prison Council," *The Walking Dead*, DVD, directed by Frank Darabont, 2010–2013, Anchor Home Entertainment, 2010–2013.

28. In the comic book series, when the survivors form the committee to share decision-making, four men are selected to the Council, to which Rick responds, "No women?" "Prison Council," *The Walking Dead*, 2013.

29. Barkman, "Women in a Zombie Apocalypse," 98.

30. Danee Pye and Peter O'Sullivan, "Dead Man's Party," in *The Walking Dead and Philosophy: Zombie Apocalypse Now*, ed. Wayne Yuen (Chicago: Open Court), 2013, 107–116.

31. Hattie McDaniel is an African American actress best known for her Academy Award-winning role of Mammy in *Gone with the Wind* (1939). "Tell It to the Frogs," *The Walking Dead*, 2010.

32. Robert Kirkman, *The Walking Dead* (comic book series) (Berkeley: Image Comics, 2003–present), Issue #3.

33. Mia Warren, "Michonne: *The Walking Dead's* Last Feminist Icon?" *Sarcasmia*, Oct. 22, 2013. Barkman defends Lori's justification for the gendered division of labor, arguing that clean clothes are important to "help distinguish the living from the dead" and to enhance a sense of wellbeing.

34. Joanna Robinson, "Does *The Walking Dead* Still Have a Woman Problem?" *Think Pieces*, Nov. 4, 2013.

35. Megan Kearns, "Nothing Can Save *The Walking Dead's* Sexist Woman Problem," *Bitch Flicks*, May 1, 2013.

36. Barkman, "Women in a Zombie Apocalypse," 99.

37. Kearns, "Nothing Can Save *The Walking Dead's* Sexist Woman Problem."

38. Lorraine Berry, "Walking Dead: Still a White Patriarchy." *Salon*, April 1, 2013.

39. For discussion of *The Walking Dead's* problematic racial politics, see Berry, "Walking Dead: Still a White Patriarchy" and Warren, "Michonne: *The Walking Dead's* Last Feminist Icon?"

40. Barkman, "Women in a Zombie Apocalypse," 106.

41. "On the other hand," Barkman adds, "Ed or the Governor are instances of men who are masculine, but because they are deeply immoral, they should never lead; in such cases, it'd be easy to agree that a woman—one who has leadership qualities—should lead." Barkman, "Women in a Zombie Apocalypse," 99.

42. Barkman, "Women in a Zombie Apocalypse," 102.

43. The DVD's special features clarifies that Dianne escapes the horde and sleeps in a tree "until the whole thing blows over." If one were to still consider this scene as part of the canon, it would mean that, once more, Dianne is much better off without men telling her how to survive.

44. A flashback shows the pair using their feminine wiles to con men out of money, and the film twice shows them tricking the unsuspecting male protagonists— the mild-mannered Columbus and tough-guy Tallahassee—out of the masculine items of both guns and vehicles.

45. Kyle Bishop, "Vacationing in *Zombieland*: The Classical Functions of the Modern Zombie Comedy," *Journal of the Fantastic in the Arts* 22.1 (2011): 11.

46. See also Leah Murray, "When They Aren't Eating Us, They Bring Us Together: Zombies and the American Social Contract," in *The Undead and Philosophy: Chicken Soup for the Soulless*, ed. Richard Greene and K. Silem Mohammad (Chicago: Open Court, 2006), 212.

47. Isaac Marion, *Warm Bodies: A Novel* (New York: Emily Bestler Books, 2011).

48. In the novel, R is initially drawn to Julie's fear and vulnerability. "She is huddled in a corner, unarmed, sobbing and screaming as M creeps toward her.... I shove M aside and snarl, 'No. Mine.'" R's command, "Mine," implies patriarchal ownership of Julie. Marion, *Warm Bodies: A Novel*, 19.

49. Lisa Eklund and Siri Tellier, "Gender and International Crisis Response: Do We Have the Data, and Does It Matter?" *Disasters* 36.4 (2012): 13.

50. J.P.W. Rivers, "Women and Children Last: An Essay On Sex Discrimination in Disasters," *Disasters* 6.4 (1982): 258.

51. Eklund and Tellier, "Gender and International Crisis Response: Do We Have the Data, and Does It Matter?" 1.

52. Amanda Ripley, *The Unthinkable: Who Survives When Disaster Strikes—and Why* (New York: Crown, 2008), 89.

53. "Women and Disaster," *Canadian Women's Health Network* 12.2 (2010): 4.

54. B.H. Morrow and B.D. Phillips, "What's Gender 'Got to Do with It'?" *Mass Emergency Disasters* 17.1 (1999): 5.

55. "Witnesses and Heroes of the Haiti Earthquake," *Oxfam International,* January 23, 2010.

56. Kerri Ritchie, "Young Girl Honoured as Tsunami Hero," *ABC News,* October 14, 2009.

57. Lucas Kavner, "Japanese Public Worker's Heroic Final Moments," *The Huffington Post.* April 5, 2011.

58. Rebecca Webber, "The Rescuers: Hurricane Katrina's Female Heroes," *Glamour* 103.12 (2005): 234.

59. "Recognize the Strength of Women and Girls in Reducing Disaster Risks: Stories from Viet Nam," edited by Miguel Coulier, Ian Wilderspin and Pernille Goodall, United Nations Entity for Gender Equality and the Empowerment of Women, 2012.

60. Bill Tiernan, "Women Emerge as Crisis Leaders in Macho Balkans," *USA Today,* March 4, 2013.

61. Tiernan, "Women Emerge as Crisis Leaders in Macho Balkans."

62. "Young and Brave: Girls Changing History," Young and Brave: Girls Changing History, http://www.nwhm.org/online-exhibits/youngandbrave/index.html, accessed August 8, 2014.

63. Katty Kay and Claire Shipman, "The Confidence Gap," *The Atlantic,* April 14, 2014.

64. Kay and Shipman, "The Confidence Gap."

65. Peter Dendle, "The Zombie as Barometer of Cultural Anxiety," in *Monsters and the Monstrous: Myths and Metaphors of Enduring Evil,* ed. Niall Scott (Amsterdam: At the Interfaces, Probing the Boundaries, 2007), 54.

66. Rinaldi, "What Feminism Has to Say about World War Z," 9.

67. Patterson, "Cannibalizing Gender and Genre," 115.

68. bell hooks, "The Oppositional Gaze: Black Female Spectators," in *Film and Theory: An Anthology,* ed. Robert Stam and Toby Miller (Malden, MA: Blackwell, 2000), 510–523.

BIBLIOGRAPHY

Barkman, Ashley. "Women in a Zombie Apocalypse." In *The Walking Dead and Philosophy: Zombie Apocalypse Now,* edited by Wayne Yuen, 97–106. Chicago: Open Court, 2012.

Berry, Lorraine. "Walking Dead: Still a White Patriarchy." *Salon,* April 1, 2013.

Bishop, Kyle. "Vacationing in *Zombieland*: The Classical Functions of the Modern Zombie Comedy." *Journal of the Fantastic in the Arts* 22.1 (2011): 24–37.

Brooks, Max. *World War Z: An Oral History of the Zombie War.* New York: Three Rivers Press, 2006.

Clover, Carol J. "Her Body, Himself." In *Critical Visions in Film Theory,* edited by Timothy Corrigan, Patricia Wright, and Meta Mazaj, 513–530. Boston: Bedford, 2011.

Creed, Barbara. "Horror and the Monstrous-Feminine: An Imaginary Abjection." In *The Dread of Difference: Gender and the Horror Film,* edited by Barry Keith Grant, 35–65. Austin: University of Texas Press, 1996.

Dendle, Peter. "The Zombie as Barometer of Cultural Anxiety." In *Monsters and the Monstrous: Myths and Metaphors of Enduring Evil,* edited by Niall Scott, 45–60. Amsterdam: At the Interfaces, Probing the Boundaries, 2007.

"Disasters and Gender Statistics." 2009. International Union for Conservation of

Nature. http://cmsdata.iucn.org/downloads/disaster_and_gender_statistics.pdf, accessed Dec. 11, 2013.

Eklund, Lisa, and Siri Tellier, "Gender and International Crisis Response: Do We Have the Data, and Does It Matter?" *Disasters* 36.4 (2012): 13.

Enarson, Elaine. 2000. "Gender and Natural Disasters." InFocus Programme on Crisis Response and Reconstruction—Working Paper. http://www.ilo.org/wcmsp5/groups/public/-ed_emp/-emp_ent/-ifp_crisis/documents/publication/wcms_116391.pdf, accessed Nov. 26, 2013.

Freeland, Cynthia A. "Feminist Frameworks for Horror Films." In *Post-Theory: Reconstructing Film Studies*, 195–218. Madison: University of Wisconsin Press, 1996.

Grant, Barry Keith. "Taking Back the *Night of the Living Dead*: George Romero, Feminism, and the Horror Film." In *The Dread of Difference: Gender and the Horror Film*, edited by Barry Keith Grant, 200–12. Austin: University of Texas Press, 1996.

Harper, Stephen. "I could kiss you, you bitch": Race, Gender, and Sexuality in *Resident Evil* and *Resident Evil 2: Apocalypse*." *Jump Cut* (2007), http://www.ejumpcut.org/archive/jc49.2007/HarperResEvil/index.html, Oct. 30, 2011.

_____. "They're Us: Representations of Women in George Romero's 'Living Dead' Series." *The Journal of Cult Media*, 2003. http://www.cult-media.com/issue3/Aharper.htm, accessed Oct. 17, 2011.

hooks, bell. "The Oppositional Gaze: Black Female Spectators." In *Film and Theory: An Anthology*, edited by Robert Stam and Toby Miller, 510–523. Malden, MA: Blackwell, 2000.

Kavner, Lucas. "Japanese Public Worker's Heroic Final Moments." *The Huffington Post*, April 5, 2011.

Kay, Katty, and Claire Shipman. "The Confidence Gap." *The Atlantic*, April 14, 2014.

Kearns, Megan. "Nothing Can Save *The Walking Dead's* Sexist Woman Problem." *Bitch Flicks*, May 1, 2013.

Kirkman, Robert. *The Walking Dead* (graphic novel). Berkeley, CA: Image Comics, 2003–present.

"List of Zombie Films." Wikipedia, 2013. http://en.wikipedia.org/wiki/List_of_zombie_films, accessed Dec. 10, 2013.

Marion, Isaac. *Warm Bodies: A Novel*. New York: Emily Bestler Books, 2011.

Matheson, Richard. *I Am Legend*. New York: Gold Medal, 1954.

Morrow, B.H., and B.D. Phillips, "What's Gender 'Got to Do with It'?" *Mass Emergency Disasters* 17.1 (1999): 5.

Mulvey, Laura. "Visual Pleasure and Narrative Cinema." *Screen* 16.3 (1975): 6–18.

Murray, Leah A. "When They Aren't Eating Us, They Bring Us Together: Zombies and the American Social Contract." In *The Undead and Philosophy: Chicken Soup for the Soulless*, edited by Richard Greene and K. Silem Mohammad, 211–20. Chicago: Open Court, 2006.

Paffenroth, Kim. "*Land of the Dead*: The Deepest Abyss of Hell and the Final Hope." In *Gospel of the Living Dead: George Romero's Visions of Hell on Earth*. Waco, TX: Baylor University Press, 115–132, 2006.

Pagano, David. "The Space of Apocalypse in Zombie Cinema." In *Zombie Culture: Autopsies of the Living Dead*," edited by Shawn McIntosh and Marc Leverette, 71–86. Lanham, MD: Scarecrow Press, 2008.

Paravisini-Gebert, Lizabeth. "Women Possessed: Eroticism and Exoticism in the Representation of Woman as Zombie." In *Sacred Possessions: Vodou, Santeria, Obeah, and the Caribbean*, edited by Margarite Fernandez Olmos and Lizabeth Paravisini-Gebert, 37–58. New Brunswick, NJ: Rutgers University Press, 1997.

Patterson, Natasha. "Becoming Zombie Grrrls On and Off Screen." In *Braaaiiinnnsss! From Academics to Zombies,* edited by Robert Smith, 226–47. Ottawa: University of Ottawa Press, 2011.

_____. "Cannibalizing Gender and Genre: A Feminist Re-Vision of George Romero's Zombie Films." In *Zombie Culture: Autopsies of the Living Dead,"* edited by Shawn McIntosh and Marc Leverette, 101–118. Lanham, MD: Scarecrow Press, 2008.

Platts, Todd K. "Locating Zombies in the Sociology of Popular Culture." *Sociology Compass* 7 (2013): 553.

Preston, John. "Prosthetic White Hyper-Masculinities and 'Disaster Education.'" *Ethnicities* 10.3 (2010): 331–343.

"Prison Council." The Walking Dead Wiki, 2013. http://walkingdead.wikia.com/wiki/Prison_Council, accessed Dec. 11, 2013.

Pye, Danee, and Peter O'Sullivan, "Dead Man's Party." In *The Walking Dead and Philosophy: Zombie Apocalypse Now,* edited by Wayne Yuen, 107–116. Chicago: Open Court, 2013.

Rinaldi, Jen. "What Feminism Has to Say About World War Z." In *Braaaiiinnnsss! From Academics to Zombies,* edited by Robert Smith, 9–19. Ottawa: University of Ottawa Press, 2011.

Ripley, Amanda. *The Unthinkable: Who Survives When Disaster Strikes—and Why.* New York: Crown, 2008.

Ritchie, Kerri. "Young Girl Honoured as Tsunami Hero." *ABC News.* October 14, 2009.

Rivers, J.P.W. "Women and Children Last: An Essay on Sex Discrimination in Disasters." *Disasters* 6.4 (1982): 258.

Robinson, Joanna. "Does *The Walking Dead* Still Have a Woman Problem?" *Think Pieces,* Nov. 4, 2013.

Rutherford, Jennifer. *Zombies.* New York: Routledge, 2013.

Shenker, Brett. "Facebook Fandom Spotlight: Zombies." *Graphic Policy,* July 17, 2013, accessed 1/4/2014.

Spines, Christine. "Horror Movies and the Women Who Love Them." *Entertainment Weekly,* 24 July 2009, http://www.ew.com/ew/article/0,,20293304,00.html, accessed Nov. 15, 2011.

Tiernan, Bill. "Women Emerge as Crisis Leaders in Macho Balkans." *USA Today,* March 4, 2013.

Tosca, Susana Pajares. "Reading Resident Evil-Code Veronica X." MelbourneDAC 01: 206–216, 2006.

Warren, Mia. "Michonne: *The Walking Dead's* Last Feminist Icon?" *Sarcasmia,* Oct. 22, 2013.

Webber, Rebecca. "The Rescuers: Hurricane Katrina's Female Heroes." *Glamour* 103.12 (2005): 234.

Williams, G. Christopher. "Birthing an Undead Family: Reification of the Mother's Role in the Gothic Landscape of *28 Days Later.*" *Gothic Studies* 9. 2 (2007): 33–44.

Williams, Linda. "When the Woman Looks." In *The Dread of Difference: Gender and the Horror Film,* ed. Barry Keith Grant, 15–34. Austin: University of Texas Press, 1996.

"Witnesses and Heroes of the Haiti Earthquake." *Oxfam International,* January 23, 2010.

"Woman and Disaster." *Canadian Women's Health Network* 12.2 (2010): 4.

"Women and Disasters." Wiki Gender. 2009. http://www.wikigender.org/index.php/Women_and_Disasters, accessed Nov. 26, 2013.

"Young and Brave: Girls Changing History." Young and Brave: Girls Changing History. http://www.nwhm.org/online-exhibits/youngandbrave/index.html, accessed August 8, 2014.

Postmortem Ethics
Personhood at the Margins of Death

CORY ANDREW LABRECQUE

Zombies in Ethics Education: Mindless Chatter, Creative Pedagogy or a Sign of the Times?

An important part of a teacher's mandate is to equip students with critical thinking skills that can go on to be cultivated in—and applied to—learning contexts that are beyond the boundaries of usual coursework and outside the walls of the academy. Although the objective, by-and-large, is for students to feel that they "own" the material they study and analyze, there is often an expectation that they leave the classroom wanting to engage—in a constructive and balanced way—ideas, perceptions, and issues that are different from the ones to which they were formally exposed. This, along with the ability to coherently integrate various streams of knowledge and to perpetually assess one's own thinking, is the mark of a robust education. In the field of bioethics—these are among the core competencies.

One of the great challenges of bioethics pedagogy is to bridge the gap between the safe, deliberated, and contained environment of the classroom and the rushed, messy, stressful life-and-death "real-world" of patients and their families, research participants, and health care providers who are often hard-pressed to make weighty decisions quickly. In class, students discuss theories of moral status; on the ground, they are grappling with how to proceed with a baby born at 24 weeks gestation. In class, students debate the difference between proportionate and disproportionate (what some might call "overzealous") treatment; on the ground, they are asked whether or not to withdraw artificial nutrition and hydration from a dying parent.

In the limitedness of human knowing, one might consider the acumen

246

of Immanuel Kant—one of the most influential thinkers in Western philosophy—who underlined the significance of moral imagination while championing the authority of reason. If it is true that "the imagination can be shown to contribute to the interpretation, as well as to the constitution, of experience," then all the more should such a faculty be fostered and focused—carefully so—in the study of ethics.[1]

The Centers for Disease Control and Prevention appears to have taken heed of this Kantian idea. In May 2011, they launched a "tongue in cheek" campaign to coach the public on how to deal with a zombie apocalypse as a creative way to address public health preparedness and response.[2] Perhaps this should come as no surprise; the CDC is in Atlanta, Georgia, which is a stone's throw away from where the award-winning hit television series *The Walking Dead* is filmed.

Many (bio)ethicists find the idea of a zombie apocalypse to be rather compelling, at least for education purposes. Since talk of zombies is an exercise in imagination and the threat does not seem immediately evident, students can approach the topic freely and without much reservation in their application of bioethical theory and principle, perhaps without the heavy-heartedness that sometimes comes with a painful real-life experience (which often, in itself, offers valuable insight for future decision-making).[3] It is, of course, when someone relates zombies and their characteristics (sometimes referred to as their "symptoms") to actual patient populations that the discussion becomes somewhat more heated. Indeed, talk of patients in light of the zombie phenomenon in bioethics education can only ever serve as a point of departure for difficult topics and should never mean to trivialize or detract from the impact that diseases have on patients and their caregivers.

If imagination, then, sheds light on experience, thoughts of a zombie apocalypse can help with the basic and preliminary task of identifying the ethical issues in a situation at hand and to frame or re-frame a number of ethical questions that are regularly examined and debated within the more familiar contexts of medicine, public health, religion, and neuroscience.[4] Here, one might ask: is the zombie dead or "just" infected? What are the implications of the answer to this question for how one ought to deal with zombies? When is a human being no longer a person? How does "zombification" impact personhood? Do memory and identity linger in the zombie? Do zombies have free will? What effect does "zombification" have on relationships? Is the zombie, for instance, still one's spouse, sibling, child, or parent? Does it all come down to the brain? What effect does "zombification" have on the dignity and sanctity of the human body? What are the ethics of isolation and quarantine? How should healthcare resources be allocated when pandemics hit? Under what circumstances, if any, do "standards" of ethical conduct become obsolete?

A zombie apocalypse presses an even more fundamental concern that has interested—and continues to interest—those who study eschatology, war, and global catastrophe: why be ethical when the end is upon us? Or, as *The Walking Dead* brings to the fore: if survival is the ultimate end, should we re-think our ethics or abandon them altogether?

Although this short chapter cannot address all of these intriguing questions, they are listed for the reader to go on and engage at greater length. For the purpose of this essay, this discussion of zombies is situated within the larger context of moral status and the concept of personhood.

Categorizing, Bestowing and Protecting Personhood

Standing under the celestial canopy, the Biblical psalmist—acknowledging his fragility—contemplates: "When I look at your heavens, the work of your fingers, the moon and the stars that you have established; what are human beings that you are mindful of them, mortals that you care for them?"[5] The personhood question is perennial and, in many ways, finds itself at the core of much bioethical inquiry. In research ethics, some ask why certain things that would never be carried out on humans are done, often without question, on non-human animals. In ethics at the beginning of life, a central concern is the moral status of the fetus at different times over the gestation period. At the end of life, the continuity of a patient's personhood as she moves through the stages of Alzheimer's disease is evaluated; theorists and practitioners alike mull over the "then self" (before the onset of disease) and the "now self" (post-onset), wondering which ought to have precedence over the course of the patient's care.[6]

James Walters, a professor of Christian bioethics, offers a working description of the two primary traditions in personhood thought that will be helpful here. In physicalism (not to be confused with the philosophical thesis sometimes denoted as "materialism"), "the essence of a person is found in his or her biological make-up. All humans are persons, ipso facto."[7] A good case in point is the Roman Catholic tradition which finds at the very root of its theological anthropology and social teaching the inherent value of human life and the understanding that from conception a human being must be recognized as a person clothed with a dignity that stems from being made in the image and likeness of God.[8] On the contrary, personalism perceives "the essence of a person as being located in one's mental capacities and ability to use these in satisfying ways. Whether one is a human is not important."[9] Adherents to this second tradition abound and their grappling with the definition of personhood is worth some attention.

In his *Foundations of the Metaphysics of Morals*, Kant argues that human beings are of unconditional worth and that one's inherent dignity, by virtue of autonomy and rationality, demands respect for all persons.[10] "Beings whose existence does not depend on our will but on nature, if they are not rational beings, have only a relative worth as means and are therefore called 'things'; on the other hand, rational beings are designated 'persons.'"[11] Only "rational nature," Kant argues, "exists as an end in itself"; it is rationality, then, that bestows unconditional worth.[12] What Kant would have argued in the context of the study in this essay is up for grabs, but it would seem that the zombie does not fit the latter interpretation. Rationality, glorified by the likes of Kant, and other eminent philosophers such as Aristotle, Plato, and Locke as *the* distinct feature that separates humankind from the rest of the non-human world, is crucial to contemporary functional definitions of personhood.

Bioethics pioneer Joseph Fletcher, the former Episcopalian priest who penned such seminal texts as *Situation Ethics* and *Humanhood*, is testimony to the influence of these philosophers. It is clear to Fletcher that

> humans without some minimum of intelligence or mental capacity are not persons, no matter how many of their organs are active, no matter how spontaneous their living processes are. If the cerebrum is gone, due to disease or accident, and only the mid-brain or brainstem is keeping autonomic functions going, they are only objects, not subjects–they are *its*, not *thous*. Just because heart, lungs, and the neurological and vascular systems persist, we cannot say a *person* exists. Noncerebral organisms are not personal.[13]

Thus, only *cerebral* organisms are persons. The category of "cerebral organisms" for Fletcher seems to refer exclusively exclusively to humans, not only in light of the fact that his manuscript dwells on the question of *human*hood and *human*ness, but also because cerebration and neocortical function constitute the fifteenth criterion in his "profile of *man*." Others, though, might also include here non-human organisms, such as primates, which exhibit higher brain function. A definition of personhood drawn primarily from the (adequate) functioning of the cerebrum and its associated anatomical structures is prototypical. References to the human person are made solely in connection to the neurological status and cognitive capacity of the organism.

Human persons, in this regard, are properly functioning machines that operate on par with, or above, the minimal standards delineated by Fletcher in his fifteen positive criteria that constitute "a profile of man": minimal intelligence (any human whose I.Q. score is less than 40 on a standard Stanford-Binet test is questionably a person; an I.Q. less than 20 is certainly not a person), self-awareness, self-control, a sense of time, a sense of futurity, a sense of the past, the capacity to relate to others, concern for others, communication, control of existence, curiosity, change and changeability, balance of rationality and feeling, idiosyncrasy, and neocortical function.[14] It is unclear which features are necessary and which are sufficient for personhood, as is often the problem with functional theorists. Fletcher seems to offer here, as Edward Keyserlingk—bioethicist and former Government of Canada Public Service Integrity Officer—rightly concludes, indicators of the "good," the "mature," or the "optimal" life.[15] Zombies, evidently, are out of the category.

Nevertheless, these criteria brought to the fore for analysis are problematic as it is evident that any "indisputably" functional human organism ceases to be a person by these standards throughout the course of daily living; not just in temperament and perspective, but also in the natural periodic suspension of consciousness such as in the regularity of sleep and in the occasional experiences of fainting, coma, and anaesthetization, for instance. A more pressing issue, perhaps, is the harmful connection Fletcher forges between disease and nonpersonhood. He argues that lack of self-awareness is pathological, that lack of concern for others is a "clinical indication of psychopathology," and that lack of control of existence characteristically applies to "severe cases of toxic and degenerative psychosis."[16] Failure to meet any one of these criteria not only risks one's status as person, but also indicates underlying disease.

In his *Abortion and Infanticide*, American philosopher Michael Tooley infers that while there are certain properties sufficient, but not necessary, for personhood (such as non-momentary interests and agency), others (such as self-consciousness, in the absence of interests, and rationality, unless defined in the weak sense of having thoughts or temporal judgment), are neither sufficient nor necessary.[17] Still, he goes on to entertain the allegation that a human being is not a person unless he or she is capable of consciousness.[18] As a personalist—that is, as a theorist who determines personhood based on whether or not a being can adequately carry out particular functions—Tooley is convinced that most people would agree that

> anything that has, and has exercised, all of the following capacities is a person, and that anything that has never had any of them is not a person: the capacity for self-consciousness; the capacity to think; the capacity for rational thought; the capacity to arrive at decisions by deliberation; the capacity to envisage a future for oneself; the capacity to remember a past involving oneself; the capacity for being a subject of non-momentary interests; the capacity to use language.[19]

Whatever the case may be, Tooley admits that, in spite of his extensive discussion of person-making characteristics, he cannot determine the precise moment at which a human being becomes a person or loses personhood given that neither contemporary science nor our working definitions of properties supposed vital for personhood are adequate to this end.[20] Tooley does assert, however, that one's right to life is directly proportional to whether he or she has consciousness because one who lacks consciousness has no desire for continued existence.[21] An organism "possesses a serious right to life only if it possesses the concept of a self as a continuing subject of experiences and other mental states, and believes that it is itself such a continuing entity."[22] A person does not forfeit this right, once acquired, if he or she temporarily loses these capacities as long as it is certain, in principle, that they will be restored. However, a person does lose his or her right to life, Tooley says, if the relevant capacities can never again be regained.[23]

This, of course, is a central theme in early seasons of *The Walking Dead* and the fear of such a loss, especially in loved ones, is skillfully captured in the character of Hershel Greene. In the beginning, the farmer-veterinarian desperately clings to the hope that the zombie state is only temporal; the viewer quickly learns that Hershel has been keeping his zombified wife in a barn on the edge of his property. He holds out on the possibility of a cure or the chance that the condition might be reversible; as such, Hershel initially believes that survivors need only wait things out and not embark on a premature killing rampage. Eventually, hope dissipates and a debate over how one is to live in a post-apocalyptic world ensues.

It is also worth making reference here to the 2013 zombie-romance film, *Warm Bodies*. The premise is very much built on the notion that traces of personhood and a lingering capacity for relationality continue to reside in the zombie. The main character, R, is a young male zombie who gradually returns to his former (pre-zombie) self because of an amorous bond that strengthens between him and Julie, one of the few human beings not infected by the zombie plague. This is brought to light here with the purpose of showing how different descriptions of what the zombie is—particularly regarding whether or not the zombie state or condition is reversible—makes talk about ethics difficult. Importantly, this problem underlines the importance of semantics. Public bioethical discourse that welcomes plurality must first be attentive to how contributors from various disciplines understand foundational concepts—such as "personhood," "consciousness," and, in this case, "zombie"—before tackling the relevant ethical questions.

Australian bioethicist Peter Singer, of *Animal Liberation* fame, reads Tooley in this way: "Continued existence cannot be in the interests of a being who *never* has had the concept of a continuing self–that is, never has been able to conceive of itself as existing over time…. To have a right to life, one must have, or at least at one time have had, the concept of having a continued existence."[24] Does this imply that a zombie, who once had a sense of existence over time, has a right to life and an interest in continued existence? Or, is this right forfeit because—when it comes to zombies (*Warm Bodies* aside)—there can be no anticipation of restored rationality, self-consciousness, or sense of continuity?

This idea of a continued self, or of the psychological continuity of self that is perpetuated by a series of psychological connections, has been made rather popular in the work on personal identity by British philosopher Derek

Parfit.[25] Jeff McMahan, an American philosopher who once studied under Parfit, brings to the fore the influential view that

> we are essentially psychological beings whose continued existence consists in psychological continuity, or the holding of certain continuities of mental life over time. A person twenty years ago and a person today are the same person if and only if the latter is related to the former by an overlapping series of psychological connections, involving memory and the persistence or gradual evolution of a particular set of desires, beliefs, intentions, dispositions of character and so on.[26]

In this vein, Australian bioethicist Helga Kuhse argues that no such continuity exists between the person before the onset of disease and the person after onset. She is convinced, for instance, that "severely demented patients, in distinction from persons, no longer have an interest in their own continued existence."[27] If one assumes that Max Brooks—author of *The Zombie Survival Guide*—has the pathogenesis and prognosis right, in the case of the zombie the "then" self is eradicated by the *Solanum* virus and, therefore, one can only speak of the surviving being as an animated nonperson that no longer has any connection to the "then" self.[28] The exploration of ethics in this essay will go on to engage this particular understanding of the zombie as incurable, apathetic to self-preservation, insatiable, and—ultimately—as the walking *dead*, which is a loaded and perceptive description.

It may be of interest to the reader to note that reactions to this quandary of defining "person" have been shaped, in large part, by arrogance; they are either preposterously simplistic or dangerously absurd. Although the U.S. Supreme Court ruled in 1857 that black people were not persons (*Dred Scott v. Sanford*) and decided in 1973 that unborn people were not persons (*Roe v. Wade*), the law did recognize entities such as corporations (the 1886 case of *Santa Clara County v. Southern Pacific Railroad Company*), partnerships, or, at times, the collection of property as "persons" entitled to the legal rights and protections that the Constitution afforded any white American male. The principle of corporate personhood slipped into history rather quietly and unattested, elevating the corporation to a status once judged unworthy of black Americans. It was only in the last century, in 1929, that Canadian women were legally declared persons (the term, up until then, was synonymous with "human male") under the British North America Act, thereby becoming eligible to appointment in the Senate. Beforehand, British Common Law claimed that women were "persons in the matter of pains and penalties, but not in the matter of rights and privileges"; hence, they had not been full-fledged members of the community of persons.[29] History exposes the selective bestowal of personhood on those whose function or utility in society was judged valuable. This exercise in zombie personhood raises two important questions: (1) who judges and by what standards? and (2) what are the practical implications of being excluded from the category of persons?

In *The Problem of Being Human*, Lloyd Averill—reading the work of Quaker economist-philosopher Kenneth Boulding—brings insight to this discussion[30]:

> Existence "on the boundary"—awareness at once of the stubbornness and of the fragility of human life, of the possibility of an expanded humanness and humaneness beyond all our dreams, and yet of the possibility, too, of a dehumanization beyond all our nightmares—this boundary-line situation is probably the most pervasive theme of our present culture.[31]

The zombie apocalypse stirs up a certain fear of this existence "on the boundary"; it brings about a discomfort with those categories of personhood/nonpersonhood and normal/abnormal, for instance, that guide moral discourse. The zombie is convoluting. On the one hand, it possesses detectable traces of the human body and begs the recall of a life once lived. On the other, the zombie is man "dehumanized beyond all our nightmares" and the want to see to its annihilation is either rooted in fear of the other or in fear of *becoming* the other.

The Moral Status of Reanimated Flesh

According to Tom Beauchamp and James Childress, authors of *Principles of Biomedical Ethics* (a pivotal text in the field that is now in its sixth edition), "[t]o have moral status is to deserve the protection afforded by moral norms" and "[s]uch protections are afforded only to entities that can be morally wronged by actions."[32] If one bases moral status on physicalism, zombies would be bereft of it because even though they were living human beings at one time, death marks the end of a mortal's existence as a human being and, therefore, a person. While it is true that contemporary medicine resuscitates human beings periodically, the patient is restored to life and is destined for death at some later time. However, the zombie, once reanimated, remains dead and goes on to experience the effects of death, such as the irreversibility of integrated function and the perpetual decay of the organism, although some underlying neural mechanism leads to a loss of satiety and its characteristic drive to consume.[33] Unlike the living human being, the zombie goes about its zombie ways without (according to Brooks) an oxygen-dependent brain, lungs that engage in pulmonary gas exchange, or working circulatory and digestive systems.[34] Living human beings must have these intact—either "naturally" or through biotechnological intervention—in order to remain "alive" as it were.

In his text, Brooks calls zombies "somnambulists" and denunciates them as "the greatest threat to humanity, other than humanity itself."[35] No longer a human being, the zombie is an embodied virus (identified as "Solanum") that, as an obligate intracellular parasite, constantly seeks a host in order to replicate. He makes this plain: "To call them predators and us prey would be inaccurate. They are a plague, and the human race their host."[36] Brooks insists that "[n]o matter who a person was in his former life, that person is gone, replaced by a mindless automaton with no instinct other than for feeding."[37] The reference here to the earlier discussion of the continuity of self as well as the appeal to Cartesian language is not lost on readers. Zombies as unconscious *automata* "possessing neither thought nor feeling nor a mental life of any kind" are devoid of personhood—certainly by the standards of functional personalism.[38]

Regardless of whether the claim to life is embedded in one's humanhood or in one's capacity for satisfactory cognitive function, physicalists and personalists alike would agree that there is no reason to account for zombies' interests in avoiding suffering (since they are not sentient), their right to life (since they are dead), or for the preservation of their dignity. One could argue, however, that even the dead are owed a certain measure of respect; many of the religions, for instance, bind on the living an obligation to pray for the dead, to appropriately treat the bodies of the dead, and to assure them a proper burial. It goes without saying that these last two tenets are particularly difficult to meet when the zombie is trying to eat the living person who chooses to continue abiding by these principles in the post-apocalyptic world.

Andrea, a primary character on *The Walking Dead*, speaks to this dilemma in the shadow of the survivors' extermination of the zombies who were kept hidden in Hershel's barn. Scanning the aftermath, she makes a rather revelatory announcement: "we bury the ones we love and burn the rest."[39] Accordingly, the human inclination to bury the dead remains somewhat intact even for certain zombies; this seems to be the case perhaps only for those with whom there had been some intimate connection.

Unlike the "unanimated" human dead, zombies are a constant threat to the living as they are governed by an uncontrollable impulse to consume; this is their *raison d'être*. Yet, while ostensibly malevolent, the zombie—bereft of free will—is not a moral agent. In this case, legitimate defense becomes not only a right—on part of the living—but a duty for the self and for one who is responsible for the lives of others.[40] If the defense of the Common Good requires that an unjust aggressor be rendered incapable of causing harm, it is usually the case that those who legitimately hold authority have the right to use arms in order to repel said aggressors against the civil community entrusted to their responsibility.[41]

Although the zombie apocalypse may very well challenge our notion of what constitutes "legitimate authority," another question must be posed. If direct and intentional killing is morally problematic and if no one can under any circumstance claim for him- or herself the right to destroy an innocent human being, can one licitly seek out zombies—zombies, which are, for instance, contained and pose no direct threat to the living—in order to exterminate them? We are reminded at this point of the third and fourth seasons of *The Walking Dead*, which largely transpire in the relative safety of an abandoned prison. Zombies are kept at bay by a fence that surrounds the compound. Even though many of them collect on the grounds and hobble around in plain view, if the zombies pose no immediate threat, the survivors leave them be (for the most part). If, however, the zombies breach the confines, the survivors kill them. This, though, does not seem to be the result of moral reasoning about the personhood status of the zombie or distress associated with the act of killing (which many—including children—have become accustomed to), but is more a strategic plan to salvage dwindling ammunition and limited resources, and to lay low lest unnecessary noise attracts others (zombie or otherwise) who might jeopardize their refuge.

Putting Death to Death

In his Pulitzer Prize winning book *The Denial of Death*, anthropologist Ernest Becker suggests that "the idea of death, the fear of it, haunts the human animal like nothing else; it is a mainspring of human activity—activity designed largely to avoid the fatality of death, to overcome it by denying in some way that it is the final destiny for man."[42] One could easily say that this discussion of personhood has very much been an exercise in distancing zombies, however reminiscent of humanhood, from living human beings. The reader will note that the conversation has moved from zombies-are-not-persons and, therefore, have no moral status to zombies-are-a-constant-threat to the self and the Common Good, and, therefore, one has a duty to eliminate them (or, at least, eliminate those which pose an immediate threat). Of course, the objective here was not to exclude zombies from personhood *in order to* relieve man from the guilt and moral accountability associated with killing zombified parents, spouses, children, or neighbors, but to identify what we are dealing with and to determine what is demanded of us as moral agents.

The death-as-enemy motif, an ever present theme in Western culture and religion, is pervasive in the zombie apocalypse when survivors must keep constant vigil with the looming specter of the walking dead.[43] "On the most elemental level," Becker explains, "the organism works actively against its own fragility by seeking to expand and perpetuate itself in living experience; instead of shrinking, it moves toward more life."[44] In this way, "the fear of death can be carefully ignored or actually absorbed in the life-expanding process."[45] Hence, the motivation to exterminate the zombie and to "kill death," as it were, comes down to the removal of a constant and global threat in order to increase the proportion of the living.

Writing in the thick of the Second World War, the late psychoanalyst Gregory Zilboorg argued that self-preservation and the ability to go about our normal functions would be largely hampered if the fear of death remained constantly at the forefront of the imagination; "it must be properly repressed," he says, "to keep us living with any modicum of comfort."[46] Zilboorg explains:

> In normal times we move about actually without ever believing in our own death, as if we fully believed in our own corporeal immortality. We are intent on mastering death. We work out medical problems of longevity; we indulge in planning for the remote future of our family, our country, humanity as a whole; we marshall all the forces which still the voice reminding us that our end must come some day, and we are suffused with the awareness that our lives will go on forever.[47]

The zombie apocalypse, however, is fear made manifest in the flesh and is also an opportunity to rise up and put death to death. This want to *control* death—already evident in the technologization of death seen in this contemporary age—seems to fall neatly in line with the trajectory that the French historian Philippe Ariès outlines in his celebrated account of Western attitudes toward death and dying.[48] Granted, controlling death by "killing" it seems to be rather oxymoronic. Instead, it is more accurate to speak here of de-animating the dead in order to bring an end to a contagion that will almost assuredly wipe out the human species.

The plight to de-animate the dead in a zombie apocalypse in order to preserve one's self, family, community—indeed, human civilization—images in many ways the context within which Zilboorg elucidates the fear of death.

> We must maintain within us the conviction that we are stronger than all those deathly dangers, and also that we, each one of us who speaks of himself in the first person singular, are exceptions whom death will not strike at all. That is why we court death in various dangerous sports. It is an effort to prove that we are fearless, and fearless to us means not afraid of death. It does not mean that we don't mind it and are ready to meet our Maker with a sense of courteous welcome and in a manner of nonchalance; it means a rather boastful, challenging bravado: "Death cannot get *me*." So strong is this propensity in us that we elevate it to the height of great virtue; we hold a "physical coward" in contempt, and the physical coward is normally ashamed to admit that he is afraid to die.[49]

In this vein, Becker reminds that "heroism is first and foremost a reflex of the terror of death."[50] Close in kind to the theater of war, the zombie apocalypse sees this terror "mastered through a more direct form of aggression—the hatred for the enemy and the enthusiasm for his defeat."[51] Zilboorg makes this plain:

The fear of death is thus transformed into a sado-masochistic combination, which on the socialized level is represented by the awareness of grim hatred of the enemy and of readiness to make great sacrifices to save the country, to win the war, to kill the enemy. The psychological emphasis is laid here on the murderous, destructive drives within us, not only because they represent the logical needs of our functioning in an atmosphere of war, but primarily because these drives are the only ones capable of maintaining the fear of death in a state of low tension.[52]

He notes that the "murderous drives" of soldiers in war "not only are sanctioned by the ethics of warfare, but enable us to feel [like] masters over life and death."[53] As many film and literary treatments of the zombie apocalypse bring to the fore, the focus on moral agency and moral accountability often shifts contexts as soon as it is understood that the threat of contagion is not a transient one. In the beginning, concern for self-preservation will beg deep and abiding moral questions about who or what the zombie is and will lead to a quick unraveling of the thou-shall-not-kill mandate that keeps most human communities intact. With the identification of the zombie as a dead, non-person, embodied virus bereft of moral status, killing will become so commonplace that one will have to frequently revisit what constitutes proper ethical conduct in the dwindling living human community that survivors will struggle to preserve. *The Walking Dead* engages this time and again; it is not always clear whether the zombies or the other survivors are the greater menace when it comes to trying to stay alive.

When the apocalypse ushers in a new clash of civilizations, living human beings will assume a new kind of mastery over death by "killing" it time and again. As this new found control over death by the living reshapes the way individuals relate to one another in what remains of the human community, one may very well discover that the longstanding fear of death will be usurped by an emerging fear of life.

Notes

1. Rudolf A. Makkreel, *Imagination and Interpretation in Kant: The Hermeneutical Import of the Critique of Judgment* (Chicago: University of Chicago Press, 1990), 1.

2. See Ali S. Khan, "Preparedness 101: Zombie Apocalypse," Centers for Disease Control and Prevention Public Health Matters Blog, May 16, 2011, http://blogs.cdc.gov/publichealthmatters/2011/05/preparedness-101-zombie-apocalypse/; see also Centers for Disease Control and Prevention, "Zombie Preparedness," 2012, http://www.cdc.gov/phpr/zombies.htm.

3. I leave treatment of various religio-cultural representations of zombies for another time, however interesting these are. As for the imminence of a zombie apocalypse and conversations about whether or not zombies are real, see also Frank Swain's *How to Make a Zombie: The Real Life (and Death) Science of Reanimation and Mind Control* (London: Oneworld, 2013) and "Why Zombies Are Real," June 18, 2013, http://www.huffingtonpost.com/frank-swain/why-zombies-are-real_b_3461580.html.

4. Karen Rommelfanger, PhD, Neuroethics Program Director at the Emory University Center for Ethics, and I organized a symposium, called "Zombies and 'Zombethics,'" at Emory University on October 31, 2012. The symposium gathered professors of philosophy, religion, psychiatry, neuroscience, ethics, sociology, history, and public health to address some of the questions listed here. See Carol Clark, "Ethicists to Wrestle with Zombies on Halloween," October 30, 2012, http://news.emory.edu/stories /2012/10/esc_zombies_and_ethics/campus.html.

5. Ps. 8.3–4.

6. See, for instance, Stephen G. Post, "Alzheimer Disease and the 'Then' Self," *Kennedy Institute of Ethics* 5.4 (1995): 307–321.

7. James W. Walters, "Is Koko a Person?" *Dialogue* 9.2 (1997), http://dialogue.adventist.org/ articles/09_2_walters_e.htm.

8. *Catechism of the Catholic Church* (Ottawa: Canadian Conference of Catholic Bishops, 1994), §2270; Gen 1.27–28.

9. Walters, "Is Koko a Person?"

10. Immanuel Kant, *Foundations of the Metaphysics of Morals and What Is Enlightenment?*, trans. Lewis White Beck (Indianapolis: Bobbs-Merrill, 1959), 46–54.

11. Ibid., 46.

12. Ibid., 47.

13. Joseph Fletcher, *Humanhood: Essays in Biomedical Ethics* (Buffalo: Prometheus, 1979), 135.

14. Ibid., 12–16.

15. Edward Keyserlingk, *Sanctity of Life or Quality of Life in the Context of Ethics, Medicine and Law*, Law Reform Commission of Canada (Ottawa: Minister of Supply and Services Canada, 1979), 98.

16. Fletcher, *Humanhood*, 12; 14–15.

17. Michael Tooley, *Abortion and Infanticide* (Oxford: Clarendon Press, 1983), 123–146.

18. Ibid., 90.

19. Ibid., 349.

20. Ibid., 421.

21. Ibid., 107–108.

22. Marshall Cohen, Thomas Nagel, and Thomas Scanlon, eds., *The Rights and Wrongs of Abortion* (Princeton: Princeton University Press, 1974), 57.

23. Tooley, *Abortion and Infanticide*, 146–157.

24. Peter Singer, *Practical Ethics*, 2d ed. (Cambridge: Cambridge University Press, 1993), 97–98.

25. Refer to Parfit's discourse on psychological continuity in his influential *Reasons and Persons* (Oxford: Clarendon Press, 1984), 204–209.

26. Jeff McMahan, "Brain Death, Cortical Death and Persistent Vegetative State," in *A Companion to Bioethics*, ed. Helga Kuhse and Peter Singer (Oxford: Blackwell, 2001), 256.

27. Helga Kuhse, "Some Reflections on the Problem of Advance Directives, Personhood, and Personal Identity," *Kennedy Institute of Ethics*, 9.4 (1999): 361.

28. Max Brooks, *The Zombie Survival Guide: Complete Protection from the Living Dead* (New York: Three Rivers Press, 2003), 2–5.

29. Heritage Community Foundation/Famous 5 Foundation, "The 'Persons' Case," Canada's Digital Collections, 2002, http://collections.ic.gc.ca/famous5/achievements/case01.html, par. 2.

30. Kenneth Boulding, *The Meaning of the Twentieth Century* (New York: Harper & Row, 1964), 192.

31. Lloyd J. Averill, *The Problem of Being Human* (Valley Forge, PA: Judson Press, 1974), 13.

32. Tom L. Beauchamp and James F. Childress, *Principles of Biomedical Ethics*, 6th ed. (New York: Oxford University Press, 2009), 66.

33. If death for a number of religious traditions is theologically defined as the separation of the body and the soul, the zombie is, then, nothing more than a physical mass. The being no longer has a soul even though it is animate (incidentally, the word "animate" is derived from the Latin word "anima" for life, breath, soul, or mind). Interestingly, especially for our discussion of zombies, the definition of death in the bioethical literature continues to spur considerable disagreement. See, for example, B.A. Manninen, "Defining Human Death: An Intersection of Bioethics and Metaphysics," *Reviews in the Neurosciences* 20.3–4 (2009): 283–92.

34. Brooks, *The Zombie Survival Guide*, 11–12.

35. Ibid., xiii.

36. Ibid., xiii.

37. Ibid., 16.

38. Peter Singer, *Animal Liberation*, 2d ed. (London: Pimlico, 1995), 10.

39. *The Walking Dead*, "Nebraska," season 2, episode 8, directed by Clark Johnson, written by Evan Reilly, AMC, 12 February 2012.

40. *Catechism of the Catholic Church*, §2265.

41. Ibid.

42. Ernest Becker, *The Denial of Death* (New York: Free Press, 1973), ix.

43. For instance, 1 Cor. 15.24–46 reads: "Then comes the end, when [Christ] hands over the kingdom to God the Father, after he has destroyed every ruler and every authority and power. For he must reign until he has put all his enemies under his feet. The last enemy to be destroyed is death."

44. Becker, *The Denial of Death*, 21.

45. Ibid., 21.

46. Gregory Zilboorg, "Fear of Death," *Psychoanalytic Quarterly* 12 (1943): 467.

47. Ibid., 468.

48. Philippe Ariès, *The Hour of Our Death*, trans. Helen Weaver (New York: Oxford University Press, 1981).

49. Zilboorg, "Fear of Death," 468.

50. Becker, *The Denial of Death*, 11.

51. Zilboorg, "Fear of Death," 472.

52. Ibid., 473.

53. Ibid., 474.

Bibliography

Ariès, Philippe. *The Hour of Our Death*. Translated by Helen Weaver. New York: Oxford University Press, 1981.

Averill, Lloyd J. *The Problem of Being Human*. Valley Forge, PA: Judson Press, 1974.

Beauchamp, Tom L., and James F. Childress. *Principles of Biomedical* Ethics, 6th ed. New York: Oxford University Press, 2009.

Becker, Ernest. *The Denial of Death*. New York: Free Press, 1973.

Boulding, Kenneth. *The Meaning of the Twentieth Century*. New York: Harper & Row, 1964.

Brooks, Max. *The Zombie Survival Guide: Complete Protection from the Living Dead.* New York: Three Rivers Press, 2003.

Catholic Church. *Catechism of the Catholic Church.* Vatican: Libreria Editrice Vaticana, 2003.

Centers for Disease Control and Prevention. "Zombie Preparedness." Last updated August 18, 2014. http://www.cdc.gov/phpr/zombies.htm.

Clark, Carol. "Ethicists to Wrestle with Zombies on Halloween." October 30, 2012. http://news. emory.edu/stories /2012/10/esc_zombies_and_ethics/campus.html.

Cohen, Marshall, Thomas Nagel, and Thomas Scanlon, eds. *The Rights and Wrongs of Abortion.* Princeton: Princeton University Press, 1974.

Fletcher, Joseph. *Humanhood: Essays in Biomedical Ethics.* Buffalo: Prometheus, 1979.

Heritage Community Foundation/Famous 5 Foundation. "The 'Persons' Case." Canada's Digital Collections. 2002. http://collections.ic.gc.ca/famous5/achievements/case01.html.

Kant, Immanuel. *Foundations of the Metaphysics of Morals and What Is Enlightenment?* Translated by Lewis White Beck. Indianapolis: Bobbs-Merrill, 1959.

Keyserlingk, Edward. *Sanctity of Life or Quality of Life in the Context of Ethics, Medicine and Law.* Law Reform Commission of Canada. Ottawa: Minister of Supply and Services Canada, 1979.

Khan, Ali S. "Preparedness 101: Zombie Apocalypse." Centers for Disease Control and Prevention Public Health Matters Blog. May 16, 2011. http://blogs.cdc.gov/publichealthmatters/2011/05/ preparedness-101-zombie-apocalypse/.

Kuhse, Helga. "Some Reflections on the Problem of Advance Directives, Personhood, and Personal Identity." *Kennedy Institute of Ethics* 9.4 (1999): 347–64.

Makkreel, Rudolf A. *Imagination and Interpretation in Kant: The Hermeneutical Import of the Critique of Judgment.* Chicago: University of Chicago Press, 1990.

Manninen, B.A. "Defining Human Death: An Intersection of Bioethics and Metaphysics." *Reviews in the Neurosciences* 20.3–4 (2009): 283–92.

McMahan, Jeff. "Brain Death, Cortical Death and Persistent Vegetative State." In *A Companion to Bioethics,* edited by Helga Kuhse and Peter Singer, 250–60. Oxford: Blackwell, 2001.

Parfit, Derek. *Reasons and Persons.* Oxford: Clarendon Press, 1984.

Post, Stephen G. "Alzheimer Disease and the 'Then' Self." *Kennedy Institute of Ethics* 5.4 (1995): 307–21.

Singer, Peter. *Animal Liberation,* 2d ed. London: Pimlico, 1995.

_____. *Practical Ethics,* 2d ed. Cambridge: Cambridge University Press, 1993.

Swain, Frank. *How to Make a Zombie: The Real Life (and Death) Science of Reanimation and Mind Control.* London: Oneworld, 2013.

_____. "Why Zombies Are Real." June 18, 2013. http://www.huffingtonpost.com/frank-swain/why-zombies-are-real_b_3461580.html.

Tooley, Michael. *Abortion and Infanticide.* Oxford: Clarendon Press, 1983.

The Walking Dead. "Nebraska." Directed by Clark Johnson.Written by Evan Reilly. AMC, 2012.

Walters, James W. "Is Koko a Person?" Accessed July 10, 2014. http://dialogue.adventist.org/articles/09_2_walters_e.htm.

Zilboorg, Gregory. "Fear of Death." *Psychoanalytic Quarterly* 12 (1943): 465–75.

Zombies in the Sunshine State

An Economic and Historical Analysis of the Zombie Apocalypse

CHRISTOPHER M. ESING

Although writer and film director George Romero sets his zombie horror classic *Day of the Dead* in post-apocalyptic Florida, his choice of setting fails to account for the sunshine state's long and varied history with catastrophic events, and the ability of Florida's citizenry to repeatedly overcome its many challenges to create one of the most populous and wealthiest states in the U.S. This essay seeks to explore how cataclysmic and apocalyptic events similar to those of a zombie pandemic have historically shaped the economic development and growth of Florida in order to better imagine the social and economic challenges faced by a modern zombie outbreak.

Although it is likely that most of society would collapse in the face of a wide scale zombie plague, pockets of survival and resistance would also expectedly emerge as individuals formed new communities, established new forms of self governance, and created varied forms of exchange via barter and monetization within a localized economy. In the event that the central government was to survive, it is possible that societies could try to revive larger governmental structures and economic trade, but history has time and again proven that to be an incredibly difficult task. While this essay focuses primarily on Florida, two recent catastrophic events help to highlight the difficult realities of dealing with modern disasters, the first being the SARS epidemic in China and the second being Hurricane Katrina in New Orleans.

Following the SARS outbreak in 2003, the Chinese government launched a counter offensive to stop the spread of the viral disease, which held just a small percentage of survival for those who contracted the contagion. The

Chinese government banned all but the most necessary travel and encouraged citizens to remain in their homes. In more extreme measures, local governments also required daily testing of all citizens using temperature guns and quarantining those individuals who had an elevated temperature. For a period of several months, almost all economic exchange had come to a complete standstill until the threat of the virus was eliminated. Even after danger had passed and restrictions had been lifted, most citizens feared returning to the streets and markets. Those who did wore masks and only bought basic necessities. These routines continued for months and decimated the Chinese economy.[1]

In order to try and revive trade, businesses collaborated with the national government and created a shopping holiday whereby prices were reduced with the government promising to help businesses recover losses. These actions worked, and within the month, the economy began to slowly recover. However, in the event that SARS had morphed into a cataclysmic plague, it is likely that the devastating effects of the outbreak would have stretched far and wide and lasted much longer and would have been compounded by shortages in foods, rations, and other necessities leading to possible starvation, widespread theft and looting, and a loss of control and confidence within the national government and the nation state.[2]

The most recent examples of this scenario can be seen in natural disasters such as Hurricane Katrina in 2005 or the earthquake in Haiti in 2010. Following the landfall of Hurricane Katrina and the unexpected breach of the levy system in New Orleans, both the local, state, and national government proved incapable of meeting the immediate needs of the city's refugees who were scattered far and wide. Inside New Orleans, the local government was largely ineffective in its ability to address the destruction due to the sheer size of the disaster that sent tidal waves surging through neighborhoods throughout the city leaving flattened and flooded homes in its path and turning the French Quarter and historic district into an inaccessible island surrounded by 12 feet of water with many of its residents stranded on the rooftops or trapped in the attics of their homes.[3]

On the state level, the widespread devastation caused by the tidal surge and winds of Hurricane Katrina inundated much of the lower lying lands along the coast causing state resources to be spread out and dispersed among all areas of the state. In this scenario, the Federal Emergency Management Agency (FEMA) should have stepped in to meet the immediate needs of the people. However, bureaucratic mismanagement under FEMA director Michael Brown largely crippled the rescue efforts and plans that had been put into place and helped to turn a natural disaster into a man-made debacle. Despite the orders for buses and boats to help remove people from the city, Brown's calls were repeatedly met with delays that prevented any form of relief for the refugees of the city. These delays turned average Americans into criminals as they looted grocery stores for food and pharmacies for medicine. With an inability to pay, people began taking what they needed, and in the melee that followed, widespread looting broke out across the city. When the police tried to halt the actions, violence ensued and turned New Orleans into a battle-ground between the police and looters.[4]

After the violence erupted, the president should have declared a state of emergency and sent in the National Guard to return order, but then-president George W. Bush failed to take any action. Over four days, the pres-ident sat by while millions of Americans watched in horror at the inability and ineffectiveness of the government to rescue its citizens in the face of a widescale natural disaster. Only after foreign countries such as Cuba promised to send international aid and workers to New Orleans did President Bush declare a state of emergency. With the help of private volunteers, many of the residents were finally evacuated but not before the lives of 1,833 Americans had been taken as a result of the events. Following the evacuation, FEMA still proved largely ineffective to meet the needs of many of the refugees who were taken in by neighboring cities and offered food and shelter.[5]

In the months that followed, FEMA promised trailers for displaced New Orleanians that many manufacturing plants worked to speedily construct only to have them sit unoccupied in empty lots because of bureaucratic mis-management. The breeched levees took months to repair and still failed to meet structural codes. At the same time, most homeowners who had housing insurance lacked flood protection and lost everything after being inundated, and many of their neighbors were in the same predicament. Large swaths of the city simply remained uninhabited with most residents unable to return and rebuild. Many were promised government aid and assistance but the money was slow to be distributed and the lack of a workforce led to con-struction delays. New work did help to revive construction, but a large con-tingent of fake contractors and migrant labor flooded within the city creating a labor pool of outsiders that failed to help bring about economic revival.[6] Although a lot of organizations helped to pitch in to revive the city, large swaths of New Orleans had yet to recover nearly nine years following the natural disaster.

These modern examples of localized and regional catastrophes help illustrate how the government might try to handle small outbreaks of a con-tagion but fail to capture what or how an ongoing epidemic of a moving and spreading contagion such as a zombie apocalypse might occur and what effect that would play on the local, national, and global economy. An examination of earlier contagious outbreaks in American history might help shed some light onto what it would be like to live with a prolonged and deadly epidemic. Of all the pre-vaccination and preventative measure diseases, Yellow Fever emerged as a perennial scourge of the American colonies as early as 1648 in Boston. Also known as Yellow Jack, the disease lasted longer than most other outbreaks and tended to render many of the afflicted as deathly reflections of themselves. The sheer physical torment of jaundiced patients and black, bloody vomit was enough to scare most people into the woods with news of an outbreak.[7]

Although eruptions of Yellow Jack were known to occur in the North during the summer, the frost line rendered Yellow Fever a primarily Southern phenomenon. Port cities remained bastions of the plague, and its continued presence in many cities ravaged local economies as economic activity frequently came to a standstill. One such city, which suffered dramatically and recurrently from Yellow Fever, was the port town of Pensacola, Florida. Yellow Fever struck repeatedly from 1867 to 1905 leaving the community economically depressed and devastated in its wake. An examination of Pensacola helps to illustrate how an ongoing outbreak would ruin a local economy and send most residents into the woods for safety, food, and survival.

Yellow Fever struck Pensacola in the summer of 1867. The fever arrived aboard the British steamer *Fair Wind*. Although the city quarantined the ship, crewmen from other ships quickly contracted the illness. Until the summer, the disease remained isolated at the port, but in July, the fever disseminated rapidly among the residential population. When news of the outbreak spread, the populace fled to the countryside and left the city abandoned. The disease initially claimed twenty-four lives and infected both of the city's doctors.[8] As it spread, the fever laid waste to many soldiers stationed at the navy yard and Fort Barrancas with more than six hundred documented cases among the troops.[9] Most of the marines contracted the disease, and the Navy was forced to close the facility until the fever had run its course. The disease finally passed during the fall, but overall it took the lives of two hundred citizens. The events marked the first of many outbreaks that would haunt the city during the second half of the 19th century.[10] The disease gripped Pensacola for the next several years leaving its citizens ravaged and weary.

A massive Yellow Fever epidemic occurred in August 1874, and the military began ordering regular quarantines of Pensacola. It blocked all traffic from the rail spur entering into the city via Montgomery, Alabama. Out of a population of 3,374, 2,000 fled into the countryside and neighboring villages. Of those that remained, most became ill and 354 died.[11] The sheer number of deaths created fear among the public, and many refused to stay long outside their homes. Most burials were makeshift. Many occurred along the beach in sand that was easy to dig. The tide eventually exposed many of the bodies making matters worse.[12] Winter brought another lull in the number of new cases reported, but in July 1875, the third straight summer of Yellow Fever outbreaks gripped the city. Starting onboard the ship *Von Moltke*, arriving out of Havana, it quickly spread among the troops at Fort Barrancas and later at Fort Pickens. With strict measures to quarantine the soldiers, the military halted the spread of the fever among the general populace, but its mere presence rattled already shaky nerves.[13]

At a time when the rest of the state was experiencing an economic boom caused by a newfound love of oranges among Northern consumers and the birth of Florida's modern tourist industry among health seekers, the continued occurrence of Yellow Fever in Pensacola prevented the city from experiencing much of the growth that was happening in the rest of Florida. Concerns over Yellow Fever became the most prominent questions asked to the state's commissioner of immigration.[14] Northern tourists and immigrants were terrified by the descriptions of yellow corpses whose clothes had been stained with the black vomit that characterized the illness.

Despite the many fears surrounding the disease, life continued as usual for most of Pensacola's inhabitants once the summer season of pestilence passed. The fever rarely disrupted the lumber industry in the city as most of the mills were located outside the town limits and continued their operations during outbreaks. While the fever did interrupt shipments, most operators waited until the winter when the disease had passed and then sent out most of their lumber. The greatest impact of Yellow Fever was upon the ability to attract Northern capital as most Northerners refused to visit Pensacola ruining most chances at new investors and curtailing any hopes of attracting the tourist industry benefiting the rest of the state.[15] Even with the setbacks, Pensacola established itself as the most important lumber export-shipping center in the South. The value of timber exported from Pensacola in 1876 alone held an estimated value of more than $3,000,000.[16]

In the face of the unremitting plague, the city grew economically doing most of its trade during the winter when the fever abated. When the first major railroad came to Pensacola, railroad and land manager William Chipley did much to try and ease fears over Yellow Fever since his rails needed cargo, passengers, and settlers along its lines. Chipley published several tracts arguing that the new quarantine procedures at the port and in the city made worries over Yellow Fever a thing of the past. He wanted to promote the city as a tourist destination and the only way for that to occur was to convince Northern travelers that Pensacola was safe.[17] Not all worked out as the railway had planned. In 1882 during construction of the railroad, the city suffered another major spate of the disease when the Spanish ship *Salita* arrived with sickened crewmembers. Despite a strict quarantine, the epidemic spread to the city, and over the summer, 2,200 people had taken ill. When the fever had run its course, it claimed the lives of more than 250 citizens.[18] Fever returned again in August 1883, but the illness was largely contained to the Navy Yard bringing a momentary reprieve.

Alabama and east Florida adopted strict quarantine measures on interstate commerce after 1882. No train car was allowed to leave West Florida until the first frost to protect passengers from what they viewed as a potential danger. Most of the empty cars were placed on the side of the railway, but the sheer number soon blocked the tracks into the city. Throughout the course of the epidemic, the city and railroad were largely forced to halt operations.[19] Because of the impact upon businesses and the new railroad the *Pensacola Commercial* pleaded with the Board of Health not to announce to the public when cases of fever were discovered in 1883. The board responded by canceling its subscription to the newspaper. The announcement threw the public into a state of frenzy, and they reacted by burning the residences of the ill. All shipments were diverted to Mobile, Alabama, and other nearby ports. The outbreak proved to be the last scourge on the city until the 1890s, but the impact proved to be impossible to surmount in particular at a time when the city and railroad were both trying to encourage tourism and new business. Northern fears of Yellow Fever largely sank both efforts and almost tanked the new railway.[20]

In August 1893, the Board of Health announced that two deaths had resulted from Yellow Fever, which led to the largest mass exodus that the city had seen in recent history. As one reporter noted, "This [news] flew like wild fire over the city and gathered wings as it went. The cheeks of timid people which but a few hours before had been radiant with smiles, suddenly paled with fear and in an incredibly short time in every part of the city preparations for departure were being made. The north bound train which left at 1:45 p.m. was filled to overflowing and hundreds of others hurried their preparations to leave on the night train."[21] With many of the workers having fled the city, commerce and business completely shut down throwing those who remained out of employment. Fearing that the fever would ruin the port, Chipley had the state health officer come and examine the bodies, and he said that the board was wrong on their assessment and that the quarantine of the city should cease immediately and commerce resumed.[22] Because of the damage caused by the scare to the city and economic recession brought about by the Panic of 1893, Chipley issued *The Pamphlet of West Florida* denouncing the presence of Yellow Fever and highlighting the industrial, commercial, and agricultural advantages of west Florida.[23]

Although Pensacola had a few minor recurrences of Yellow Fever, the last major scourge of Yellow Jack occurred in 1905. The troubles for the city began in July of that year when a number of local families had gone to New Orleans to visit with relatives and friends. When they returned, it was announced that Yellow Fever had struck the Crescent City, and three Pensacolains had contracted the disease. The news led to another exodus of most of the city's 25,000 inhabitants despite the knowledge that mosquitoes were now the known carriers of the disease. Mayor Bliss ordered the first extensive fumigation of any city in the South and ordered all breeding grounds to be flushed out. In fear that infected persons would leave, armed guards surrounded and blocked all roads and a water patrol blocked the entrance of the harbor. By October, the fever continued to affect new citizens, so the State Board of Health sent an army of 100 men to again fumigate the city and surrounding village. The measures largely worked, and by November the city recovered from its final battle with the plague.[24]

Yellow Fever had wreaked havoc upon the small port city of Pensacola throughout the nineteenth and early twentieth centuries. With each outbreak, businessmen and wealthy inhabitants consistently fled the struggling community leaving many of the poorer inhabitants to fend for themselves undercutting all economic activities for weeks and months at a time. More than any other city in Florida, Pensacola suffered from more Yellow Fever epidemics marking it as the most undesirable port and city in which to engage in business. The repeated epidemics scared away many of the tourists, businessmen, and settlers that flocked to the state after the Civil War and hindered almost all development.

These events help illustrate some of the potential scenarios of what could occur should a zombie apocalypse transpire. If widespread, most citizens would likely flee into the country to seek refuge from an ongoing, unknown, and frightful menace. With cities abandoned, many would be forced to rely upon their own instincts for survival. Unlike Yellow Fever, which experienced a seasonal lull, citizens would be afflicted by the knowledge that the contagion had no visible end, and would in all likelihood lead to the looting and the burning of cities with zombie outbreaks similar to what occurred to the homes of those afflicted with Yellow Fever in Pensacola. With cities abandoned and destroyed, money would lose all value on a local level and fear of contact with strangers would result in the loss of an exchange based economy altogether. Trying to imagine some of these outcomes, film director George Romero addresses some of these possibilities in *Dawn of the Dead* depicting the military working with rural governments to try and eradicate the threat of zombies as those same communities fell to the onslaught of a zombie horde.

This scenario is rooted in the notion of a widescale zombie apocalypse, but if the contagion spread slowly and emerged in one city, it is more likely that the national government or private citizens would establish military blockades to prevent or halt the spread of the outbreak. While it seems logical that the government would direct these activities, questions over whether zombies could be cured or were still human might halt or delay any major military maneuvers. In that event, it seems likely that private armed militias and vigilante groups would move to prevent the spread of the plague by surrounding their own cities with a physical blockade. An example of these activities can be seen with the Yellow Fever epidemic of 1888 in Jacksonville, Florida.

When Yellow Fever struck Jacksonville in 1888, the plague shut down the city, scared away much of the tourist for several seasons, and caused the population to flee en mass costing the city thousands of dollars in commercial losses.[25] During the outbreak, steamers and rail lines were forced to stop operations to Florida. The Clyde Line of steamships turned back to Charleston, South Carolina, and refused to send passengers to Jacksonville, while railroads stopped most of their railcars in Savannah, Georgia, or Waycross, Georgia. When citizens from Jacksonville tried to exit the city by rail, the inhabitants of Waycross threatened to tear up the tracks if refugees were permitted to pass their way.[26] At the same time, armed locals surrounded nearly every community and village in Florida and southern Georgia."[27] The quarantine of the city forced many of the businesses to close for several months, and the population plummeted overnight causing a permanent drop from 25,000 people in 1888 to 17,201 in 1890.[28] The economic impact stretched far and wide across the state causing a devaluation of properties across Florida. In the city of Sanford, Henry Sanford's Land Colonization Company recorded its first losses of any kind with international land sales falling from 5,961 pounds [British pounds sterling] in 1888 to 4,333 pounds in 1889."[29] Three years elapsed before the populace and commerce returned to its previous levels, but the city never fully recovered as tourists failed to return and instead chose locales further South that had not been gripped by the plague.

In the Jacksonville example, private citizens moved to protect their own vested interests by blockading the source of the outbreak in order to guarantee both their physical and economic survival. It seems logical, especially in the context of a post Katrina and 9/11 America where the government failed to protect its citizens, that private corporations and individuals would band together or at least act individually to protect their homes, families, and communities. A collective movement of private groups and individuals to create a safe haven also suggests the possibility that contagion-free cities would emerge in the midst of a wide scale zombie apocalypse. These localities would be free from the outbreak and develop their own localized economies and means of protection and production to restart life. Many would likely become thriving and flourishing cities helping take in refugees as they sought out protection.

The state of Florida has historically served as one of the largest examples of an area emerging as a sanctuary for those seeking refuge from war, economic disaster, and disease. Following the Civil War, soldiers seeking escape from physical and mental battle wounds and ailments, along with factory workers unable to cope with the new pace of a streamlined machine life, arthritic and pneumatic patients seeking reprieve from the cold weather, invalids looking to enjoy the healthful rays of the sunshine, and wealthy businessmen and their families looking to escape the dreary cities of the North, all came to Florida to find self-rejuvenation and recuperation helping to make Florida into the nation's leading sanatorium.

One of the great boosters of this movement was famed *Uncle Tom's Cabin* novelist Harriet Beecher Stowe who moved to Florida shortly after the Civil War. She rented a home in the village of Mandarin for her son Frederick who had been wounded in the head during the war and experienced severe alcoholism and delirium.[30] Stowe chose Florida because of the large number of tracts and pamphlets discussing the state as a healthy place to recuperate from neurasthenia and other nervous conditions. During her time in Florida, Stowe wrote *Palmetto Leaves* as a narrative of her spiritual and physical rejuvenation while living in Florida. In the work, she proclaimed, "my visit here has been like sunshine and spring to a frost bitten plant. I have had more life, more rest, more appetite, more conscious pleasure in existence than I have had for years in New England. Here must be my future home, for at least half of the year, if I am to live and do anything."[31]

Palmetto Leaves reflected Stowe's belief in Florida's regenerative possibilities. Whether emblematic of her hopes for her son or her own physical transformation upon arrival, Stowe believed that Florida was suitable for a wide array of "nervously-organized dyspeptics who require a great deal of open, out-door life."[32] Along with her work, hundreds of dispatches from newspapermen and soldiers depicted accounts of exotic beaches, rivers, lakes, animals, and the flora and fauna of the state. Enchanted by the landscape and atmosphere, few writers could resist contrasting the warmth and sunshine of Florida in the winter to the ice and snow of the north. These reports portrayed Florida as a new American Eden and a tropical paradise.[33] For many Americans, these writings reflected the growing romanticism of nature among a society increasingly distraught with industrialization, urbanization, and an uncertainty of the effectiveness of modernization and development to solve the woes of society.

The period of Reconstruction and the rise of a new industrial revolution witnessed a dramatic increase in the number of individuals seeking escape into the wilderness to find renewal.[34] For this reason, the earliest tracts that highlighted Florida's scenery frequently emphasized the healthfulness of its climate. T.F. Smith's *Florida and Texas* explained that "as respects to health, the climate of Florida stands preeminent. The peninsular climate of Florida is much more salubrious than that of any other state in the Union. The general healthfulness of many parts of Florida, particularly on its coast, is proverbial. The average annual mortality of the whole peninsula, is found to be 2.06 percent, while the other portions of the United States (previous to the war with Mexico) is 3.05 percent."[35] These statistics offered comfort and security to immigrants who held fears of swamp miasmas and Malaria but desired a return of their vigor.

Following the war, climate and health statistics addressed growing number of medical concerns. The *Florida Settlers and Immigrants Guide of 1873* claimed, "the climate of the State is well adapted for the cure of rheumatism; in fact it may be regarded as a specific for this disease."[36] For wounded veterans who had returned home with broken bones and amputated limbs, these passages promised seasonal and permanent relief from arthritic and swollen joints that caused agony and discomfort.[37] In the same year, Robert Spier's *Going South for the Winter with Hints to Consumptives* promised reinvigorated "life and health" for those who head South for the winter in particular in Jacksonville and Florida.[38] By 1875, the healing virtues of Florida had become so well established within the medical field that Rambler's *Guide to Florida* proclaimed, "The wonderful salubrity of the climate of Florida is its greatest attraction, and is destined to make it to America what the South of France and Italy are to Europe, -the refuge of those who seek to escape the rigor of a Northern winter. So well convinced are our physicians of this fact, that they now advise their patients to seek health in Florida."[39]

These pronouncements sparked a large movement of invalids into the state primarily to Jacksonville, St. Augustine, and the St. Johns River. Frederick D. Lente's *Florida as a Health Resort* sought to guide invalids along their path in Florida while giving comfort to locals who addressed concerns at the number of mentally and physically ill making their way in the state. Lente claimed that most invalids exemplified no traits to the general observer and fall under the growing category of nervous prostration. He observed that "certain forms of dyspepsia, which is, like other nervous affections, are becoming more and more common—which is merely one of the many symptoms of modern "wear and tear," and when various other treatment has failed—are permanently relieved by a winter's residence in Florida."[40] Lente encouraged most invalids to journey to Florida, and directed everyone to take advantage of Florida's outdoor lifestyle.[41]

James A. Hensall's *Camping and Cruising in Florida* reflects this sentiment with its account of a doctor's journey with his patients on a camping excursion through Florida. As Hensall explains, what makes Florida special is that "one can live in the open air during the winter without discomfort, and therein lies the great and lasting benefit to the invalid who requires the open air life and nature's great restorers, air, sunshine, exercise, and refreshing sleep."[42] These conditions offered "the poor broken down man of business and the nervous wife and mother wearied and worn from household chores, a rejuvenating balm well calculated to restore nerve action to its healthy condition."[43] With prescriptions in hand, consumptives and invalids turned Florida and Jacksonville into a burgeoning tourist destination as invalids in need of respite spent their winters in Florida and returned home each spring. The seasonal influx served as the basis for the new industry.

In order to house the growing number of visitors to the state, northern investors were the first to take advantage of the influx building a large number of hotels and resorts. The St. James Hotel opened in Jacksonville on January 1, 1869, and marked the beginning of a construction boom as northerners financed hotels across North Florida.[44] In particular, sulphur and mineral springs served as locations for larger resorts. Doctors widely held that "the waters from sulphur springs were efficacious in curing all forms of consumption, scrofula, jaundice, and other bilious affections, chronic dysentery, diarrhea, disease of the uterus, chronic rheumatism, gout, dropsy, gravel, neuralgia, tremor, ringworm, and itch."[45] Along with hotels, "private boarding houses kept by northern people, with pleasant surroundings, and quite as inviting as the hotels to those seeking but a temporary home" emerged to fill the need."[46]

A growing number of northerners also began purchasing winter homes. Stowe's *Palmetto Leaves* first suggested that "Florida is peculiarly adapted to the needs of people who can afford two houses, and want a refuge from the drain that winter makes on the health."[47] The wealthy plotted houses from Jacksonville to St. Augustine and all up and down the banks of the St. Johns River. With neurasthenia growing in prevalence among the industrial classes, many began to build houses in which to perennially spend each winter. The numbers of second homebuilders increased exponentially after the start of the Franco-Prussian war blocked almost all travel in and out of Europe sending America's wealthiest citizens in search of a new tourist playground.

Beyond the American elite, a number of middle-class workers in the North started receiving a week long paid vacation during the year.[48] Interested in showing off their newly acquired status and escaping the extremities of city life, a tide of tourists and travelers began to peel back the Florida frontier. By 1872, George Washington Olney could proclaim that "fifty thousand people visited Florida last winter, of whom, about ¼ were invalids."[49] A large contingent of businesses stepped forward to take advantage of the tourism trade. Foremost among them were the steamboat and steamship lines, and preeminent among the boat owners was Colonel Hubbard L. Hart. The colonel operated the first steamships on the St. Johns after the war and developed the largest and most lucrative fleets along the river.[50]

Because of the number of northern tourists, migrants, and settlers, most promotional tracts referred to Jacksonville and Florida as a northern colony. George Barber called Jacksonville and Florida a "New England Garden."[51] Edward King noted that Jacksonville and neighboring Palatka had been rebuilt "according to the New England pattern," and estimated that half its resident population was from the Northern states."[52] Northern physicians, dentists, and lawyers began to outnumber the ministers and teachers who arrived after the war highlighting the economic turn that immigration began to take.[53] Tourist brochures remarked, "This element, although by no means the largest, is yet by far the most important, and to it is due all the prosperity which is now spreading over every portion of the state."[54] Drawing attention to northern businessmen, many guidebooks exclaimed, "Northerners are developing the true resources and capabilities of the State, and who are engaged in all the enterprises of private and public benefit. They are building churches, schools, erecting saw mills, building new hotels, in fact, civilizing the entire region."[55]

Jacksonville's and Florida's growth following the Civil War serves as a historical example of how not only sanctuaries against disease can emerge but also how they can become thriving economic centers. Although the state was not under continuous threat of attack by a wandering army of zombies, a defensible city could easily emerge as an asylum from such a threat. Exploring this possibility, George Romero depicts this idea in *Land of the Dead* imagining post-apocalyptic Pittsburgh, Pennsylvania, as a new feudal sanctuary from a zombie onslaught. Should a number of cities emerge as Romero mentions in *Land of the Dead*, it is also likely that a larger national and international economy could revive with the use of air transportation to move goods from one region to another. These safe zones would serve as the only real buffer against a complete destruction of modern civilization.

In the wake of recent events such as Hurricane Katrina, it seems unlikely that national or international governments would be capable of meeting a large-scale outbreak of a zombie plague leading most private citizens to fend for themselves. Those with money could flee at first, but the longer and more widespread that the outbreak occurred, money would lose its inherent value. Those with money would be better to spend their wealth creating and investing in defensible cities and regrow local economies where their capital could be reinvested and their influence reestablished.

Of all the places and regions likely to meet these criteria and emerge as a new refuge against a zombie apocalypse, Florida appears to be most prepared to handle the onslaught of an epidemic of the living dead. Florida has a long and storied past with zombies from those created by Voodoo to those produced by plague. Florida author Zora Neal Hurston explores both themes in number of her works and studies on the state. Hurston's work in Haiti and among Haitian immigrants follows and traces accounts of Voodoo zombies throughout Florida and the Caribbean in hopes of understanding how priests not only brought the dead back to life but also turned zombies into contemporary slaves. At the same time, Hurston's character Teacake in *Their Eyes Were Watching God* embodies an early archetype for a modern plague zombie after he contracts rabies from an alligator. Mentally ravaged by the disease,

Teacake looses basic sensibilities and repeatedly tries to attack and kill his wife Janey. In order to save her life, Janey is forced shoot her husband to free them both from his madness.[56]

Both of the examples help illustrate some of the state's early dealing with zombie like individuals, but Florida has also had a number of contemporary zombie attacks that have helped prepare its populace for a future zombie apocalypse. Fueled by drug-induced trances at least two recent cases in 2012 have been labeled by the press and media as zombie assaults. The first occurred on May 26, 2012, when Rudy Eugene, a homeless man in Miami, Florida, attacked another homeless man by beating him, stripping him of his clothes, and eating half of the victim's face. Unresponsive to police officers demands to cease the assault, the attack ended when police fatally shot Eugene.[57] In a similar incident on June 21, 2012, Charles Baker of Palmetto, Florida, stripped himself of his clothes and in a frenzied craze barged into his girlfriend's home and when confronted by another man in the house attacked the individual by biting and eating the flesh off his arm. Police arrived soon afterwards and tasered Baker twice before he could be restrained.[58] These two attacks promoted state senator Dwight Bullard to introduce the so-called "Zombie Apocalypse Amendment" on April 30, 2014. The Bill would allow anyone to carry a concealed weapon during a state of emergency without the threat of prosecution in order to defend themselves from attack and protect fellow citizens.[59]

While Florida tries to set new parameters to allow its citizens to guard themselves and their property against a threat of a zombie outbreak, the state's geographic position as both a peninsula and the southern most landmass in the eastern U.S. offers its citizens a secure and strategic location from a zombie invasion assuming that most of the living dead can not swim or would sink and be eaten by ocean predators if they tried to float with the tide. With only a single defensible border across its northern frontier, the geography of Florida allows the state's government and citizens to direct almost all of their military and economic resources towards preventing a zombie incursion from the North with the ability to fall further South in the peninsula should a break in its defenses occur. Besides these tactical benefits, Florida also contains some of the best agricultural lands in the United States, has vast open defensible stretches of glades and prairie lands, and is also home to many of the wealthiest individuals in the U.S. With all of these advantages, Florida seems positioned to meet many of the challenges presented by a zombie pandemic. Not only has it made allowances for its citizens to publicly arm themselves to confront a zombie threat, its wealthier citizens also have the means to support the defense of the state in the event that both the national and state government fail. Moreover, its large agricultural farms and industries also give the state's farmers the ability to meet the nutritional and dietary needs of inhabitants. Under such conditions, Florida's economy would likely continue unabated throughout the pandemic, and its defense and resources would help it emerge as a refuge and safe haven from a Zombie apocalypse.

NOTES

1. Laura Wang, interview by Christopher Esing, Chengde, China, April 25, 2004.

2. Laura Wang, interview by Christopher Esing, Chengde, China, April 28, 2004.

3. *Surviving Katrina*, DVD, directed by Johnathan Dent, 2006, Discovery Channel, 2007.

4. *Surviving Katrina*.

5. *Surviving Katrina*.

6. *Surviving Katrina*.

7. Seth French, Commissioner of the Bureau of Immigration, *Semi-Tropical Florida: Its Climate, Soil, and Productions, with a Sketch of Its History* (Chicago: Rand, McNally, 1879), 44.

8. George Pearce, "Torment of Pestilence," *Pensacola Florida Historical Quarterly* 56, no. 4 (April 1978), 456.

9. William Otto Robinson, *A History of Yellow Fever with Specific Emphasis on the Pensacola Epidemics* (Pensacola: Robinson, 1991), 34–36.

10. John Matthew Brackett, "The Naples of American, Pensacola During the Civil War and Reconstruction," MA thesis, Florida State University, 2005, 46–49.

11. Robert B.S. Hargis, "Correspondence, Pensacola, Nov. 9, 1873. Yellow Fever," *New Orleans Medical and Surgical Journal*, M.S. O1 (March 1874), 781–785. Robinson, *A History of Yellow Fever*, 36. Eirlys Mair Barker, "Seasons of Pestilence: Tampa and Yellow Fever, 1824–1905," MA. Thesis, University of South Florida, 1984, 3.

12. Pearce, *Torment of Pestilence*, 456.

13. Dianne Catherine Dusevitch, "Fort Barrancas," *Pensacola Historical Society Quarterly* 7, no. 3. (Summer 1974), 31–34.

14. Seth French, Commissioner of the Bureau of Immigration, *Semi-Tropical Florida: Its Climate, Soil, and Productions, with a Sketch of Its History* (Chicago: Rand, McNally, 1879), 44.

15. Brackett, 73.

16. T.C. Rigby, *Dr. Rigby's papers on Florida giving a general view of every portion of the state, its climate, resources, statistics, society, crops, trade, &c.* (Cincinnati: E. Mendenhall, 1876), 37–38.

17. Chipley, *A New Light on Florida: Pensacola: The Naples of America and Its Surroundings Illustrated. September 1877. New Orleans, Mobile, and the Resorts of the Gulf Coast* (Louisville, KY: Courier-Journal Press, 1877), 9–13.

18. Hargis, 781–785. Occie Clubbs, "Yellow Pine Lumber," *Pensacola in Retrospect: 1870–1890, The Florida Historical Quarterly* 37, no. 03 (January 1959), 382.

19. Pearce, *Torment of Pestilence*, 462.

20. Ibid., 463–466.

21. Ibid., 66–67.

22. Ibid., 79.

23. W.D. Chipley, *This Pamphlet on Western Florida Is Issued by the Louisville and Nashville R.R. Pensacola, Florida. In Co-operation with the Counties of Escambia, Santa Rose, Walton, and Holmes. And the Young Men's Business Association of Pensacola* (Louisville, KY: L&N R.R., 1895), 3–7.

24. Pearce, *Torment of Pestilence*, 469–47.

25. Carolina Rawls, *The Jacksonville Story: A Pictorial Record of a Florida City* (Jacksonville: Jacksonville's Fifty Years of Progress Association, 1950), 10.

26. Charles S. Adams, *Report of the Jacksonville Auxiliary Sanitary Association of Jacksonville, Florida. Covering the work of the association during the Yellow Fever Epidemic, 1888* (Jacksonville: Times-Union Print, 1889), 24.

27. Thomas Frederick Davis, *History of Jacksonville, Florida and Vicinity 1513 to 1924* (Gainesville: University Press of Florida, 1925, 1964), 180.

28. Glenn Emery, *Urban Boosterism in Florida: Tallahassee and Jacksonville, 1865–1917* (thesis, University of Florida, 1987), 86.

29. Richard J. Amundson, "The Florida Land and Colonization Company," *Florida Historical Quarterly* 44, no. 3 (January 1966), 166.

30. John T. Foster, Jr., *Beechers, Stowes, and Yankee Strangers: the Transformation of Florida* (Gainesville: University Press of Florida, 1999), 55.

31. Florida Land Agency, *Florida: Its soil, climate, health, productions, resources with a sketch of its history. A manual of reliable information. Concerning the resources of the State and Inducements to Immigration* (Jacksonville: Florida Land Agency, 1875). 15.

32. Harriet Beecher Stowe, *Palmetto Leaves* (Gainesville: University Press of Florida, 1873, 1968), 124–127.

33. Rambler, *Guide to Florida* (Gainesville: University of Florida Press, 1875, 1964), ix.

34. Claudia Bell and John Lyall, *The Accelerated Sublime: Landscape, Tourism, and Identity* (Westport, CT: Praeger, 2002), 174.

35. T.F. Smith, *Florida and Texas: A series of Letters, Comparing the Soil, Climate, and Productions of these states, setting forth many advantages that East and South Florida offers to Emigrants* (Ocala: Printed at the East Florida Banner Office, 1866), 5.

36. Dennis Egan, *The Florida Settler or Immigrants Guide; A Complete Manual of Information Concerning the Climate, Soil, Productions, and Resources of the State* (Tallahassee: Office of the Floridian, 1873), 12.

37. Henry Martyn Field, *Bright Skies and Dark Shadows.* (New York: Scribner's, 1890), 76.

38. Robert Speir, *Going South for the Winter with Hints to Consumptives* (New York: Edward O. Jenkins, 1873), 9.

39. Rambler, *Guide to Florida*, 63.

40. Frederic D. Lente, M.D., *Florida as a Health Resort. Reprinted from the New York Medical Journal. November 1876* (New York: D. Appleton, 1876), 17.

41. Ibid., 22.

42. Bates (C.F.) & Company, *Florida tourist and southern investor's guide* (Cedar Rapids: Bates [C.F.], 1898), 23.

43. James R. Nichols, *From Snow banks to Orange Orchards: An essay read before the Essex North Massachusetts Medical Society May, 1878, and published in the Boston Journal of Chemistry* (Haverhill: Press of Franklin P. Stiles, 1878), 4.

44. Foster, 55.

45. Rambler, xiii.

46. Ledyard Bill, *A winter in Florida, or, Observations on the soil, climate, and products of our semi- tropical state* (New York: Wood & Holbrook, 1869), 83.

47. Stowe, 38.

48. Nina Silber, *The Romance of Reunion: Northerners and the South 1865–1900* (Chapel Hill: University of North Carolina Press, 1999), 68.

49. George Washington Olney, *A Guide to Florida, "The Land of Flowers": Containing an Historical Sketch, geographical, agricultural, and climactic statistics, routes of travel by land and sea, and general information invaluable to the invalid, tourist, or immigrant* (New York: Crushing, Bermuda, 1872), 25. Eric E. Elliot, *Paddle Wheels on the St. Johns: An Analysis of the Impact of Steamboat Technology on a Southeastern Region of the United States*, Ph.d. dissertation, Carnegie Mellon University, 1987, 215.

50. Edward A. Meuller, *Along the St. Johns and Ocklawaha Rivers* (Charleston: Arcadia, 1999), 10. Elliot, *Paddle Wheels on the St. Johns*, 158.

51. George Barbour, *Florida for Tourists, Invalids, and Settlers* (New York: D. Appleton, 1882), 226.

52. Benjamin F. Rogers, "Florida as Seen Through the Eyes of Nineteenth Century Travelers" *Florida Historical Quarterly* 34, no. 2 (October 1955), 28–29.

53. Maurice M. Vance, "Northerners and Their Professions in Jacksonville." *The Florida Historical Quarterly* Vol. 38 No. 1(July 1959), 11.

54. Frank Simpson, *A Trip Through Northern and Central Florida, During March and April, 1882* (East Orange, N.J.: East Orange Gazette Print, 1882), 23.

55. Barbour, 295.

56. Zora Neale Hurston, *Their Eyes Were Watching God* (New York: Perennial Library, 1990), Ch. 19.

57. Richard Luscombe, "Miami Man Shot Dead Eating a Man's Face May Have Been on LSD Like Drug." *The Guardian*, May 29, 2012.

58. Philip Caulfield, "Another Florida 'Zombie' Attack? Naked Man Storms Girlfriends House, Bites Chunk Out of Man's Arm." *New York Daily News*, June 21, 2012

59. Greg Seals, "Florida Is Actually Preparing for a Zombie Apocalypse." *The Daily Dot*, May 1, 2014. Amy Hubbard, "Florida Senate Bill Amended to Include Zombie Apocalypse," *Los Angeles Times*. April 30, 2014.

BIBLIOGRAPHY

Adams, Charles S. *Report of the Jacksonville Auxiliary Sanitary Association of Jacksonville, Florida. Covering the work of the association during the Yellow Fever Epidemic.* Jacksonville: Times-Union Print, 1889.

Amundson, Richard J. "The Florida Land and Colonization Company." *Florida Historical Quarterly* 44, no. 3 (January 1966).

Barbour, George M. *Florida for Tourists, Invalids, and Settlers: Containing Practical Information Regarding Climate, Soil, and Productions, Cities, Towns, and People, the Culture of the Orange and Other Tropical Fruits, Farming and Gardening, Scenery and Resorts, Sports, Routes of Travel, Etc.* New York: D. Appleton, 1882.

Barker, Eirlys Mair. *Seasons of Pestilence: Tampa and Yellow Fever, 1824–1905.* MA thesis, University of South Florida, August 1984.

Bell, Claudia, and John Lyall. *The Accelerated Sublime: Landscape, Tourism, and Identity.* Westport, CT: Praeger, 2002.

Bill, Ledyard. *A winter in Florida, or, Observations on the soil, climate, and products of our semi-tropical state with sketches of the principal towns and cities in eastern Florida: to which is added a brief historical summary together with hints to the tourist, invalid, and sportsman.* New York: Wood & Holbrook, 1869.

Brackett, John Matthew. *The Naples of American Pensacola During the Civil War and Reconstruction.* MA thesis, Florida State University, 2005.

C.F. Bates & Company. *Florida Tourist and Southern Investor's Guide.* Cedar Rapids, 1898.

Caulfield, Philip. "Another Florida 'Zombie' Attack? Naked Man Storms Girlfriends House, Bites Chunk out of Man's Arm." *New York Daily News*, June 21, 2012.

Chipley, William D. *A New Light on Florida: Pensacola: The Naples of America and Its Surroundings Illustrated. September 1877. New Orleans, Mobile, and the Resorts of the Gulf Coast.* Louisville, KY: Courier-Journal Press, 1877.

_____. *This Pamphlet on Western Florida is issued by W.D. Chipley. General Land Commissioner Louisville and Nashville R.R. Pensacola, Florida. In Cooperation with the Counties of Escambia, Santa Rose, Walton, and Holmes. And the Young Men's Business Association of Pensacola.* Louisville, KY: Courier-Journal, 1895.

Clubbs, Occie. "Pensacola in Retrospect: 1870–1890." *Florida Historical Quarterly* 37, no. 3 (January 1959): 377–396.

Davis, Thomas Frederick. *History of Jacksonville Florida and Vicinity, 1513–1924.* Jacksonville: Florida Historical Society, 1925.

Dent, Johnathan, dir. *Surviving Katrina.* DVD, 2006. Discovery Channel, 2007.

Dusevitch, Dianne Catherine "Fort Barrancas." *Pensacola Historical Society Quarterly* 7, no. 3 (Summer 1974).

Eagan, Dennis. *The Florida settler: or Immigrants' guide; a complete manual of information concerning the climate, soil products and resources of the state.* Tallahassee: Printed at the office of the Floridian, 1873.

Elliot, Eric E. *Paddle Wheels on the St. Johns: An Analysis of the Impact of Steamboat Technology on a Southeastern Region of the United States.* Ph.D. dissertation, Carnegie Mellon University, 1987.

Emery, Glenn. *Urban Boosterism in Florida: Tallahassee and Jacksonville, 1865–1917.* MA thesis, University of Florida, 1987.

Field, Henry Martyn. *Bright Skies and Dark Shadows.* New York: Scribner's, 1890.

Florida Land Agency. *Florida, its soil, climate, health, productions, resources, and advantages: with a sketch of its history. A manual of reliable information concerning the resources of the state and inducements to immigration.* Jacksonville: Florida Land Agency, 1875.

Fosters, Mark S. *Castles in the Sand: The Life and Times of Carl Grahm Fisher.* Gainesville: University Press of Florida, 2000.

French, Seth. *Semi-tropical Florida: Its climate, soil, and productions with a sketch of its history, natural features and social condition.* Chicago: Rand McNally, 1879.

Hargis, Robert B.S. "Correspondence, Pensacola, Nov. 9, 1873. Yellow Fever." *New Orleans Medical and Surgical Journal*, M.S. O1 (March 1874).

Hubbard, Amy. "Florida Senate Bill Amended to Include Zombie Apocalypse." *Los Angeles Times*, April 30, 2014.

Hurston, Zora Neal. *Their Eyes Were Watching God.* New York: Perennial Library, 1990.

Lente, Frederick Divoux. *Florida as a Health-Resort.* Reprinted from the *New York Medical Journal*, November 1876. New York: D. Appleton, 1876.

Luscombe, Richard. "Miami Man Shot Dead Eating a Man's Face May Have Been on LSD Like Drug." *The Guardian*, May 29, 2012.

Meuller, Edward A. *Along the St. Johns and Ocklawaha Rivers.* Charleston: Arcadia, 1999.

Nichols, James R. *From Snow banks to Orange orchards: An essay read before the Essex North Massachusetts Medical Society May, 1878, and published in the Boston Journal of Chemistry.* Haverhill: Press of Franklin P. Stiles, 1878.

Olney, George Washington. *A Guide to Florida: "the land of flowers," containing an historical sketch, geographical, agricultural and climatic statistics, routes of travel by land and sea, and general information invaluable to the invalid, tourist or emigrant.* New York: Cushing, Bardua, 1872.

Pearce, George. "Torment of Pestilence." *Pensacola Florida Historical Quarterly*, 56 (April 1978).

"Rambler," pseud. *Guide to Florida*. Gainesville: University of Florida Press, 1875, 1964.

Rawls, Carolina. *The Jacksonville Story: A Pictorial Record of a Florida City*. Jacksonville: Jacksonville's Fifty Years of Progress Association, 1950.

Rigby, T.C. *Dr. Rigby's papers on Florida giving a general view of every portion of the state, its climate, resources, statistics, society, crops, trade, &c*. Cincinnati: E. Mendenhall, 1876.

Robinson, William Otto. *A History of Yellow Fever with Specific Emphasis on the Pensacola Epidemics*. Pensacola: Robinson, 1991.

Rogers, Ben F. "Florida Seen Through the Eyes of Nineteenth Century Travellers." *Florida Historical Quarterly* 34, no. 2 (October 1955): 177–189.

Seals, Greg. "Florida Is Actually Preparing for a Zombie Apocalypse." *The Daily Dot*, May 1, 2014.

Silber, Nina. *The Romance of Reunion: Northerners and the South 1865–1900*. Chapel Hill: University of North Carolina Press, 1999.

Simpson, Frank. *A Trip Through Northern and Central Florida, During March and April, 1882*. East Orange, N.J.: East Orange Gazette Print, 1882.

Smith, T.F. *Florida and Texas: A series of Letters, Comparing the Soil, Climate, and Productions of these states, setting forth many advantages that East and South Florida offers to Emigrants*. Ocala: Printed at the East Florida Banner Office, 1866.

Speir, Robert. *Going South for the Winter with Hints to Consumptives*. New York: Edward O. Jenkins, 1873.

Stowe, Harriet Beecher. *Palmetto-Leaves*. Gainesville: University of Florida Press, 1873, 1968.

Wang, Laura. Interview by Christopher Esing. Personal interview. Chengde, China, April 28, 2004.

About the Contributors

Diem-My T. **Bui** is a clinical assistant professor in the Department of Communication at the University of Illinois, Chicago. Her research interests include transnational feminist media studies, critical cultural studies, ethnic studies, popular culture, and film. She focuses on cultural production, cultural memory, and embodiments of difference in racialized and sexualized representations in popular culture.

Christopher M. **Esing** is a writer and a social and economic historian. His works range from children's books to historical essays and monographs. His areas of specialization include modernization, the New South, the modern United States, race, and Florida history.

Cory Andrew **Labrecque** is the Raymond F. Schinazi Scholar in Bioethics and Religious Thought and the director of the Master of Arts program at the Emory University Center for Ethics. He also serves as co-director of Catholic Studies. His research lies at the intersection of religion, medicine, biotechnology, environment, and ethics.

Jennifer M. **Lankford** is an attorney with Thompson Burton, PLLC, in Franklin, Tennessee. She has extensive experience litigating employment disputes within the federal and state courts of Tennessee and is a member of the Nashville Bar Foundation Leadership Forum as well as the American, Tennessee, and Nashville Bar Associations.

Scott **Mirabile** is a developmental psychologist and assistant professor of psychology at St. Mary's College of Maryland. In his research he investigates how parents and teachers help children learn about and cope with emotions. An avid fan of the zombie genre, he is the faculty mentor to the St. Mary's Humans vs. Zombies Club and has given public lectures on human adaptation to a zombie apocalypse.

Jeff **Moehlis** is a professor in the Department of Mechanical Engineering at the University of California, Santa Barbara. His research includes the application of dynamical systems and control techniques to neuroscience, cardiac dynamics, energy harvesting, fluid dynamics, and collective behavior. He has received a Sloan Fellowship in mathematics, a National Science Foundation CAREER Award, and a Northrup Grumman Excellence in Teaching Award.

Nick **Proctor** is a professor of history at Simpson College. He is the author of a historical monograph on hunting in the Old South and has written a number of games for the Reacting to the Past series of immersive historical role-playing games. To aid other authors in the series, he wrote the *Reacting to the Past Game Designer's Handbook.*

LuAnne **Roth** teaches in the English department at the University of Missouri, where her courses involve the topics of folklore, foodways, contemporary legend, and critical theory. Roth's current research focuses on media representations of food and culture (including zombie foodways), and she maintains a digital archive of scenes related to food/culture, legend/rumor, and the undead.

Steve **Schlozman**, M.D., is an assistant professor of psychiatry at Harvard Medical School and the former director of Medical Student Education in Psychiatry for Harvard and associate director for The Clay Center for Young Healthy Minds at Massachusetts General Hospital. He is the author of a novel, *The Zombie Autopsies: Secret Notebooks from the Apocalypse*, which was optioned and adapted for film by George Romero. He has also discussed this work for the History Channel, the Discovery Channel, and in three documentary films.

Kate **Shoults** has a degree in English and is currently in the secondary education language arts program at the University of Missouri, Columbia. Shoults serves as a substitute teacher in the Columbia Public Schools district and is an educational outreach assistant for the True/False Film Festival. She presented on feminism and zombie films at Mizzou's Zombie Week conference in 2013.

David F. **Steele** is an associate professor in the Department of Sociology at Austin Peay State University. He received a Ph.D. from the University of Tennessee, Knoxville. His research interests are in the areas of the sociology of popular culture, environmental sociology, and the sociology of teaching.

Amy L. **Thompson** is an associate professor in the Department of Biology and director of Pre-Professional Health Programs at Austin Peay State University. In addition to zombies, her research focuses on brown recluse spider bite treatment. She was recognized in 2014 as one of the American Society for Clinical Pathologists' 40 Under 40.

Antonio S. **Thompson** is an associate professor at Austin Peay State University. In 2009–2010 he taught on fellowship at the United States Military Academy at West Point. He is the author of books on POWs in America during World War II and German POWs in Kentucky between 1942 and 1946.

James F. **Thompson** is a semi-retired professor of biology at Austin Peay State University. He is also a former medical technologist who worked in blood banking for eleven years.

Linda W. **Thompson** is a professor in the School of Nursing at Austin Peay State University. Her clinical area is psychiatric mental health nursing. Her research interests include the image of nursing, nursing ethics and long-term mental health issues.

Jason W. **Warren** is a strategist at the U.S. Army War College and was an assistant professor of military history at West Point from 2009 to 2012. His research focuses on warfare in early colonial America and he recently published a book on the Great Narragansett War. He also has published on modern military affairs and teaches ancient military history.

Jeremy **Youde** is an associate professor of political science at the University of Minnesota Duluth. He received a Ph.D. from the University of Iowa and previously taught at San Diego State University and Grinnell College. His research focuses on global health politics and African politics. He has published more than 20 peer-reviewed articles and is the author of three books.

Index

www.ingramcontent.com/pod-product-compliance
Lightning Source LLC
Chambersburg PA
CBHW021136090426
42740CB00008B/804